CRC Series in Naturally Occurring Pesticides
Series Editor-in-Chief
N. Bhushan Mandava

Handbook of Natural Pesticides: Methods
Volume I: Theory, Practice, and Detection
Volume II: Isolation and Identification
Editor
N. Bhushan Mandava

Handbook of Natural Pesticides
Volume III: Insect Growth Regulators
Volume IV: Pheromones
Editors
E. David Morgan
N. Bhushan Mandava

Handbook of Natural Pesticides
Volume V: Microbial Insecticides
Editor
Carlo M. Ignoffo

Handbook of Natural Pesticides
Volume VI: Insect Attractants and Repellents
Editors
E. David Morgan
N. Bhushan Mandava

CRC
Handbook
of
Natural Pesticides

Volume VI
Insect Attractants
and
Repellents

Editors

E. David Morgan, D.Phil.
Reader
Department of Chemistry
University of Keele
Staffordshire, England

N. Bhushan Mandava, Ph.D.
Senior Partner
Todhunter, Mandava and Associates
Washington, D.C.

CRC Series in Naturally Occurring Pesticides

Series Editor-in-Chief
N. Bhushan Mandava, Ph.D.

CRC Press
Taylor & Francis Group
Boca Raton London New York

CRC Press is an imprint of the
Taylor & Francis Group, an **informa** business

First published 1990 by CRC Press
Taylor & Francis Group
6000 Broken Sound Parkway NW, Suite 300
Boca Raton, FL 33487-2742

Reissued 2018 by CRC Press

© 1990 by Taylor & Francis
CRC Press is an imprint of Taylor & Francis Group, an Informa business

No claim to original U.S. Government works

Publisher's Note
The publisher has gone to great lengths to ensure the quality of this reprint but points out that some imperfections in the original copies may be apparent.

Disclaimer
The publisher has made every effort to trace copyright holders and welcomes correspondence from those they have been unable to contact.

ISBN 13: 978-1-315-89359-4 (hbk)
ISBN 13: 978-1-351-07269-4 (ebk)

Visit the Taylor & Francis Web site at http://www.taylorandfrancis.com and the
CRC Press Web site at http://www.crcpress.com

INTRODUCTION

The United States has been blessed with high quality, dependable supplies of low cost food and fiber, but few people are aware of the never-ending battle that makes this possible. There are at present approximately 1,100,000 species of animals, many of them very simple forms, and 350,000 species of plants that currently inhabit the planet Earth. In the U.S., there are an estimated 10,000 species of insects and related acarinids which at sometime or other cause significant agricultural damage. Of these, about 200 species are serious pests which require control or suppression every year. World-wide, the total number of insect pests is about ten times greater. The annual losses of crops, livestock, agricultural products, and forests caused by insect pests in the U.S. have been estimated to aggregate about 12% of the total crop production and to represent a value of about $4 billion (1984 dollars). On a world-wide basis, the insect pests annually damage or destroy about 15% of total potential crop production, with a value of more than $35 billion, enough food to feed more than the population of a country like India. Thus, both the losses caused by pests and the costs of their control are considerably high. Insect control is a complex problem for there are more than 200 insects that are or have been subsisting on our main crops, livestock, forests, and aquatic resources. Today, in the U.S., conventional insecticides are needed to control more than half of the insect problems affecting agriculture and public health. If the use of pesticides were to be completely banned, crop losses would soar and food prices would also increase dramatically.

About 1 billion pounds of pesticides are used annually in the U.S. for pest control. The benefits of pesticides have been estimated at about $4/$1 cost. In other words, chemical pest control in U.S. crop production costs an estimated $2.2 billion and yields a gross return of $8.7 billion annually.

Another contributing factor for increased crop production is the effective control of weeds, nematodes, and plant diseases. Crop losses due to unwanted weed species are very high. Of the total losses caused by pests, weeds alone count for about 10% of the agricultural production losses valued at more than $12 billion annually. Farmers spend more than $6.2 billion each year to control weeds. Today, nearly all major crops grown in the U.S. are treated with herbicides. As in insect pest and weed control programs, several chemicals are used in the disease programs. Chemical compounds (e.g., fungicides, bactericides, nematicides, and viracides) that are toxic to pathogens, are used for controlling plant diseases. Several million dollars are spent annually by American farmers to control the diseases of major crops such as cotton and soybeans.

Another aspect for improved crop efficiency and production is the use of plant growth regulators. These chemicals that regulate the growth and development of plants are used by farmers in the U.S. on a modest scale. The annual sale of growth regulators is about $130 million. The plant growth regulator market is made up of two distinct entities, growth regulators and harvest aids. Growth regulators are used to increase crop yield or quality. Harvest aids are used at the end of the crop cycle. For instance, harvest aids defoliate cotton before picking or desiccate potatoes before digging.

The use of modern pesticides has accounted for astonishing gains in agricultural production as pesticides have reduced the hidden toll exacted by the aggregate attack of insect pests, weeds, and diseases, and also improved the health of humans and livestock as they control parasites and other microorganisms. However, the same chemicals have allegedly posed some serious problems to health and environmental safety because of their high toxicity and severe persistence, and have become a grave public concern in the last two decades. Since the general public is very much concerned about their hazards, the U.S. Environmental

Protection Agency enforced strong regulations for the use, application, and handling of the pesticides. Moreover, such toxic pesticides as DDT, 2,4,5-T, and toxaphene were either completely banned or approved for limited use. They were, however, replaced with less dangerous chemicals for insect control. Newer approaches for pest control are continuously sought, and several of them look very promising.

According to a recent study by the National Academy of Sciences, pesticides of several kinds will be widely used in the foreseeable future. However, newer selective and biodegradable compounds must replace older highly toxic persistent chemicals. The pest control methods that are being tested or used on different insects and weeds include: (1) use of natural predators, parasites, and pathogens, (2) breeding of resistant varieties of species, (3) genetic sterilization techniques, (4) use of mating and feeding attractants, (5) use of traps, (6) development of hormones to interfere with life cycles, (7) improvement of cultural practices, and (8) development of better biodegradable insecticides and growth regulators that will effectively combat the target species without doing damage to beneficial insects, wildlife, or man. Many leads are now available, such as the hormone mimics of the insect juvenile and molting hormones. Synthetic pyretheroids are now replacing the conventional insectides. These insecticides, which are a synthesized version of the extract of the pyrethrum flower, are much more attractive biologically than the traditional insecticides. Thus, the application rates are much lower, in some cases one tenth the rates of more traditional insecticides such as organophosphorus pesticides. The pyrethroids are found to be very specific for killing insects and apparently exhibit no negative effects on plants, livestock, or humans. The use of these compounds is now widely accepted for use on cotton, field corn, soybeans, and vegetable crops.

For the long term, integrated pest management (IPM) will have tremendous impact on pest control for crop improvement and efficiency. Under this concept, all types of pest control — cultural, chemical, inbred, and biological — are integrated to control all types of pests and weeds. Chemical control includes all of the traditional pesticides. Cultural controls consist of cultivation, crop rotation, optimum planting dates, and sanitation. Inbred plant resistance involves the use of varieties and hybrids that are resistant to certain pests. Finally, biological control involves encouraging natural predators, parasites, and microbials. Under this system, pest-detection scouts measure pest populations and determine the best time for applying pesticides. If properly practiced, IPM could reduce pesticide use up to 75% on some crops.

Naturally occurring pesticides appear to have a prominent role for the development of future commercial pesticides, not only for agricultural crop productivity, but also for the safety of the environment and public health. They are produced by plants, insects, and several microorganisms, which utilize them for survival and maintenance of defense mechanisms, as well as for growth and development. They are easily biodegradable, oftentimes species-specific, and also sometimes less toxic (or nontoxic) to other nontarget organisms or species — an important consideration for alternate approaches to pest control. Several of the compounds, especially those produced by crop plants and other organisms, are consumed by humans and livestock, and yet appear to have no detrimental effects. They appear to be safe and will not contaminate the environment. Hence, they will be readily accepted for use in pest control by the public and the regulatory agencies. These natural compounds occur in nature only in trace amounts and require very low dosage for pesticide use. It is hoped that the knowledge gained by studying these compounds is helpful for the development of new pest control methods such as their use for interference with hormonal life cycles and trapping insects with pheromones, and also for the development of safe and biodegradable chemicals (e.g., pyrethroid insecticides). Undoubtedly, the costs are very high as compared to the presently used pesticides. But hopefully, these costs would be compensated for by the benefits derived through these natural pesticides from the lower volume of pesticide use and reduction of risks. Furthermore, the indirect or external costs resulting from pesticide poisoning, fatalities, livestock losses, and increased control expenses (due to the destruction of natural enemies and beneficial insects as well as the environmental contamination and pollution from chlorinated, organophosphorus, and carbamate pesticides) could be assessed against benefits vs. risks. The development and use of such naturally occurring chemicals could become an integral part of IPM strategies.

As long as they remain endogenously, several of the natural products presented in this handbook series serve as hormones, growth regulators, and sensory compounds for growth, development, and reproduction of insects, plants, and microorganisms. Others are useful for defense or attack against other species or organisms. Once these chemicals or their analogs and derivatives are applied by external means to the same (where produced) or different species, they come under the label "pesticides" because they contaminate the environment. Therefore, they are subject to regulatory requirements, in the same way the other pesticides are handled before they are used commercially. However, it is anticipated that the naturally occurring pesticides would easily meet the regulatory and environmental requirements for their safe and effective use in pest control programs.

A vast body of literature has been accumulated on natural pesticides during the last two or three decades; we have been assembling this information in these handbooks. We have limited our attempts to chemical and a few biological aspects concerned with biochemistry and physiology. Wherever possible, we tried to focus attention on the application of these compounds for pesticidal use. We hope that the first two volumes which dealt with theory and practice served as introductory volumes and will be useful to everyone interested in learning about the current technology that is being adapted from compound identification to field trials. The subsequent volumes deal with the chemical, biochemical, and physiological aspects of naturally occurring compounds, grouped under such titles as insect growth regulators, plant growth regulators, etc.

In a handbook series of this type with diversified subjects dealing with plant, insect, and microbial compounds, it is very difficult to achieve either uniformity or complete coverage while putting the subject matter together. This goal was achieved to a large extent with the understanding and full cooperation of chapter contributors, who deserve my sincere appreciation.

The editors of the individual handbooks relentlessly sought to meet the deadlines and, more importantly, to bring a balanced coverage of the subject matter; however, that seems to be an unattainable goal. Therefore, they bear full responsibility for any pitfalls and deficiencies. We invite comments and criticisms from readers and users as they will greatly help to update future editions. It is hoped that these Handbooks will serve as a source for chemists, biochemists, physiologists, and other biologists alike — those engaged in active research as well as those interested in different areas of natural products that affect the growth and development of plants, insects, and other organisms.

The editors wish to acknowledge their sincere thanks to the members of the Advisory Board for their helpful suggestions and comments. Their appreciation is extended to the publishing staff, especially Amy Skallerup, Melanie Mortellaro, and Sandy Pearlman for their ready cooperation and unlimited support from the initiation to the completion of this project.

N. Bhushan Mandava
Editor-in-Chief

FOREWORD

Pests of crops and livestock annually account for multi-billion dollar losses in agricultural productivity and control costs. Insects alone are responsible for more than 50% of these losses.

For the past 40 years, the principal weapons used against these troublesome insects have been chemical insecticides. The majority of such materials used during this period have been synthetic organic chemicals discovered, synthesized, developed, and marketed by commercial industry. In recent years, environmental concerns, regulatory restraints, and problems of pest resistance to insecticides have combined to reduce the number of materials available for use in agriculture. Replacement materials reaching the marketplace have been relatively few due to increased costs of development and the general lack of knowledge about new classes of chemicals having selective insecticidal activity.

In response to these trends, it is gratifying to note that scientists in both the public and private sectors have given significant attention to the discovery and evaluation of natural products as fertile sources of new insecticidal agents. Not only are these materials directly useful as insect control agents, but they also serve as models for new classes of chemicals with novel modes of action to attack selective target sites in pest species. Such new control agents may also be less susceptible to the cross resistance difficulties encountered with most classes of currently used synthetic pesticide chemicals to which insects have developed immunity.

Natural products originating in plants, animals, and microorganisms are providing a vast source of bioactive substances. The rapid development and application of powerful analytical instrumentation, such as mass spectrometry, nuclear magnetic resonance spectroscopy, gas chromatography, high-performance liquid chromatography, and immuno- and other bioassays have greatly facilitated the identification of miniscule amounts of active biological chemicals isolated from natural sources. These new scientific approaches and tools are addressed and reviewed extensively in these volumes.

Some excellent examples of success in this research involve the discovery of insect growth regulators, especially the so-called juvenoids, which are responsible for control of insect metamorphosis, reproduction, and behavior. Pheromones, which play essential roles in insect communication, feeding, and sexual behavior, represent another important class of natural products holding great promise for new pest insect control technology. All of these are discussed in detail in volumes dealing with insects.

It is hoped that the scientific information provided in these volumes will serve researchers in industry, government, and academia, and stimulate them to continue to seek even more useful natural materials that produce effective, safe, and environmentally acceptable materials for use against insect pests affecting agriculture and mankind.

Orville G. Bentley
Assistant Secretary
Science and Education
U.S. Department of Agriculture

PREFACE TO VOLUME VI

Since we began to assemble this series of Handbooks on Natural Pesticides, the climate of opinion towards environmentally acceptable pesticides has changed enormously. What was once the pursuit of a few, has become the demand of the many. As our knowledge of the effects of stable, persistant chemicals in the environment has grown, an aversion to the use of such substances has formed in the minds of the general public. Independently of that, concern has spread among agriculturalists at the rapid development of resistance to some of the newer pesticides in many of the more serious agricultural pests.

If the depredations of insects on our crops, outlined in the general preface to this series, are to be contained, let alone reduced, then new insecticides must be found. The general public will demand that these new insecticides must be environmentally acceptable too, and preferably selective towards only those harmful insects. These conditions are more easily stated than fulfilled. From where are the ideas for these new insecticides to come?

In the belief that the natural physiology and communication of these insects will be the best source of ideas, we have assembled these volumes. We mean communication in the broadest sense. Communication between individuals of the same species, between plants and insects, between predator and prey, between host and parasite, symbiont or inquiline, is the subject of allelochemics, of chemical signals between species, whether these be plants or animals.

In this volume, we draw together some of the well-recognized allelochemical interactions of insects. To help us to understand how these allelochemical interactions are detected, Prof. Louis Schoonhoven has written an introductory chapter "Insects in a Chemical World". Insects can frequently be observed to have preferences for some plants and avoidance of others. In the hands of man, this can be turned into a potent weapon. The recognition in plants of feeding deterrents of anti-feedants (Chapter 2) has led to a great expansion of research in the area, and isolation of many active compounds. The types and origins of these deterrent compounds are very diverse, and there is little or no understanding yet of the character that creates deterrency, so this chapter is largely a compilation of tables of data. Table 6 in this chapter, which lists number of known species and number of species examined in each plant family emphasizes how we have only scratched the surface in our examination of the plant world. While I have a personal interest in the exploitation of one of the most active and broad-spectrum feeding deterrents, I also believe that a compound with a narrow spectrum but high activity against an important pest, may provide protection against that pest and win acceptance as attitudes change.

As well as the specific feeding deterence for plant-eating insects, there is recognized a more general repellency towards any group of insects, which is covered in the next chapter. Likewise, there are specific attractants for insects which might also be turned to our own purposes. These are covered in Chapter 5 by Waage and Hedin, but we also begin to recognize specific types of behavior, pupation, and other kinds of behavior still to be discovered. Of these, only ovipositional attraction and deterency has yet reached a state of knowledge to warrant a section (Chapter 4) to itself.

We must also recognize that some insects have overcome the toxic or deterrent substances of plants, and others will adapt themselves to try to overcome deterrents that we may apply against them. Therefore, a treatment of how insects detoxify plant chemicals is important in this strategy (Chapter 6). We must be always one jump ahead of the pest and be considering new methods of control and new methods of incorporating or disseminating them. Genetic engineering offers a great prospect here. How much more efficient it would be if we can induce the plant itself to produce an insect deterrent or repellent. That is much more efficient than producing, isolating, and applying it ourselves. The application may yet seem far away, but a consideration of present knowledge and potential of genetic manipulation for insect control is important, and concludes this volume.

We hope the combined efforts of our authors will bring hope and inspiration to those entrusted with the task of protecting our food supply, our timber, and the beauty of our surroundings.

Once again I must thank all our contributors and our secretaries at Keele, Christine Owen and Ann Billington for their help and patience, especially with some of the lengthy and difficult tables.

E. David Morgan, Ph.D.
Keele, England

THE EDITORS

E. David Morgan, D.Phil., is a Chartered Chemist, a Fellow of the Royal Society of Chemistry, and a Fellow of the Royal Entomological Society of London. He received his scientific training in Canada and England, and has worked for the National Research Council of Canada, Ottawa; The National Institute for Medical Research, London; the Shell Group of Companies; and is now Reader in Chemistry at the University of Keele, Staffordshire, England. He is co-author of a textbook on aliphatic chemistry with the Nobel prizewinner, Sir Robert Robinson, and with him is a co-inventor of a number of patents. Dr. Morgan has contributed to over 150 papers, most of them on aspects of insect chemistry, and has written a number of reviews on insect hormones, pheromones, and anti-feedants.

N. Bhushan Mandava, holds B.S., M.S., and Ph.D. degrees in Chemistry and has published over 150 papers including two patents, several monographs and reviews, and books in the areas of pesticides and plant growth regulators and other natural products. As editorial advisor, he has edited three special issues on countercurrent chromatography for the *Journal of Liquid Chromatography*. He has also edited a book on countercurrent chromatography and is now a consultant in pesticides and drugs. Formerly, he was associated with the U.S. Department of Agriculture and the Environmental Protection Agency as Senior Chemist. He has been active in several professional organizations, was President of the Chemical Society of Washington, and serves as Councillor of the American Chemical Society.

CONTRIBUTORS

Andrew F. Cockburn, Ph.D.
Research Geneticist
Insects Affecting Man and Animals
  Laboratory
Agricultural Research Service
U.S. Department of Agriculture
Gainesville, Florida

Patrick F. Dowd, Ph.D
Research Entomologist
Northern Regional Research Center
Agricultural Research Service
U.S. Department of Agriculture
Peoria, Illinois

Paul A. Hedin, Ph.D.
Research Chemist
Crop Science Research Laboratory
Agricultural Research Service
U.S. Department of Agriculture
Mississippi State University, Mississippi

E. David Morgan, D. Phil.
Reader
Department of Chemistry
University of Keele
Staffordshire, England

Dale M. Norris, Ph.D.
Department of Entomology and Graduate
  Neurosciences Program
University of Wisconsin
Madison, Wisconsin

J. A. A. Renwick, Ph.D.
Chemist
Boyce Thompson Institute
Ithaca, New York

L. M. Schoonhoven, Ph.D.
Professor
Department of Entomology
Agricultural University
Wageningen, The Netherlands

Jack A. Seawright, Ph.D.
Research Leader
Insects Affecting Man and Animals
  Laboratory
Agricultural Research Service
U.S. Department of Agriculture
Gainesville, Florida

Susan K. Waage, Ph.D.
Postdoctoral Research Fellow
School of Forestry
Auburn University
Auburn, Alabama

J. David Warthen, Jr., Ph.D.
Research Chemist
Agricultural Research Service
U.S. Department of Agriculture
Beltsville, Maryland

TABLE OF CONTENTS

INSECTS IN A CHEMICAL WORLD

L. M. Schoonhoven

Dedicated to the memory of Professor J. de Wilde (1916-1983)

The world in which we live impinges upon us primarily by visual and auditory stimuli. To most animals chemical signals are often more important than vision and hearing to appraise their environment. Lower animals especially, strongly depend on olfactory and taste stimuli. Plants and animals, during millions of years of living together, have developed a large array of chemicals which play a role in intricate ecological interactions between myriads of species. Not only interspecific relationships, but also various kinds of interactions between individuals of the same kind are often performed by way of chemical signals. The effectiveness of chemical communication depends, of course, on the quality of the chemoreceptory system of the recipient. The chemical sense represents a universal feature of living systems. Bacteria, for instance, are known to possess a fairly detailed chemoreceptive system, which is used to guide them toward a chemically more suitable environment or to avoid detrimental substances.[1] The primordial chemo-reactivity as exhibited by prokaryotes has developed during evolution into the acute chemical senses of present-day animals. The senses of smell and taste have reached a high degree of perfection with respect to specificity as well as sensitivity. The insect world provides many examples of a full exploitation of the possibilities of chemical communication within the limits set by physical laws.

Insects, like all animals, are born to grow and to reproduce before dying. For growth they may feed on a large variety of food sources, including decaying organisms, plants, and animals. Almost any organic material on the earth's surface is vulnerable to insect attack. Some insects excavate pine needles, others tunnel in the skin of elephants, some grow on dung, and larvae of the moth *Tinea cloacella* flourish on corks in wine cellars. This does not mean that insects are unselective feeders; on the contrary. Different from many vertebrates, most insect species are highly specialized with regard to their food requirements. Selective feeding habits demand a finely tuned sensory system to locate acceptable food sources and to recognize their quality. The ultimate goal in an insect's life, the production of offspring, requires finding a mating partner, small-sized like itself, and often at a relatively large distance. Once inseminated, females in many species search for an oviposition site on or near a food source felicitous to her neonate offspring. These kinds of behaviors depend, like food selection, to a large extent on chemical cues.

The more we learn about communication in social and gregarious insects like honey bees, cockroaches, termites, bark beetles, and aphids, the more it appears that here also chemical signals are involved. But even in nonsocial insects there is rapidly growing evidence for the presence of chemical markers which are used to communicate with conspecifics. Thus the parasitic wasp *Pseudeucoila bochei*, after depositing an egg in its host, a banana fly larva, also leaves a marking substance which deters conspecific females from ovipositing in the same larva. In this way superparasitization and the consequential competition for food is prevented.[2] The more we learn about the factors which govern insect behavior, the more the overwhelming importance of chemical cues becomes evident. Indeed, it seems that the outside world appears to an insect largely as a world full of chemicals. Insects, therefore, live in a chemical world, which man can appreciate only dimly. This does not mean, of course, that insects are devoid of receptors for physical stimuli, such as visual, magnetic, mechanical, or temperature factors. The fact that apple maggot flies are lured to red wooden spheres, especially when they exceed the size of their host fruits,[3] illustrates the importance

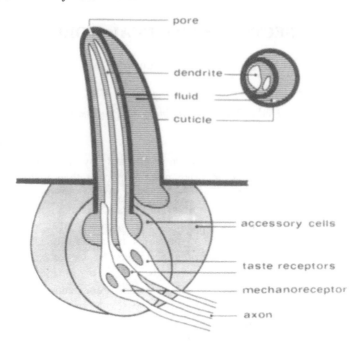

FIGURE 1. Schematic drawing of a longitudinal and a transverse section of an insect taste hair. The sensillum is innervated by two chemoreceptors and one mechanoreceptor. (Courtesy of Dr. F. W. Maes, Groningen State University, The Netherlands.)

of vision in fast-flying insects. Zebras derive protection from tsetse fly attack by their silhouette-disrupting stripes, and many flowers attract pollinating insects because of their colors. These facts demonstrate that vision, especially in diurnal insects, is often of great significance. But even in insects with a highly developed visual system, chemical signals remain to be of paramount importance.[4]

CHEMICAL SENSES

In a number of species, the arthropods represent by far the largest phylum in the animal kingdom. Insects excel not only in number of species and number of individuals, but also in diversity of both structure and behavior. The great variety in feeding habits and the extreme diversity in habitats occupied is reflected in the structure and physiological properties of their chemoreceptors. In insects all chemoreceptors are cuticular structures in which the internal communication system, i.e. the nervous system via its neural extensions, makes direct contact with the animal's outside world.[5,6] As such the chemical senses represent the outcome of two conflicting principles. On the one hand, the cuticle is made to protect the insect's inside against desiccation, toxic compounds, and microorganisms. On the other hand, neurons, created to perceive chemicals in the environment, can only be stimulated if there is a free, albeit very small, passage to them.

Taste sensilla often have the shape of a hair (sensillum trichodeum) or a small peg (sensillum basiconicum). Usually they are innervated by a few neurons. Their number is rarely less than three or more than eight. The cell bodies of these receptors lie in a sub-hypodermal position beneath the hair. The axons of these neurons are believed to extend, without synapsing, to the central nervous system. Their dendrites, on the other side of the cell body, extend to the tip of the sensillum and terminate close to the single apical pore of the hair (Figure 1). Olfactory sensilla are essentially cuticular structures perforated with a

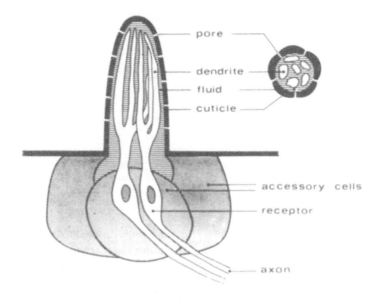

FIGURE 2. Schematic drawing of a longitudinal and a transverse section of an insect olfactory hair. The sensillum is innervated by two bipolar neurons. (Courtesy of Dr. F. W. Maes, Groningen State University, The Netherlands.)

large number of minute pore canals. They often have the shape of a thin circular or oval plate (sensillum placodeum), a short cone (sensillum basiconicum), or a blunt hair (sensillum trichodeum). Olfactory sensilla (Figure 2) are in most cases innervated by a few to several bipolar neurons, although plate organs may have much bigger numbers of neurons, occasionally over 100. The total number of sensilla which an insect bears may vary enormously, depending on species and developmental stage. Thus antennae of drones of the honeybee contain over 300,000 neurons, whereas lepidopterous larvae have on their head only 30 sensilla, innervated by 188 receptory cells in total. It has been suggested that the reduced numbers of sensilla in, e.g., Endopterygota, is related to increased food specialization.[7]

Chemoreceptors have a dual function. They inform the insect about the quality of certain chemicals in its environment, and, in addition, they measure the concentration of the stimulating compounds. To determine the quality of chemical stimuli, insects employ two strategies. One makes use of very specific receptors which ideally react to only one compound. Many pheromone receptors fit into this category of specialized receptors. An extreme specificity is manifested in chemoreceptors which discriminate between different enantiomers of the same compound. Bark beetle *(Ips paraconfusus)* receptors, for instance, show strong reactions to one of its pheromonal components, (S)-(−)-ipsenol, but they are essentially insensitive to its antipode, (R)-(+)-ipsenol (Figure 3).[8] An alternative neural coding strategy is based on receptors with fairly wide stimulus spectra. When different receptors have different though overlapping spectra, a particular stimulus may activate some receptors, whereas another stimulus excites only some of them and/or other neurons. Each stimulus is coded by neural activity in some receptors. This "across-fiber-pattern" coding allows a large number of stimuli to be coded with a limited number of receptors.

The sensitivity of the sense of smell is sometimes very high. Male silk moths, for instance, smell the female sex attractant at a concentration of 200 molecules per ml of air, i.e., one odor molecule among 10^{17} molecules of air. The high sensitivity in this case is due to a large number of specialized receptors on the antennae (34,000) in combination with a very low threshold. Thus only one pheromone molecule suffices, after reaching the dendritic surface, to elicit a nerve impulse.[9] Of course, not all (pheromone) receptors systems show

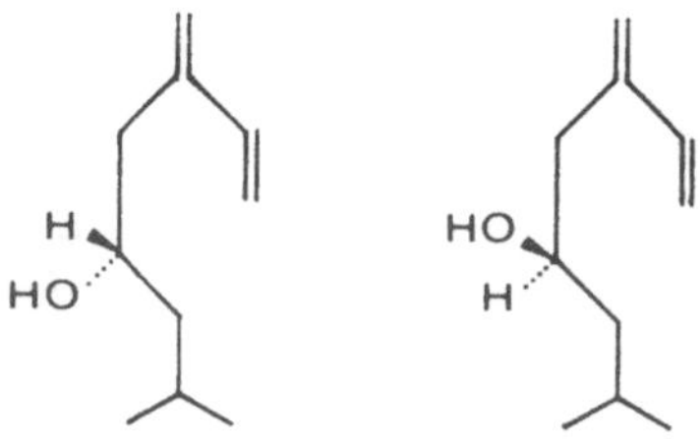

FIGURE 3.   (S)-(−)-ipsenol (left), a component of the aggregation pheromone of *Ips paraconfusus*, and its inactive antipodal enantiomer, (R)-(+)-ipsenol (right).

the same degree of sensitivity. Chemoreceptor thresholds undoubtedly vary in order to function optimally by retrieving only meaningful information from the environment. Too high a sensitivity for certain compounds may be just as bad as too low a sensitivity.

FOOD FINDING

The instinct to secure the right type of food depends in most insects to a large extent on the chemical senses. Only when the chemoreceptors detect a particular configuration of chemical stimuli in the external world which sufficiently matches a central response profile located in the central nervous system, certain elements of feeding behavior may be triggered. Distance perception and orientation towards food sources is often facilitated by olfactory cues. Mosquitoes are attracted by mixtures of lactic acid, carbon dioxide, and other unknown components of smell of the human skin,[10] and starved Colorado potato beetles orient themselves upwind in an airstream containing a combination of leaf-alcohols such as is produced by potato leaves.[11] Odorous compounds, as well as chemicals perceived by contact chemoreceptors, may incite biting movements or other preparatory feeding behavior. Thus when the tarsal receptors of a walking blowfly come into contact with sugar particles, locomotion is arrested and the proboscis is extended, ready to suck up nutritive solutions.[12] Not only the initial phase of food intake, but also continued feeding requires adequate sensory input from taste receptors located on external and internal mouthparts.[13] It has been suggested that receptors in the suboral cavity govern food swallowing, e.g., in lepidopterous larvae.[14]

Many insects show remarkably strong food specializations, putting high demands on the senses involved in food recognition. An analysis of the food range of 4895 phytophagous insects in Great Britain reveals that 52% feed on one plant genus only. Twenty percent are restricted to one plant family (not including grasses), 9% feed on Gramineae and/or Cyperaceae, and 19% of the species accept more than one plant family.[7] These figures, of course, give only a general picture. Certain insect groups are more catholic eaters than others. From the 21 species of British Orthoptera, 41% live on grasses and/or sedges, and the remaining 59% are polyphagous. Among the British aphids, however, 76% of the species are strictly monophagous, and the remaining 24% are predominantly oligophagous, including only 6% which are polyphagous.[15] Food specialization occurs also in other insect groups. For example, 53% out of 528 species which live as external parasites on mammals and birds, are restricted to one host species. Ichneumonid wasps, which parasitize mainly lepidopteran caterpillars, show monophagy in 53% of the 14,800 species known.[16]

Food recognition no doubt depends to a large extent on chemicals. About half of the total number of insect species feed on plants, which synthesize as primary producers the essential nutrients for animals. The diversity of green plants is astounding, especially to the chemist, since they possess a biochemical wealth which can only be grasped in cursory

TABLE 1
Number of Secondary Plant Substances Involved in Plant-Animal Interactions

Alkaloids	5500
Phenolics	1700
Terpenoids	3850
Other	1250
Known compounds	12,300

Note: Data from Harborne, J. B., *Introduction to Ecological Biochemistry*, 2nd ed., Academic Press, Orlando, FL, 1982, 3.

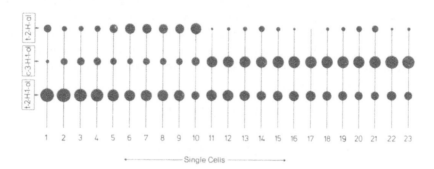

FIGURE 4.   Intensity of neural activity in 23 olfactory receptor cells in the antenna of the Colorado beetle, in response to three volatiles emitted by its host plant. The intensity of the responses to the three individual components, trans-2-hexen-1-ol, cis-3-hexen-1-ol, and trans-2-hexenal, are represented by the sizes of the circles. Different cells show different responses to certain host chemicals. (From Visser, J. H., Differential sensory perceptions of plant compounds by insects, in *Plant Resistance to Insects*, Hedin, P. A., Ed., ACS Symposium Series 208, 1983, ch. 12. With permission.)

terms. The 12,000 or so secondary plant substances presently known (Table 1) is guessed to represent only 3 to 10% of all plant substances occurring in nature. Many of these compounds, either alone or in complex combinations, are used as chemical stimuli, which help insects to differentiate acceptable from unacceptable plants. This does not mean that there is one insect species which is able to perceive all existing plant chemicals. Insects, on the contrary, have developed during evolution a combination of receptors allowing them to distinguish optimally between food plants and nonfood plants. This is accomplished by combining information from receptors with fairly broad sensitivity spectra, such as many olfactory receptors. In the Colorado beetle, plant odor quality is encoded by response patterning across a number of olfactory receptor types (Figure 4).[18] A large number of contact chemoreceptors on the palps of locusts also show limited specificity and there does not seem to be a single peripheral code either for acceptance or rejection.[19] In other insects, however, receptors with a narrow action spectrum have been found. Some gustatory receptors are tuned to certain chemicals, which specifically occur only in the insect's food plant. Thus, larvae of the cabbage white butterfly have in their maxillary taste hairs one neuron which reacts only to glucosinolates, compounds which are typical of their food plants. Stimulation of this particular cell type evokes feeding behavior in a hungry caterpillar, especially when at the same time sugar-sensitive cells and a receptor for amino acids are activated. After biting, the insect sees a chemical profile which should match, at least to a certain degree, a reference pattern in the insect's brain to evoke continued feeding.[20,21] So-called deterrent receptors are stimulated by a number of secondary plant compounds which occur in unacceptable plants. Some deterrents manifest their presence in another way, namely by distorting the normal neural input from feeding stimulants.[22]

Knowledge of the functioning of chemoreceptors in nonherbivorous insects is far less advanced than in phytophages. The information available, however, suggests that in some cases fairly specific sense cells play a role in checking food quality, whereas in other instances the sensory image is coded by the increase (or occasionally decrease) of nerve activity in a large number of receptors.

Narrow-spectrum olfactory cells are abundant on the antennae of adult carrion beetles. Two cell types with overlapping reaction spectra may be distinguished. One is the carrion receptor, which strongly responds to the putrid odor of carrion. It reacts also with excitation to some other compounds, but is inhibited by the ketone cyclo-heptanone. The latter compound, however, excites the other cell type.[23] Lactic acid seems to be the natural stimulus for olfactory receptors in mosquitoes.[24] Several blood-sucking insects respond[25] to low concentrations of ATP and related nucleotides.[26] These substances stimulate receptors on the labellar lobes of tsetse flies.[27] Presumably, these receptors control piercing activity, thus guiding the labium toward ATP-rich sources, e.g., blood cells. Mammalian red blood cells are known to contain appreciable quantities of adenine nucleotides, mostly in the form of ATP. Thus, some chemicals may attract an insect to its food; others incite biting or piercing, and stimulate swallowing or sucking.

REPRODUCTION

Mating and oviposition, items of overwhelming importance on the life agenda of all insects, unfold themselves faultlessly only in the setting of a large array of chemicals. Many small animal species like insects, which often occur at sparse population densities, have perfected their mating partner location by employing pheromones. Their long-range effectiveness and specificity are unexcelled.

The idea that insects produce odors to attract the opposite sex was proposed almost 150 years ago by the zoologist T. von Siebold, but it was not until 1959 that Butenandt and co-workers identified the first insect pheromones, obtained from 500,000 virgin female silkworm moths. Since then sex attractants from hundreds of species have been elucidated at an ever increasing rate, as each issue of the *Journal of Chemical Ecology* shows. General interest in sex pheromones was stimulated because these chemical signals may attract males sometimes over very large distances (several kilometers), although they are produced in small quantities (some nanograms). Moreover, they are highly species-specific. Applied entomologists realized that because of their specificity it might be possible to disturb this communication system and thus control certain pest insects. Therefore, efforts to identify sex pheromones were concentrated especially on noxious insect species.[28]

Although the large distance attraction fascinates our imagination, recent experiments under natural conditions have indicated that calling females normally attract males from ranges of some meters up to a few hundred meters. Pheromones are usually a precise blend of some components. Behavioral studies showed that certain compounds in the attractant mixture may affect precopulatory behavior, such as landing and other close-range elements in the male response sequence.[29] During courtship other species-specific pheromones may be produced, a fact which raises the specificity of the communication channel. Males may produce aphrodisiacs in glandular scent scales or extrusible organs in the abdomen. A male grayling butterfly, for example, clasps the female's antennae between his wings at a certain moment during courtship, thus bringing them into contact with a group of scent scales (Figure 5). Specific behavior patterns, in which species-characteristic chemicals play their crucial role, help to prevent interspecific matings.

The identification of many sex pheromones has made use, in many cases, of behavioral tests of various design. Although such bioassays are in principle the most natural and therefore the most reliable tests, they often require much time, many insects, and a thorough knowledge

FIGURE 5. Grayling male (right) bowing so that the female's antennae come in contact with the scent organ on the forewings of the male. (From Tinbergen, N., Meeuse, B. J. D., Boerema, L. K., and Varossieau, W. W., Die Balz des Samtfalters, *Eumenis* (= *Satyrus*) *semele* (L.), *Z. Tierpsychol.*, 5, 182, 1942.)

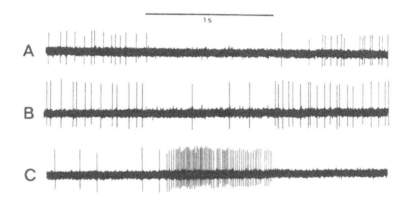

FIGURE 6. Action potentials from an olfactory sensillum on the antenna of a female *Yponomeuta rorellus* on stimulation with three plant odors. All stimulations lasted one second during the period indicated by the upper trace. A: hexanal, B: benzaldehyde, C: linalool. Hexanal completely suppresses nerve activity in the unstimulated sensillum, whereas linalool strongly activates the sensillum. (Modified from Van der Pers, J. N. C., Thomas, G., and Den Otter, C. J., Interactions between plant odors and pheromone reception in small ermine moths (Lepidoptera: Yponomeutidae), *Chem. Senses*, 5, 367, 1980.)

of the insect's behavior under natural conditions. Electrophysiological techniques have come to our aid and proved to be powerful tools which can replace, at least partly, elaborate behavioral tests. The electrophysiological tests make use of the fact that the olfactory receptors attuned to pheromone signals are often located in large numbers on the antennae. When a pheromone molecule enters an olfactory sensillum and interacts with the dendrite of a chemosensitive neuron an electrical signal is generated, which is transmitted to the central nervous system. When a microelectrode is implanted near the sensillum, the electrical signals of individual neurons can be recorded (Figure 6). With this "single-cell" technique, information can be obtained on the specificity of the receptors as well as their sensitivity. This method was successfully used to show that different components of the pheromone may stimulate different cell types.[30] Using the same delicate method it was found that the bark beetle, *Ips pini*, has two types of receptor cells, each of them reacting to a different enantiomer of the aggregation pheromone ipsdienol.[31]

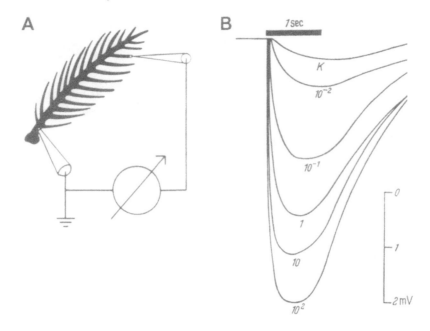

FIGURE 7.   One electrode inserted near the tip and another electrode at the base of an antenna of a male *Bombyx mori* moth; (A) record EAG's elicited by odor puffs. The EAG-responses (B) to an air puff passing a control paper (K) and to a series of papers holding bombykol of increasing concentration (log 10 values in μg) are shown. The stimulus lasts 1 second. (Modified from Boeckh, J., Kaissling, K. E., and Schneider, D., Insect Olfactory receptors, *Cold Spring Harbor Symp. Quant. Biol.*, 30, 263, 1965).

Another electrophysiological method, discovered and developed by J. Boeckh and collaborators,[32] measures the electric potential change between the base and the tip of the antenna when an odor which stimulates a number of receptors is blown over the antenna. The size of the electroantennogram (EAG) is a measure of the stimulating capacity of the odor (Figure 7). Because of its simplicity this method allows rapid and routine determinations, which was found to be of great help identifying compounds with biological activity. Mayer et al.[33] recently made a critical examination of EAG signals with regard to methodology and quantitative aspects. They clarified some of the relationships between the EAG, single cell responses, and behavior.

The ultimate objective of a gravid female is finding a suitable oviposition site. Since neonate larvae usually are unable to move over any considerable distance, it is of utmost importance that oviposition is done in the immediate vicinity of the larval food source. In general, the behavioral chain of events leading to oviposition closely parallels that used in food location. Some phytophagous insects take a test bite before oviposition or even lay their eggs during feeding, as in the case of the potato leafhopper *Empoasca fabae*. But in those species in which the adults do not feed at all or derive their energy from nectar sucked from flowers, such as many lepidopterans and dipterans, host suitability can only be tested on the basis of chemicals present on the host surface or, when the eggs are inserted into the host, by the ovipositor. Olfactory cues are used to select habitats and to locate hosts from a distance. Once a potential host has been found, contact chemoreceptors are used to test its suitability as an oviposition substrate. Female cabbage butterflies, for example, show after landing on a green plant characteristic drumming movements with their forelegs. When their tarsal sensilla (Figure 8) during this behavior are stimulated by glucosinolate molecules, which typically occur in cruciferous plants, the butterfly starts egg laying.[34] Parasitoids likewise use olfactory and contact chemical stimuli to locate hosts for oviposition. The

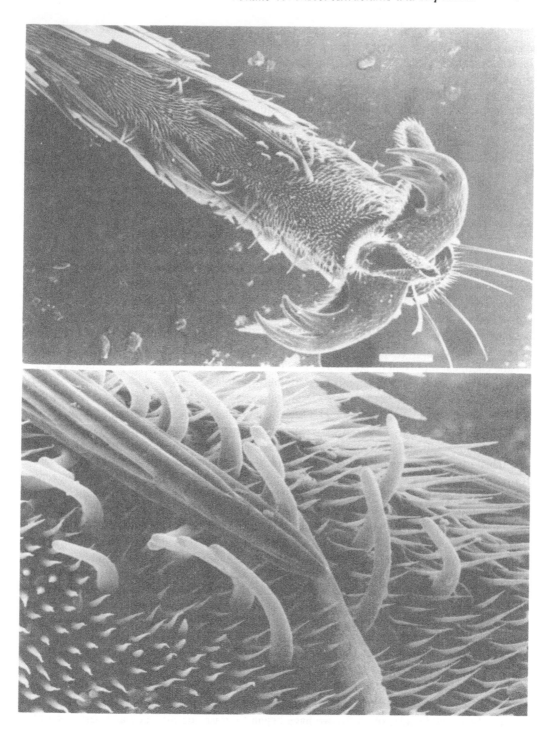

FIGURE 8. Scanning electron micrographs of ventral side of the foretarsi of a female *Pieris brassicae* butterfly. The slightly curved hairs with a blunt tip are contact chemoreceptors which touch the substrate when the insect has landed on a plant leaf. White bar indicates 100 μm for the upper picture and 16 μm for the lower picture.

tachinid *Eucelatoria bryani* is attracted by volatile chemicals emitted by the food plant of its insect host. Subsequently, certain host-specific chemicals act as contact stimulants and have an arrestant effect. Still other substances serve as oviposition stimulants.[35] The ovipositors of a number of insects bear sensilla with a chemoreceptive function. They seem to occur especially in those species which lay their eggs inside their host, such as hymenopteran parasitoids[36] and tephritid fruitflies.[37] These sensilla collect information on host quality, and in the case of some parasitoids, on the presence or absence of other parasitoids.[36]

Thus, chemicals play a crucial role in all elements of reproductive behaviors. Distance, location, and short-range orientation to mating partners often depend on pheromones, and courtship behavior may involve male- or female-produced chemical signals. Search for an oviposition substrate, be it a living organism or dead material, again depends in most cases on the presence of adequate distance as well as contact chemical cues.

COMMUNICATION

Communication is a basic element in the behavior of all Metazoa. Burghardt,[38] after reviewing many definitions of communication and rejecting most of them, arrived at the following statement: "Communication is the phenomenon of one organism producing a signal that, when responded to by another organism, confers some advantage . . . to the signaler or his group."[11] This broad definition not only includes interactions between members of the same species, but encompasses also many animal secretions which are specialized for the chemical repulsion of predators, or plant substances functioning as attractants (e.g., flower scents), or feeding deterrents (e.g., bitter principles).

The chemical senses, as most universal and most supreme senses, play a leading role in communication. Especially in social insects, many intricate behavioral and physiological interactions have recently been shown to depend on a large array of chemical signals.[39] Male bumblebees mark their flightroutes by scent-beacons, using compounds from their mandibular and labial glands. Caterpillars,[40] ants,[41] and termites[42] deposit chemical trails which can last for hours. Honeybees mark the flowers from which they collect nectar. Thereby they prevent other bees from visiting in vain flowers which are temporarily depleted of nectar.[43] Caste determination is regulated by pheromones; alarm and defense strategies in ants and termites are governed by a variety of pheromones, produced in a number of exocrine glands. Assembly and recruitment communication in many social insects is almost entirely chemoreceptive in nature. Fire ants, for example, find one another by carbon dioxide gradients, and honey bee workers, when attempting to attract lost foragers home to the hive or during swarming to a new nesting site, emit from their Nasanov glands a complex pheromone which attracts other workers from considerable distances[44] (Figure 9). Although bark beetles cannot be classified as truly social insects, they do aggregate in large numbers after some pioneer insects have found a susceptible host. The beetles then start to produce a pheromone which, often in combination with volatiles from the tree itself, causes the approach of others of both sexes and from large distances.[45,46]

The development of social life in insects could only arise by the creation of a system of chemical messengers which is of great subtleness and at the same time very efficient. After only a few decades of study, we have begun to grasp the intricacy and perfection of this chemical communication system, based upon the production by the signaler of combinations of chemicals with proper volatilities and, at the receiving side, receptors finely tuned to variations in composite pheromones. Cases are known, for instance, in which a pheromone is a mixture of a few chemicals which vary in volatility. As a consequence of volatility differences, the composition of the pheromone changes with distance (Figure 10). The behavioral responses of the receiving insects vary accordingly.[41]

In nonsocial insects communication between individuals of one species is more common

FIGURE 9. A honeybee worker with its abdomen in a slanted position, showing a recruitment calling behavior with the Nasanov gland exposed and fanning with the wing.

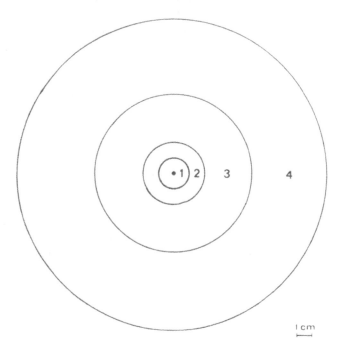

FIGURE 10. When at the central dot the mandibular gland secretion of major workers of the African weaver ant is deposited, the active spaces of 4 of its components after 20 s show different sizes. Component 1 (2-butyl-2-octenal) releases biting responses of conspecifics, component 2 (3-undecanone) releases attraction and biting, component 3 (1-hexanol) is an attractment, and component 4 (hexanal) has an alerting function. (From Bradshaw, J. W. S. and Howse, P. E., Sociochemicals of ants, in *Chemical Ecology of Insects*, Bell, W. J. and Cardé, R. T., Eds., Chapman & Hall, London, 1984, chap. 15.)

than might be expected. We have already referred to communication between mating partners, which serves the achievement of a pairing between two conspecifics. In its extension one finds some cases in which the female pheromone repels other females, and male insects produce a pheromone which is repulsive to other males. Dispersal effects are still more striking in the case of ovipositing females which mark their oviposition sites with a chemical that repels conspecific females from ovipositing at the same place. The use of these so-called epideictic pheromones is well-known in, e.g., parasitoids and phytophages. By acting to discourage or prevent overcrowding of any unit of resource, epideictic pheromones promote a uniformity in spacing among available resources, and thus reduce, for instance, food competition.[47,48] Likewise, some predators leave their chemical marks in areas which have been searched for prey, which helps to optimize foraging behavior.[49]

Territorial behavior in insects is less spectacular than in many vertebrates. Yet every facet of territoriality shown by vertebrates is also found in insects, including scentmarking.[50,51]

The production of defensive secretions[52] with a warning function falls also under Burhardt's broad definition of communication. Interestingly, the borderline between defensive chemicals and pheromones is, indeed, not sharp. In several social Hymenoptera and gregarious Heteroptera, for instance, defensive secretions act as alarm pheromones to conspecifics, causing dispersion or releasing aggressive behavior. In higher hymenopterans they may also alert recruits. An interesting case of tuning in at another's species warning system is exemplified by wild potato plants which emit volatiles which are identical to alarm pheromones of some aphids. These plants thus protect themselves to aphid settlement.[53] Defensive chemicals emitted by larvae of the leafbeetle *Gastrophysa viridula* are repellent to adults of the same species. In this case, the defensive secretion has also an epideictic pheromone function, which helps to prevent overcrowding on single plants.[54]

From the examples given it is clear that specific chemicals may have more than one function. When dealing with combinations of signal chemicals in varying ratios, as no doubt is the case in natural situations, very complicated communication patterns with several possibilities of "crosstalk" will arise. One example of the complexity of such a "communication web" is found in a study by Dixon and Payne.[55] These workers used six volatiles known from the pheromone of the southern pine beetle and its host in various combinations in traps, placed in a beetle-infested wood. Fifteen insect species, known as parasites or predators of the southern pine beetle, were caught along with 13 other associated species (food competitors, mycetophages, scavengers). This "communication web" thus encompasses 29 species (at least), indicating that information networks in nature may be very complex.

An outstanding advantage of chemical signals is that they not only transmit specific information through darkness and around obstacles, but that they also have great energetic efficiency. Most pheromones are chemically fairly simple compounds with molecular weights under 150. They are energetically cheap to synthesize, and usually are produced in minute quantities. Less than a microgram of a moderately simple compound can produce a signal that lasts for hours or even days. A female gypsy moth emits its pheromone at a rate of 13 μg/min for a 30 min period only once in its lifetime. When chemicals are produced in larger quantities, such as defensive secretions in insects or protective allelochemics in plants, the costs involved are, of course, correspondingly higher.

VARIABILITY OF BEHAVIORAL RESPONSES

Chemical and physical stimuli from the external world govern insect behavior. Insects have often been regarded as little automatons which must be stimulated by the right stimulus at the right time to perform certain activities. Behavior is based upon discrete sequences of events: a specific stimulus activates a receptor; nerve impulses travel to the central nervous system (CNS) and trigger a fixed motor reaction pattern, which leads to highly stereotyped behavior. When certain behavioral responses show little or no flexibility, it may be concluded that all elements in the insect's reaction (i.e., the receptors, the CNS, and the motor system) are fixed and unchangeable. Descartes' view that animals are automatic, thoughtless machines may be interpreted as essentially all their behavioral responses are predictable. This position was disputed as long ago as 1906 by Jennings,[56] who wrote: "Such accounts have given rise to a widespread impression that behavior in lower animals differs from that of higher forms in that it is of a fixed, stereotyped character, occurring invariably in the same way under the same external conditions. This impression is in a high degree erroneous . . . In lower as well as in higher animals varied internal conditions and changes are of the greatest importance in determining behavior, the animal by no means behaving always in the same way under the same external conditions." According to this view, differences in behavior under identical stimulus situations have to be attributed to intrinsic differences between individuals. Indeed, systematic variations are known to occur in receptor sensitivity. Also, the CNS may process sensory input in different ways, according to the insects' antecedents, resulting in different behavior in response to the same stimuli. These two sources of variation may influence the reactions of an insect to chemical stimuli.

Larvae of *Manduca sexta* show changes in the sensitivity of specific taste cells when the chemical composition of a semisynthetic diet or of their natural food is varied. The dietary receptor changes develop relatively slowly, occurring over a day or more (of continuous food access), and may reverse on a similar time course.[57] Olfactory receptors, likewise, show alterations in their sensitivity. Antennal receptors of mosquitoes responding to the host-attractant lactic acid show a decreased sensitivity after a blood meal. This change correlates with an inhibition of host-seeking behavior. At the same time olfactory receptors which detect oviposition site attractants show an increased sensitivity after a meal. Apparently, sensory input may change drastically in this insect depending on its nutritional state. Presumably, the altered behavior in a satiated mosquito may be explained, at least partially, to changes in sensory input to the CNS.[24] Some other reports scattered in the literature point to other factors, such as age, food deprivation, diapause, and light-regime, which may influence the sensitivity of chemoreceptors in insects. Some authors hypothesize that humoral factors, possibly hormones, are responsible for the changes observed.[58] Whatever the mechanism causing this phenomenon may be, the studies cited reveal that the primary chemosensory input of an insect is not constant, but may rather vary with the animal's physiological state.

Since Von Frisch's classical experiments with honeybees it is common knowledge that insects can learn. Many students using different insects have shown that behavioral responses may be changed, due to previous experience. Because chemical stimuli constitute the most detailed image of an insect's external world they logically are an important element in learning behavior. Several examples of habituation, the simplest and most universal form of learning, are known. When newly emerged *Drosophila* flies are exposed to an airstream bearing the odor of cedarwood oil, their natural aversion to this odor is transformed into a high degree of tolerance.[59] Some orthopterous and lepidopterous larvae show a gradually increasing acceptance of certain feeding deterrents after repeated exposure. The waning of the response to the inhibitory chemical can properly be regarded as habituation.[60]

Classical conditioning, a phenomenon inseparable from the name of the great Russian physiologist I. P. Pavlov, represents a higher form of learning. No doubt insects of most orders are capable of classical conditioning, albeit that learning ability appears to be relatively poor in, e.g., Diptera in contrast to Hymenoptera. Flies can be conditioned to extend their proboscis when stimulated with certain volatiles, if they are rewarded concomitantly with a sugar meal.[61] Gustatory conditioning too can be induced in flies, and has been used as a behavioral method to determine their capability to discriminate various contact chemicals.[62] Also, under more natural conditions than exposure to single chemicals, learning can be demonstrated. Several hymenopterous parasitoids with a wide host range often prefer for oviposition the host species from which the female has been reared. Such conditioning to host odors can take place either in the preimaginal state, i.e., when the parasitoid larvae is feeding inside the host, or in the adult stage immediately on emergence.[63] Many phytophagous insects develop a predilection for that foodplant on which they have been fed for some time over others previously equally, or more acceptable. When, on the other hand, lepidopterous larvae are made ill by sublethal doses of insecticides they may after recovery show the behaviorally reverse process, i.e., a dislike for the plant on which they were feeding before the treatment. This aversion-learning is probably advantageous to polyphagous insects which may sample potentially toxic plants.[64] In all cases of preference induction or aversion learning in phytophagous insects studied so far the conditioning expresses itself only within the same behavioral activity, i.e., feeding. Oviposition preferences are not influenced by previous feeding experiences. In this respect there seems to exist an hitherto unexplained contrast to parasitoids, where oviposition behavior may be affected by the nutritional history of the insect.

Although behavior is basically a matter of stimulus and response, this does not imply that behavior, even in lower animals, can be explained as an outcome solely or even primarily of simple reflexes. Antecedent events can modify behavioral reactions due to changes at the receptor level, as well as alterations in the processing of sensory input in the CNS.

EVOLUTION

The enormous diversity of life forms on earth is paralleled by a vast diversity of chemicals associated with life. It is true, the unifying concepts of cell biology have recently brought to our attention the common biochemical heritage of all living organisms. At the same time, however, it has become evident that around a number of common chemical processes a wealth of biochemicals has arisen which forms the basis of ecosystems and the cause of competition and symbiosis; it also plays a role in coevolution. Green plants as primary producers especially have evolved the capacity to synthesize many kinds of biochemicals. Animals, directly or indirectly dependent on plants, are confronted with all these chemicals, which often have adverse effects. Animals also protect themselves with chemicals, albeit their creativity in this respect is less impressive than in the case of plants.

No chemosensory system can cope with the whole variety of chemicals present in the biosphere. The small size of insects limits their number of chemoreceptors and the space for integrative units in the brain. Within these confines each species has evolved a (physiologically) unique chemosensory setup tuned to recognizing, for example, its specific range of plant or animal hosts in its particular biotype and the communicative chemicals employed for information exchange with conspecifics. This explains the striking finding that no two insect species seem to possess identical receptor systems. Although from a morphological point of view chemosensory structures often show a high degree of uniformity, especially in related species, their functional features may differ widely.[65] This is nicely illustrated in eight closely related *Yponomeuta* species, all of which have narrow and often different food preferences. These groups are considered to be in the phase of differentation, some members already being taxonomically more diverged than others. All species have different receptors, though some overlap occurs. Adaptation to compounds typical of their host plant often occurs. Thus, all species living on rosaceous host plants have sorbitol-sensitive receptors, in contrast to the other *Yponomeuta* species specialized on nonrosaceous plants.[66]

The evolution of a chemoreceptor may also be studied within one species when different strains are available. One chemoreceptory cell in the maxilla of larvae of *Mamestra brassicae*

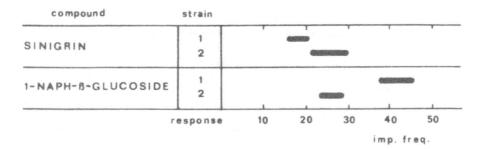

FIGURE 11.   Sensitivity (in impulses per second) of deterrent receptor to standard concentrations of two glycosides in two strains of *Mamestra brassicae*. (Data from Wieczorek, H., The glycoside receptor of the larvae of *Mamestra brassicae* L. (Lepidoptera: Noctuidae), *J. Comp. Physiol.*, A106, 153, 1976.)

detects the presence of a number of glycosides in the larval food. This cell reacts in two strains of this insect to the same compounds, but the responses differ quantitatively. Some glycosides, e.g., sinigrin, stimulate strain I more intensively than strain II, whereas some other glycosides, among them 1-naphtyl-β-glucoside, elicit reversed reactions (Figure 11). This observation may be interpreted by assuming that this glycoside receptor has two types of reaction sites and that the number of each differ for the two strains.[67] It could be conjectured that an evolutionary change of a chemoreceptor proceeds via a gradual change of relative frequencies of certain types of receptor sites. The *Mamestra* receptor exemplifies, in this view, a transient phase. Unfortunately, it is unknown whether the differences between the glycoside receptors of the two strains manifest themselves in behavioral differences. In another case, behavioral difference between two geographically isolated populations of the oriental fruit fly was used to investigate changes in olfactory receptors. It was concluded that perceptible evolution of the receptor protein has occurred during a period of 50 years.[68] The fact that European corn borer moths show polymorphism in pheromone composition and responses also indicates the presence of a fairly flexible chemoreceptory system.[69] Thus, it is not uncommon to find differences in sensory response intensities when individuals obtained from different populations are studied. EAGs from the Colorado beetle, which reflect total responses of the insect's olfactory receptors, differ when American and European insects are compared to each other.[70]

Plants, to herbivores a source of food, to predators and parasitoids places of shelter and hunting-grounds, produce by virtue of the ingenuity of their chemical synthesis apparatus a bewildering array of chemicals. Insects, aided by a great evolutionary flexibility of their chemoreceptors, have evolved into a very species-rich group. Each species developed a receptor system specialized to detect only those chemicals which are relevant for its maintenance. Each insect species lives in its own chemical world. And, in view of the high degree of insect adaptability, the chemical world of a species is not a fixed and unchangeable one, but may rather alter fairly rapidly.

SEMIOCHEMICALS — TERMINOLOGY

Chemical technology has developed during the past decades sufficiently to reveal the existence of numerous chemical messengers which guide the behaviors of insects. When analyzing this chemical world it is helpful to systematize signal compounds or "semiochemicals" (term proposed by Law and Regnier[71]) according to their function in the behavior of the producer and/or the receiving organism. Substances used in communication between members of the same species have been recognized by Bethe,[72] who designated them as "ectohormones". It was not until 1959 when ectohormones were rebaptized with the currently common term "pheromones".[73] These were defined as "substances that are secreted

TABLE 2
Some Definitions of Chemical Signals

Semiochemicals:	A chemical involved in the interaction between individuals of the same or different species.
Pheromone:	An externally secreted substance that produces a specific reaction in a receiving individual of the same species.
Allelochemic:	A substance that affects organisms of a species different from the emitting organism, for reasons other than food as such.
Allomone:	A substance produced by an organism that evokes in the receiving organism of another species a reaction that is favorable to the emitter but not to the receiver.
Kairomone:	A substance produced by an organism that evokes in the receiving organism of another species a reaction that is favorable to the receiver, but not to the emitter.

From Nordlund, D. A., *Semiochemicals: Their Role in Pest Control*, Nordlund, D. A., Jones, R. L., and Lewis, W. J., Eds., John Wiley & Sons, New York, 1981, chap. 2.

by an individual to the outside and cause a specific reaction in a receiving individual of the same species, for example, a release of certain behavior or a determination of physiologic development.'' Subclassification may be done on the basis of the type of response elicited, such as sex pheromone, alarm pheromone, trail pheromone, aggregation pheromone, or epideictic pheromone. Chemical communication of a high degree of sophistication also occurs in exchanges between species that are closely adapted to each other, for instance between symbionts or between predators and their prey (Table 2). Semiochemicals which evoke reactions in another species, and because of this are of adaptive advantage to the organism sending them, have been called "allomones". "Kairomones", on the other hand, are of adaptive value to the organism receiving them. Thus, for instance, allomones serve plants as defense substances against phytophages and protect insects from predators. A cunning adaptation is found in insects which use (or rather misuse) an organism's semiochemicals as cues to recognize the emitter as an acceptable or even preferred food source. Glucosinolates in cruciferous plants have been evolved as protecting chemicals against insect attack and infestation by microorganisms. The cabbage aphid, however, is stimulated to feed when tasting glucosinolates.[74] The bark beetle sex pheromone brevicomin is used by a predatory insect species to locate its prey.[75,76] These examples show that allomones and sex pheromones may at the same time function as kairomones. Because of this overlap of terms it has been suggested in the literature that the term kairomone be abandoned.[77] Disposal of the term, however, does not remove the overlap in functions. This complexity illustrates the fact that chemical signals often form elements in an intricate communication web involving many species of plants, insects, and other animals. We are truly dealing with chemical ecology with many participants, each in its turn being the winner or loser.

The term pheromone was coined at the time when the structure of bombykol, the sex attractant of the silkworm moth, was discovered. As in the case of hormones, insect pheromones were in the beginning presumed to be single compounds. It soon became clear that a pheromone may consist of several compounds at fairly specific ratios. In some cases they even appear to be very complex blends which may vary between individuals. For example, the mandibular gland of the weaver ant contains at least 33 chemicals in amounts greater than 5 ng and at least 4 of these are used in alarm communication behavior.[78] Of course, the mixtures may vary between individuals. The question arose whether the term pheromone should be restricted to pure chemicals, or that each natural (i.e., variable) blend should be considered as a pheromone. It is now generally accepted that the term pheromone relates to the natural blend.[79]

A succinct review of the terminology of semiochemicals is given by Nordlund.[80]

SEMIOCHEMICALS AND PEST CONTROL

The decoding of the chemical communication system in insects began in 1959 with the identification of the first sex pheromone. During the following relatively short period of 25 years, not only the structures of many hundreds of semiochemicals were resolved, but other properties as well. Thus it appears that pheromones and other semiochemicals are usually multicomponent systems which may show geographical and temporal variations in composition and production rate. They may interact with other signals, also affect other species, etc.

Now that we have started to unravel the insect's communication network, one may logically expect that this knowledge will be of considerable help in the control of pest species. Nature provides numerous examples of insects which tune in on some other species' pheromonal communication channel, or use protective allomones in plants to unmask them as adequate food sources. When insects during a long period of evolution have found ways to interfere with the intended function of a communication system and have learned to use some chemicals to their own advantage, we must also be able to intrude the system and to manipulate the insect's behavior with chemicals mimicking the signals from its natural environment.

Although a generalized concept of the role of chemical communication has been attained, this does not mean that straightforward and simple solutions are available to handle specific problems. This is due to the fact that the great diversity of form and life habits of insects is based on an equally great diversity of relationships, communication systems, and variations in behavior. Therefore, each specific case requires a thorough biological and chemical analysis before control measures based on disruption of chemical communication channels are to be attempted. Some examples of the successful application of our cognition of the crucial role of chemical signals seem promising for important developments. In Florida the Mediterranean fruit fly was eradicated by employing baits which contained food attractants laced with a quick-acting insecticide.[81] In California a study was conducted to test potentialities of sex pheromones. Here synthetic pheromones attractive to the western pine beetle were placed in traps distributed in an area of twenty-five square miles. They captured approximately 730,000 beetles. It had been predicted on the basis of extensive population studies and aerial photographic surveys that 750,000 beetles would develop that year. From these figures it was concluded that the pheromone traps had caught 97 percent of the population.[82] In addition to mass trapping as performed with western pine beetles, sex attractants have successfully been used to monitor the appearance and growth of populations. In another type of application, an insect population is confronted with unnaturally high concentrations of their synthetic pheromone which interfere with the normal process of female localization by males. This method is known as confusing or mating disruption technique.

The use of oviposition-deterring pheromones is another promising example. When some cherry trees were sprayed in a dry year with the oviposition-deterring pheromone of the cherry fruit fly, larval infestation was reduced by 85 to 90%.[83]

Host pheromones are important cues to parasitoids. Lewis and co-workers could inhibit the drive to disperse, which laboratory reared parasitoids show after their release, by exposing them during the release procedure to host chemicals. Also, search activity in parasitoids present in the wild can be stimulated by exposing them (under field conditions) to slow-release sources of synthetic chemicals mimicking chemicals present in the ovipositor gland of their hosts. As a result, parasitization rates could be raised significantly.[84]

The examples given allow a moderate optimism about the potentialities of these methods. Many recent books profess this optimism, e.g., those by Mitchell,[85] Nordlund et al.,[86] Kydonieus and Beroza,[87] and Whitehead and Bowers.[88] There is general agreement, however, that many fundamental biological questions still await elucidation, and that many technical problems are to be solved. For example, pheromones are generally produced by the insect only when needed and they are often very volatile. Long-lasting interferences will require methods allowing the controlled release of our chemicals.

The near past has shown that insects may rapidly adapt to the presence of insecticides and become resistant. Is the same to be expected to control measures based on manipulations with signal substances? When insects are considered as stimulus-bound reflex machines, their adaptability is by definition small. But, although individuals show fixed reaction patterns, the flexibility of the species is considerable. Insects have in the past often been found (though not always) to adapt to our control measures. On the other hand, it may be inferred that it is much more difficult (though certainly not *a priori* impossible) for an insect to counteract manipulations with natural or species-specific chemicals than to escape, e.g., poisoning by man-made insecticides. The development of crop plants resistant to attack by insects proves that insects, at least for some time, may be influenced by alterations in the chemical signals of plants, and change their behavior accordingly.

Antifeedant substances derived from plants unedible to most insects offer several potential crop protectants, and scientists in various parts of the world investigate substances produced by the neem tree[89,90] and several other plant species.[22]

Entomologists are ready to accept the challenge and to apply our insights on the function of semiochemicals in nature to control insects. In the words of Lewis[91]: "The challenge of deciphering (the role of chemical mediators) and devising effective approaches for their employment will require zeal, innovativeness, persistence, and long-term monetary investments. The result will be precise, specific, ecologically sound, and effective techniques essential for dealing with the pest control problems of the future." We may add that this type of strategic research should be fed by a constant input of basic knowledge and acumen to be gained by the concerted efforts of biologists and chemists.

REFERENCES

1. **Koshland, D. E.**, *Bacterial Chemotaxis as a Model Behavioral System*, Raven Press, New York, 1980.
2. **Van Lenteren, J. C.**, The development of host discrimination and the prevention of superparasitism in the parasite *Pseudeucoila bochei* (Hym., Cynipidae), *Neth. J. Zool.*, 26, 1, 1976.
3. **Prokopy, R. J.**, Attraction of *Rhagoletis* flies (Diptera; Tephritidae) to red spheres of different sizes, *Can. Entomol.*, 109, 593, 1977.
4. **Bell, W. J. and Cardé, R. T., Eds.**, *Chemical Ecology of Insects*, Chapman & Hall, London, 1984.
5. **Altner, H. and Prillinger, L.**, Ultrastructure of invertebrate chemo-, thermo-, and hygroreceptors and its functional significance, *Int. Rev. Cytol.*, 67, 69, 1980.
6. **Zacharuk, R. Y.**, Ultrastructure and function of insect chemosensilla, *Annu. Rev. Entomol.*, 25, 27, 1980.
7. **Chapman, R. F.**, Chemoreception: The significance of sensillum numbers, *Adv. Insect Physiol.*, 16, 247, 1982.
8. **Light, D. M.**, Sensitivity of antennae of male and female *Ips paraconfusus* (Coleoptera: Scolytidae) to their natural aggregation pheromone and its enantiomeric components, *J. Chem. Ecol.*, 9, 561, 1983.
9. **Kaissling, K. E. and Priesner, E.**, Die Riechschwelle des Seidenspinners, *Naturwissenschaften*, 57, 23, 1970.
10. **Schreck, C. E., Smith, N., Carlson, D. A., Price, G. D., Haile, D., and Godwin, D. R.**, A material isolated from human hands that attracts female mosquitoes, *J. Chem. Ecol.*, 8, 429, 1981.
11. **Visser, J. H. and Avé, D. A.**, General green leaf volatiles in the olfactory orientation of the Colorado beetle, *Leptinotarsa decemlineata*, *Entomol. Exp. Appl.*, 24, 738, 1978.
12. **Dethier, V. G.**, *The Hungry Fly*, Harvard University Press, Cambridge, 1975, chap. 13.
13. **Städler, E.**, Contact chemoreception, in *Chemical Ecology of Insects*, Bell, W. J. and Cardé, R. T., Eds., Chapman & Hall, London, 1984, chap. 1.
14. **Ma, W. C.**, Dynamics of feeding responses in *Pieris brassicae* (Linn.) as a function of chemosensory input: a behavioural, ultrastructural and electrophysiological study, *Meded. Landbouwhogesch. Wageningen*, 72, 1, 1972.

15. **Hodkinson, I. D. and Hughes, M. K.,** *Insect Herbivory,* Chapman & Hall, London, 1982, chap. 1.

16. **Price, P. W.,** *Parasites,* Princeton University Press, Princeton, 1980, chap. 5.

17. **Harborne, J. B.,** *Introduction to Ecological Biochemistry,* 2nd ed., Academic Press, Orlando, FL, 1982, 3.

18. **Ma, W. C. and Visser, J. H.,** Single unit analysis of odour quality coding by the olfactory antennal receptor system of the Colorado beetle, *Entomol. Exp. Appl.,* 24, 521, 1978.

19. **Blaney, W. M.,** Chemoreception and food selection by locusts, *Olfaction and Taste,* 7, 127, 1980.

20. **Schoonhoven, L. M.,** Plant recognition by lepidopterous larvae, *Symp. R. Entomol. Soc. London,* 6, 87, 1973.

21. **Schoonhoven, L. M.,** On the individuality of insect feeding behaviour, *Proc. K. Ned. Akad. Wet.,* Amsterdam, C80, 341, 1977.

22. **Schoonhoven, L. M.,** Biological aspects of antifeedants, *Entomol. Exp. Appl.,* 31, 57, 1982.

23. **Boeckh. J.,** Elektrophysiologische Untersuchungen an einzelnen Geruchsrezeptoren auf der Antenne des Totengräbers (*Necrophorus,* Coleoptera), *Z.Vgl. Physiol.,* 46, 212, 1962.

24. **Davis, E. E. and Takahashi, F. T.,** Humoral alteration of chemoreceptor sensitivity in the mosquito, *Olfaction and Taste,* 7, 139, 1980.

25. **Holscher, K. H., Barker, R. W., and Staats, J. J.,** Selected bibliography of sensory systems in hematophagous arthropods, using electrophysiological methods, *Bull. Entomol. Soc. Am.,* 28, 27, 1982.

26. **Friend, W. G. and Stoffolano, J. G.,** Feeding responses of the horsefly, *Tabanus nigrovittatus,* to phagostimulants, *Physiol. Entomol.,* 8, 377, 1983.

27. **Mitchell, B. K.,** Physiology of an ATP receptor in labellar sensilla of the tsetse fly *Glossina morsitans morsitans* Westwood (Diptera: Glossinidae), *J. Exp. Biol.,* 65, 259, 1976.

28. **Minks, A. K.,** Attractants and pheromones of noxious insects (selected references), *SROP/WPRS Bull.,* 1, 1, 1984.

29. **Baker, T. C. and Cardé, R. T.,** Analysis of pheromone-mediated behaviors in male *Grapholitha molesta,* the Oriental fruit moth (*Lepidoptera: Tortricidae*), *Environ. Entomol.,* 9, 956, 1979.

30. **Den Otter, C. J.,** Single sensillum responses in the male moth *Adoxophyes orana* (F.v.R.) to female sex pheromone components and their geometrical isomers, *J. Comp. Physiol.,* 121, 205, 1977.

31. **Mustaparta, H.,** Olfaction, in *Chemical Ecology of Insects,* Bell, W. J. and Cardé, R. T., Eds., Chapman & Hall, London, 1984, chap. 2.

32. **Boeckh, J., Kaissling, K. E., and Schneider, D.,** Insect olfactory receptors, *Cold Spring Harbor Symp. Quant. Biol.,* 30, 263, 1965.

33. **Mayer, M. S., Mankin, R. W., and Lemire, G. F.,** Quantitation of the insect electro-antennogram: measurement of sensillar contributions, elimination of background potentials, and relationship to olfactory sensation, *J. Insect Physiol.,* 30, 757, 1984.

34. **Ma, W. C. and Schoonhoven, L. M.,** Tarsal contact chemosensory hairs of the large white butterfly, *Pieris brassicae* and their possible role in oviposition behaviour, *Entomol. Exp. Appl.,* 16, 343, 1973.

35. **Nettles, W. C.,** Contact stimulants from *Heliothis virescens* that influence the behavior of females of the tachinid, *Eucelatoria bryani, J. Chem. Ecol.,* 8, 1183, 1982.

36. **Van Lenteren, J. C.,** Host discrimination by parasitoids, in *Semiochemicals, Their Role in Pest Control,* Nordlund, D. A., Jones, R. L., and Lewis, W. J., Eds., John Wiley & Sons, New York, 1981, chap. 9.

37. **Szentesi, A., Greany, P. D., and Chambers, D. L.,** Oviposition behavior of laboratory-reared and wild Caribbean fruit flies (*Anastrepha suspensa*; Diptera: Tephritidae): I. Selected chemical influences, *Entomol. Exp. Appl.,* 25, 227, 1979.

38. **Burghardt, G. M.,** Defining "communication", in *Communication by Chemical Signals,* Johnston, J. W., Moulton, D. G., and Turk, A., Eds., Appleton-Century-Crofts, New York, 1970, chap. 2.

39. **Hölldobler, B.,** Communication in social Hymenoptera, in *How Animals Communicate,* Sebeok, T. A., Ed., Indiana University Press, Bloomington, 1977, chap. 19.

40. **Fitzgerald, T. F. and Peterson, S. C.,** Elective recruitment by the Eastern tent caterpillar (*Malacosoma americanum*), *Anim. Behav.,* 31, 417, 1983.

41. **Bradshaw, J. W. S. and Howse, P. E.,** Sociochemicals of ants, in *Chemical Ecology of Insects,* Bell, W. J. and Cardé, R. T., Eds., Chapman & Hall, London, 1984, chap. 15.

42. **MacFarlane, J.,** Observations on trail pheromone, trail laying and longevity of natural trails in the termite, *Macrotermes michaelseni, Insect Sci. Appl.,* 4, 309, 1983.

43. **Wilhelm, K. T. and Pflumm, W. W.,** Ueber den Einfluss blumenhafter Düfte auf das Duftmarkieren der Sammelbiene, *Apidologie,* 14, 183, 1983.

44. **Wilson, E. O.,** *Sociobiology, the New Synthesis,* Harvard University Press, Cambridge, 1975, chap. 20.

45. **Borden, J. H.,** Behavioral responses of Coleoptera to pheromones, allomones, and kairomones, in *Chemical Control of Insect Behavior,* Shorey, H. H. and McKelvey, J. J., Eds., Wiley & Sons, New York, 1977, chap. 11.

46. **Birch, M. C.,** Aggregation in bark beetles, in *Chemical Ecology of Insects,* Bell, W. J. and Cardé, R. T., Eds., Chapman & Hall, London, 1984, chap. 12.

47. **Prokopy, R. J.,** Epideictic pheromones that influence spacing patterns of phytophagous insects, in *Semiochemicals: Their Role in Pest Control,* Nordlund, D. A., Jones, R. L., and Lewis, W. J., Eds., John Wiley & Sons, New York, 1981, chap. 10.

48. **Roitberg, B. D., Cairl, R. S., and Prokopy, R. J.,** Oviposition deterring pheromone influences dispersal distance in tephritid fruit flies, *Entomol. Exp. Appl.,* 35, 217, 1984.

49. **Hislop, R. G. and Prokopy, R. J.,** Mite predator responses to prey and predator: emitted stimuli, *J. Chem. Ecol.,* 7, 895, 1981.

50. **Baker, R. R.,** Insect territoriality, *Annu. Rev. Entomol.,* 28, 65, 1983.

51. **Vinson, S. B.,** Parasitoid-host relationship, in *Chemical Ecology of Insects,* Bell, W. J. and Cardé, R. T., Eds., Chapman & Hall, London, 1984, chap. 8.

52. **Prestwich, G. D.,** Defense mechanisms of termites, *Annu. Rev. Entomol,* 29, 201, 1984.

53. **Gibson, R. W. and Pickett, J. A.,** Wild potato repels aphids by release of aphid alarm pheromone, *Nature,* 302, 608, 1983.

54. **Pasteels, J. M., Grégoire, J. C., and Rowell-Rahier, M.,** The chemical ecology of defense in arthropods, *Annu. Rev. Entomol.,* 28, 263, 1983.

55. **Dixon, W. N. and Payne, T. L.,** Attraction of entomophagous and associate insects of the Southern pine beetle to beetle- and host tree-produced volatiles, *J. Georgia Entomol. Soc.,* 15, 378, 1980.

56. **Jennings, H. S.,** *Behavior of the Lower Organisms,* 2nd Ed., Indiana University Press, Bloomington, 1962, 237.

57. **Schoonhoven, L. M.,** Sensitivity changes in some insect chemoreceptors and their effect on food selection behaviour, *Proc. K. Ned. Akad. Wet.,* Amsterdam, C72, 491, 1969.

58. **Angioy, A. M., Liscia, A., Crnjar, R., and Pietra, P.,** An endocrine control mechanism for chemosensillar activity in the blowfly, *Experientia,* 39, 545, 1983.

59. **Thorpe, W. H.,** Further experiments on pre-imaginal conditioning in insects, *Proc. R. Soc.,* B127, 424, 1939.

60. **Jermy, T., Bernays, E. A., and Szentesi, A.,** The effect of repeated exposure to feeding deterrents on their acceptability to phytophagous insects, in *Proc. 5th Int. Symp. Insect-Plant Relationships,* Visser, J. H. and Minks, A. K., Eds., Pudoc, Wageningen, The Netherlands, 1982, 25.

61. **Fukushi, T.,** Properties of olfactory conditioning in the housefly, *Musca domestica, J. Insect Physiol.,* 25, 155, 1979.

62. **Maes, F. W. and Bijpost, S. C. A.,** Classical conditioning reveals discrimination of salt taste quality in the blowfly *Calliphora vicina, J. Comp. Physiol.* A133, 53, 1979.

63. **Vinson, S. B.,** Host selection by insect parasitoids, *Annu. Rev. Entomol.,* 21, 109, 1976.

64. **Dethier, V. G.,** Food-aversion learning in two polyphagous caterpillars, *Diacrisia virginica* and *Estigmene congrua, Physiol. Entomol.* 5, 321, 1980.

65. **Schoonhoven, L. M.,** Chemical mediators between plants and phytophagous insects, in *Semiochemicals, Their Role in Pest Control,* Nordlund, D. A., Jones, R. L. and Lewis, W. J., Eds., John Wiley & Sons, New York, 1981, chap. 3.

66. **Van Drongelen, W.,** Contact chemoreception of host plant specific chemicals in the larvae of various *Yponomeuta* species (Lepidoptera), *J. Comp. Physiol.,* A134, 265, 1979.

67. **Wieczorek, H.,** The glycoside receptor of the larvae of *Mamestra brassicae* L. (Lepidoptera: Noctuidae), *J. Comp. Physiol.* A106, 153, 1976.

68. **Metcalf, R. L., Metcalf, E. R., Mitchell, W. C., and Lee, L. W. Y.,** Evolution of olfactory receptor in oriental fruit fly *Dacus dorsalis, Proc. Natl. Acad. Sci. U.S.A.,* 76, 1561, 1979.

69. **Anglade, P., Stockel, J., and I. W. G. O. Cooperators,** intraspecific sex-pheromone variability in the European corn borer, *Ostrinia nubilalis* Hbn. (Lepidoptera, Pyralidae), *Agronomie,* 4, 183, 1984.

70. **Visser, J. H.,** Differential sensory perceptions of plant compounds by insects, in *Plant Resistance to Insects,* Hedin, P. A., Ed., *ACS Symp.,* 208, 1983, 215.

71. **Law, J. H. and Regnier, F. W.,** Pheromones, *Annu. Rev. Biochem.,* 40, 533, 1971.

72. **Bethe, A.,** Vernachlassigte Hormone, *Naturwissenschaften,* 20, 177, 1932.

73. **Karlson, P. and Butenandt, A.,** Pheromones (ectohormones) in insects, *Annu. Rev. Entomol.,* 4, 39, 1959.

74. **Wensler, R.,** Mode of host selection by an aphid, *Nature,* 195, 830, 1962.

75. **Vité, J. P. and Williamson, D. L.,** *Thanasimus dubius:* prey perception, *J. Insect Physiol,* 16, 233, 1970.

76. **Hansen, K.,** Reception of bark beetle pheromone in the predaceous clerid beetle, *Thanasimus formicarius* (Coleoptera: Cleridae), *J. Comp. Physiol.,* A153, 371, 1983.

77. **Pasteels, J. M.,** Is kairomone a valid and useful term?, *J. Chem. Ecol.,* 8, 1079, 1982.

78. **Bradshaw, J. W. S., Baker, R., and Howse, P. E.,** Multicomponent alarm pheromones in the mandibular glands of major workers of the African Weaver ant, *Oecophylla longinoda, Physiol. Entomol.,* 4, 15, 1979.

79. **Katz, R. H. and Shorey, H. H.,** In defense of the term ''pheromone'', *J. Chem. Ecol.,* 5, 299, 1979.

80. **Nordlund, D. A.,** Semiochemicals: a review of the terminology, in *Semiochemicals: Their Role in Pest Control,* Nordlund, D. A., Jones, R. L., and Lewis, W. J., Eds., John Wiley & Sons, New York, 1981, chap. 2.

81. **Steiner, L. F., Rohwer, G. G., Ayers, E. L., and Christenson, L. D.,** The role of attractants in the recent Mediterranean fruit fly eradication program in Florida, *J. Econ. Entomol.* 54, 30, 1961.

82. **Bedard, W. D. and Wood, D. L.,** Programs utilizing pheromones in survey and control: Bark beetles — the western pine beetle, in *Pheromones,* M. C. Birch, Ed., North-Holland, Amsterdam, 1974.

83. **Katsoyannos, B. I. and Boller, E. F.,** First field application of oviposition-deterring pheromone of European cherry fruit fly, *Rhagoletis cerasi, Environ. Entomol.,* 5, 151, 1976.

84. **Lewis, W. J., Nordlund, D. A., and Gueldner, R. C.,** Semiochemicals influencing behaviour of entomophages: roles and strategies for their employment in pest control, in *Les Mediateurs Chimiques,* INRA Publ., 225, 1982.

85. **Mitchell, E. R., Ed.,** *Management of Insect Pests with Semiochemicals,* Plenum Press, New York, 1981.

86. **Norlund, D. A., Jones, R. L., and Lewis, W. J., Eds.,** *Semiochemicals: Their Role in Pest Control,* John Wiley & Sons, New York, 1981.

87. **Kydonieus, A. F. and Beroza, M., Eds.,** *Insect Suppression with Controlled Release Pheromone Systems,* 2 Vols., CRC Press, Boca Raton, FL, 1982.

88. **Whitehead, D. L. and Bowers, W. S., Eds.,** *Natural Products for Innovative Pest Management,* Pergamon Press, Oxford, 1983.

89. **Schmutter, H., Ascher, K. R. S., and Rembold, H., Eds.,** *Natural Pesticides from the Neem Tree* (Azadirachta Indica *A. Juss*), Proc. 1st Int. Neem Conf., German Agency for Technical Cooperation (GTZ), Eschborn, 1981.

90. **Schmutter, H. and Ascher, K. R. S.,** Natural Pesticides from the Neem tree and other Tropical Plants, *Proc. 2nd Int. Neem Conf.,* German Agency for Technical Cooperation (GTZ), Eschborn, 1984. 1984 (in press).

91. **Lewis, W. J.,** Semiochemicals: their role with changing approaches to pest control, in *Semiochemicals: Their Role in Pest Control,* Norlund, D. A., Jones, R. L., and Lewis, W. J., Eds., John Wiley & Sons, New York, 1981, 11.

INSECT FEEDING DETERRENTS

J. David Warthen, Jr. and E. David Morgan

INTRODUCTION TO PARTS A AND B

Insect feeding deterrents appear to be one of three groups of compounds from natural sources that have immediate prospects of application for insect pest management; the other two groups are the volatile sexual and aggregating pheromones that already have limited commercial application. Recognizing the prospects, scientists have carried out a great deal of exploratory work on insect feeding deterrents in the last two decades. The kinds of biologically active substances investigated are as varied as the types of substances elaborated by higher plants, from simple monoterpenes to complex glucosides and alkaloids.

Because of the size of the subject and the difficulty in summarizing it by similar groups of compounds, we have attempted a comprehensive listing by first dividing it into two time periods: 1976-1980 and 1980-1987. In these time periods, we have listed the deterrents in Tables 1 and 4 with biological data, where possible, and chemical structures of active substances, and then cross referenced the latter two tables in Tables 2 and 5 by insect order and scientific name. In Table 3, we list spectroscopic and molecular data with references up to 1980; from 1980 onward, these data are nearly always found in the references quoted in Table 4 or in the 5th edition of *Dictionary of Organic Compounds* and its annual supplementary volumes. In Table 6, we list the plant division and family from which the compounds in Tables 1 and 4 originated.

We have tried, where possible in Tables 1 and 4, to give some indication of the level of biological activity with either a natural or synthetic diet. The type of test may vary between reports but typically a choice test between treated and control leaves, paper discs, or blocks of synthetic diet is used; sometimes an absolute measure of amount of feeding in a no-choice test is recorded.

In recent years, the activity and discoveries in the area of insect feeding deterrents have continued, though surprisingly, the expected advance towards practical use of some of these substances by agrochemical companies, more accustomed to the economics, selling, and practical application of conventional synthetic pesticides, have been slow and very cautious. However, some hopeful indicators were emerging as this survey of the literature was being completed.

In almost every case, the name of a compound when it appears in all of the Tables, is followed by an alphanumeric code (A1, A2, etc.) which refers to the chemical structure of that compound. These structures can be found following Tables 1 and 4.

PART A: INSECT FEEDING DETERRENTS (1976-1980)

J. David Warthen, Jr.

Many natural products from plants possess insect feeding deterrent activity. These feeding deterrents for insects appear to be a possible alternative, when used alone or as part of an integrated pest management system, to the use of insecticides for control of many insect pests.

Natural deterrents probably play a role as resistance factors within the plant itself.[1] Such chemicals that give an adaptive advantage to the organism producing them are called allomones. These allomones or allelochemicals may act against insects in many ways, particularly as (1) repellents, actively repelling insects away from the plant; (2) suppressants, inhibiting the initiation of feeding upon the plant; (3) deterrents, deterring the continuation of feeding; (4) antibiotics, interfering metabolically with normal growth and development; and (5) anorexigenics, producing a loss of appetite.[2] Repellents and suppressants are quite often classified as rejectants. To add to the confusion, repellents may be olfactory or gustatory. Olfactory repellents in the vapor phase stimulate olfactory receptors and drive the insect away from the treated material. Gustatory repellents act upon receptors that are not normally receptive to vapors but are sensitive to food stimuli.[3]

The first three categories in the literature — namely repellents, suppressants, and deterrents — often describe compounds that cause feeding inhibition or antifeedant activity. However, until a particular active compound is assigned to one of the five categories through elaborate bioassays, it is usually referred to as a feeding deterrent or an antifeedant. An antifeedant was described by Munakata[4] as a chemical that inhibits feeding but does not kill the insect directly. The insect remains near the treated plant and dies from starvation. Kubo and Nakanishi[5] state that the feeding inhibition caused by an antifeedant may be temporary or permanent. Perhaps the best definition of deterrent is that proposed by Dethier:[6] a chemical which inhibits feeding or oviposition when present in a place where insects would, in its absence, feed or oviposit.

Host selection by phytophagous insects is governed by the presence of attractants and feeding deterrents in plants.[4] Quite often, when pure deterrent compounds are bioassayed individually at the same concentration at which they occur in a plant, they are less active than the plant itself. Such substances are always deterrent, but at high doses. However, when all individual components are combined, there is an additive effect to produce a deterrence comparable to the effect produced by the plant itself.[7]

I have attempted to cover the literature of 1976-1980 in this review on insect feeding deterrents. Prior to 1976, one must consider the reviews of Chapman and Bernays,[8] Munakata,[9] Chapman,[10] Hedin et al.,[11] and Wright.[12] One must also not overlook the more recent work of Schoonhoven and Jermy[13] on *"A Behavioural and Electrophysiological Analysis of Insect Feeding Deterrents"*, and the work of Ascher[14] on *"Fifteen Years (1963-1978) of Organotin Anti-feedants — A Chronological Bibliography."*

Insect feeding deterrents, plant sources, insect scientific names, and doses for deterrent action are listed in Table 1; selected deterrents have alphanumeric codes for structures in Figure 1. For cross reference, insect orders and scientific names, insect common names, feeding deterrents, and alphanumeric codes for structures in Figure 1 are listed in Table 2. Some of the natural feeding deterrents in the tables have no specified plant source in the literature. Other deterrents in the tables are synthetics or not of plant origin and also have no plant source. Physical constants and sources of spectral data of selected feeding deterrents appear in Table 3.

Of all the feeding deterrents listed in Table 1, warburganal, muzigadial, and azadirachtin generally appear to be the most potent insect antifeedants known.[15,16,17] Warburganal and muzigadial from *Warburgia* species are active at 0.1 ppm for *Spodoptera* species; whereas azadirachtin from the *Azadirachta (Melia)* species is active on a host of different insects down to 0.35 ppm for *Spodoptera frugiperda* (J. E. Smith). However, ungandensidial and polygodial from *Warburgia* species are also active on *Spodoptera* species at 0.1 ppm; aristolochic acid is active on *Locusta migratoria migratorioides* (Reiche and Fairmaire) at 0.000001% (0.01 ppm), pilocarpine is active on *Bombyx mori* (L.) at 2×10^{-7} M (0.04 ppm), salicin is active on *B. mori* also at 2×10^{-7} M (0.06 ppm), and strychnine is active also on *B. mori* at 10^{-7} M (0.03 ppm). It is extremely difficult, however, to compare potency when comparing different deterrents on different insects, since some potent deterrents have no effect whatsoever on certain insects.[3]

TABLE 1
Insect Feeding Deterrents

Feeding deterrent	Plant source	Insect scientific name, conc for deterrent activity	Ref.
Absinthin: A1[a]	—	*Spodoptera litura* (F.)	1
6-Acetoxytoonacilin: A2	*Cedrela toona* Roxb. ex Willd. var. *australis*	*Epilachna varivestis* Mulsant	18
(E)-Aconitic acid: A4	*Echinochloa crusgalli* var. *oryzicola* (Vasing) Ohwi	*Nilaparvata lugens* (Stal), 0.25—0.5% @ pH 7	19
Aesculetin: A5	*Aesculus octandra* Marsh. (yellow buckeye)	*Scolytus multistriatus* (Marsham), 0.002 *M*	20
Ajugarin I: A6, II: A7, III: A8	*Ajuga remota* Benth.	*Spodoptera exempta* (Walker), I—100 ppm, II—100 ppm	17, 21, 22, 23
		Spodoptera littoralis (Boisduval), I—300 ppm, II—300 ppm	22
L-Alanine	—	*Spodoptera littoralis* (Boisduval), 0.125 *M*	24
Alanine + serine + aminobutyric acid	—	"Grasshoppers, no sp. specified," 0.5 *M*	25
Alantolactone: A9	*Inula helenium* L.	*Tribolium confusum* Jacquelin duVal, 3—7%	26,27
Albizziin: A10	*Albizia julibrissin* Durazz.	*Anacridium melanorhodon arabafrum* Dirsh, 5%	27, 28
		Locusta migratoria (L.), 1%	7, 27
		Locusta migratoria migratorioides (Reiche & Fairmaire), 1%	27, 28
2-Amino-3-acetylamino-propionic acid (ADAP)	—	*Anacridium melanorhodon arabafrum* Dirsh, 1.0%	28
Amino acids (non protein)	—	*Chrotoicetes terminifera* (Walker), 0.1M	29
		Locusta migratoria migratorioides (Reiche & Fairmaire), 0.1 M	29
		Schistocerca americana gregaria (Dirsh), 0.1 *M*	29
DL-α-Aminobutyric acid	—	*Spodoptera littoralis* (Boisduval), 0.125 *M*	24
2-Amino-4-(oxalylamino)butyric acid (ODAB)	—	*Anacridium melanorhodon arabafrum* Dirsh, 5%	28
2-Amino-3-(oxalylamino) propionic acid (ODAP)	—	*Locusta migratoria migratorioides* (Reiche & Fairmaire), 10%	28
		Anacridium melanorhodon arabafrum Dirsh, 0.5—1.0%	28
		Locusta migratoria migratorioides (Reiche & Fairmaire), 1—10%	28

TABLE 1 (continued)
Insect Feeding Deterrents

Feeding deterrent	Plant source	Insect scientific name, conc for deterrent activity	Ref.
Amitraz (Mitac®) and *Bacillus thuringiensis* var *alesti* Berliner, HD-1 strain (Dipel®)	—	*Heliothis armigera* (Hübner), 0.53% in 1:1 mixture	30
Ammonium nitrate	*Melilotus infesta* Guss. (sweetclover)	*Sitona cylindricollis* Fahraeus	1, 31, 32
Amphetamine: A11		*Periplaneta americana* (L.), 50 µg/insect	33
Angelicin: A12	*Angelica archangelica* L.	*Spodoptera litura* (F.), 500—1000 ppm	27, 34
Arbutin: A13	—	*Locusta migratoria* (L.), 0.2%	7
		Locusta migratoria migratorioides (Reiche & Fairmaire), 1%	35
Aristolochic acid: A14	*Aristolochia* spp.	*Locusta migratoria migratorioides* (Reiche & Fairmaire), 0.000001%	8, 27, 35
Asarone: A15	*Acorus calamus* L.	"Mosquitoes, moths, fleas, no sp. specified"	27, 36
		Musca domestica L.	27, 36
Ascorbic acid	—	*Trichoplusia ni* (Hübner)	32
Atropine: A16	Various	*Melanoplus bivittatus* (Say)	31
		Pieris brassicae (L.)	31
		"Chrysomelidae (Fam.), no sp. specified"	32
		"Pieridae (Fam.) no sp. specified"	32
Aucubin: A17	Various	*Locusta migratoria migratorioides* (Reiche & Fairmaire), 0.1%	27, 35
Azadirachtin: A18*	*Azadirachta indica* A. Juss. (*Melia azadirachta* L. and *M. indica Brandis*), *Melia azedarach* L.	*Acalymma vittatum* (F.), 0.1—0.25%	3, 37, 38
		Diabrotica undecimpunctata howardii Barber	37, 38
		Earias insulana (Boisduval), 0.005%	39
		Galleria mellonella (L.)	3
		Heliothis virescens (F.), 0.035—0.1 mg/36 cm² leaf	3, 40
		Heliothis zea (Boddie)	31
		Hypsipyla grandella (Zeller), 0.002 *M*	3, 13
		Leptinotarsa decemlineata (Say), 10^{-3} *M*	3,13
		Locusta migratoria (L.), 0.04%	1, 7, 23

* Corrected structure, see A30.

Compound	Source	Test organism (concentration)	Ref.
Bacillus thuringiensis var *alesti* Berliner, HD-1 strain (Dipel®)	—	*Locusta migratoria migratorioides* (Reiche & Fairmaire), 0.01%	8, 17, 35, 41
		Plutella xylostella (L.), 0.025—0.1 mg/36 cm leaf	3
		Popillia japonica Newman, 0.25%	3, 40
		Reticulitermes sp.	3
		Schistocerca gregaria Forsk, 70 µg/L	1, 3, 31, 32, 42
		Spodoptera frugiperda (J. E. Smith), 0.35 ppm	3, 31, 43
		Spodoptera littoralis (Boisduval), 0.001%	39
		Heliothis armigera (Hübner)	30
Benomyl: B1	—	*Myzus persicae* (Sulzer), 0.5 g/L	44
		"Weevils, no sp. specified"	44
p-Benzoquinone: B2	—	*Scolytus multistriatus* (Marsham)	31, 45
2-Benzoxazolinone: B3	*Zea mays* L. subsp. *mays*	*Heliothis zea* (Boddie)	36
		Ostrinia nubilalis (Hübner)	36
Benzyl alcohol: B4	Small grains	*Schizaphis graminum* (Rondani), 100 ppm	32, 46
Berberine: B5	Various	*Bombyx mori* (L.)	27, 31
		Pieris brassicae (L.)	27, 31
Bergapten: B6	*Orixa japonica* Thunberg	*Blattella germanica* (L.), 67 ppm	34
		Neostylopyga rhombifolia (Stoll), 67 ppm	34
		Periplaneta americana (L.), 67 ppm	34
		Spodoptera litura (F.), 5—100 ppm	1, 4, 34, 36, 47
Betaine: B7	—	*Danaus plexippus* (L.), 1.0 M	48
N,3-Bis(4-chlorophenyl)-4,5-dihydro-1H-pyrazole-1-carboxamide (TH 6041)	—	*Ceramica picta* (Harris), 0.25—0.5 lb/acre	49
		Leptinotarsa decemlineata (Say), 0.25—0.5 lb/acre	49
Borneol: B8	*Pinus silvestris* L.	*Dendrolimus pini* (L.), 0.3%	50
Borneol acetate: B9	*Pinus silvestris* L.	*Dendrolimus pini* (L.), 0.3%	50
Brucine: B10	*Strychnos* spp.	*Bombyx mori* (L.)	27, 31
		Danaus plexippus (L.)	27, 48
		Pieris brassicae (L.)	27, 31
Bufadienolides: B11	Toads	*Phaedon cochleariae* F.	27, 51
Caffeic acid: C1	*Sorghum bicolor* (L.) Moench	*Locusta migratoria* (L.), 0.07%	7
Caffeine: C2	Various	*Danaus plexippus* (L.)	27, 48
		Manduca sexta (L.)	27, 31
L-Canavanine: C3	*Canavalia ensiformis* (L.) DC.	*Manduca sexta* (L.)	27, 52

TABLE 1 (continued)
Insect Feeding Deterrents

Feeding deterrent	Plant source	Insect scientific name, conc for deterrent activity	Ref.
Cantharidin: C4	Insects	*Epicauta* spp., 10^{-5} M	27, 32, 53
Capsaicin: C5	Various	"Chrysomelidae (Fam.), no sp. specified"	25, 32
		"Pieridae (Fam.), no sp. specified"	32
Cardenolides	—	*Phaedon cochleariae* F.	51
Cardol	*Anacardium occidentale* L.	"mosquitoes, no sp. specified"	36
		"*Musca domestica* L."	36
Δ³-Carene: C6	*Pinus silvestris* L.	*Dendrolimus pini* (L.), 0.3%	50
Carvone: C7	Various plants	*Locusta migratoria migratorioides* (Reiche & Fairmaire), 0.1%	27, 35
Caryophyllene: C8	Various plants	*Locusta migratoria migratorioides* (Reiche & Fairmaire), 0.01%	27, 35
Caryoptin: C9	*Caryopteris divaricata* Maxim.	*Spodoptera litura* (F.), 200 ppm	4, 5, 23, 47, 54, 55
Caryoptin hemiacetal: C13	*Caryopteris divaricata* Maxim.	*Spodoptera litura* (F.), 200 ppm	1, 4
Caryoptionol: C10	*Caryopteris divaricata* Maxim.	*Spodoptera litura* (F.), 200 ppm	1, 4
Chlordimeform: C19	—	*Calopilos miranda* (Butler)	56
		Hyadaphis erysimi (Kaltenback), 10—1000 ppm	56
		Laodelphax striatellus (Fallen), 10—1000 ppm	56
		Myzus persicae (Sulzer), 10 ppm	56
		Nephotettix cincticeps (Uhler), 10—1000 ppm	56
		Nilaparvata lugens (Stal), 10—1000 ppm	56
		Pryeria sinica Moore	56
		Spodoptera litura (F.), 5 ppm	56
Chlordimeform HCl (Fundal™)	—	*Plathypena scabra* (F.)	57
		Bombyx mori (L.), 10 ppm	58
		Heliothis armigera (Hübner), 0.144%	30
		Periplaneta americana (L.), 1 μg/insect	33
Chlordimeform HCl (Fundal™) and *Bacillus thuringiensis* var alesti Berliner HD-1 strain (Dipel®)		*Heliothis armiger* (Hübner), 0.063%	30
Cholesterol acetate: C20	Corn, others	*Diatraea grandiosella* (Dyar), 0.2—0.32%	32, 59
Cholesterol myristate: C21	Corn, others	*Diatraea grandiosella* (Dyar), 0.2—0.32%	32, 59

Compound	Plant source	Insect (activity)	Ref.
Cholesterol oleate: C22	Corn, others	*Diatraea grandiosella* (Dyar), 0.2—0.32%	32, 59
Chrysin: C23	*Pinus lambertiana* Dougl.	*Chryptotermes brevis* (Walker), 0.5%	27, 36, 60
1,4-Cineole: C24	*Eucalyptus* spp.	"Mosquitoes, no sp. specified"	36
1,8-Cineole: C25	*Eucalyptus* spp.	*Musca domestica* L., 1—5%	36, 61
		Locusta migratoria (L.), 0.01%	4
		Locusta migratoria migratorioides (Reiche & Fairmaire), 0.005%	35
		"Mosquitoes, no sp. specified"	36
		Musca domestica L., 1—5%	36, 61
Cinnamic acid: C26	—	*Locusta migratoria migratorioides* (Reiche & Fairmaire)	8
Citral: C27	—	*Locusta migratoria migratorioides* (Reiche & Fairmaire), 0.001%	35
Citronellal: C28	—	*Locusta migratoria migratorioides* (Reiche & Fairmaire), 0.001%	35
(+)-Citronellene: C30	—	*Reticulitermes lucifugus santonensis* (Feytaud), + EAG	36, 62
(−)-Citronellol: C29	—	*Reticulitermes lucifugus santonensis* (Feytaud), 0.1 ng/insect	36, 62
Clerodendrin-A: C31	*Clerodendrum cryptophyllum* Turcz., *Clerodendrum trichotomum* Thunberg	*Calospilos miranda* (Butler), 5000 ppm	1
		Euproctis subflava (Bremer), 1000 ppm	1
		Ostrinia nubilalis (Hübner), 5000 ppm	1
		Spodoptera litura (F.), 300 ppm	4, 32, 36, 47, 63
Clerodendrin-B: C32	*Clerodendrum trichotomum* Thunberg	*Calospilos miranda* (Butler), 5000 ppm	1
		Euproctis subflava (Bremer), 1000 ppm	1
		Ostrinia nubilalis (Hübner), 5000 ppm	1
		Spodoptera litura (F.), 200 ppm	4, 36, 47, 63
Clerodin: C11	*Caryopteris divaricata* Maxim., *Clerodendrum infortunatum* Gaertn.	*Spodoptera litura* (F.), 50 ppm	1, 4, 36, 47, 63, 64
Clerodin hemiacetal: C14	*Caryopteris divaricata* Maxim.	*Spodoptera litura* (F.), 50 ppm	1, 4, 65
Cocculolidine: C33	*Cocculus trilobus* (Thunb.) DC.	*Callosobruchus chinensis* (Lucas)	66
		Nephotettix cincteps (Uhler)	66
Colchicine: C34	*Colchicum autumnale* L.	*Locusta migratoria migratorioides* (Reiche & Fairmaire), 0.001%	27, 35
Condensed tannin	—	*Locusta migratoria migratorioides* (Reiche & Fairmaire), 4.0%	35

TABLE 1 (continued)
Insect Feeding Deterrents

Feeding deterrent	Plant source	Insect scientific name, conc for deterrent activity	Ref.
o-Coumaric acid: C35	*Sorghum bicolor* (L.) Moench	*Locusta migratoria* (L.), 0.01%	7
		Scolytus multistriatus (Marsham) 0.002 *M*	20
p-Coumaric acid: C36	*Sorghum bicolor* (L.) Moench	*Locusta migratoria* (L.), 0.02%	7
		Scolytus multistriatus (Marsham) 0.002 *M*	20
(Z)-o-Coumaric acid glucoside	*Melilotus infesta* Guss. (sweetclover)	*Epicauta fabricii* (Leconte)	32
		Epicauta pestifera Werner	32
		Epicauta vittata (F.)	32
Coumarin: C37	*Melilotus officinalis* Lam. (sweetclover)	*Epicauta fabricii* (Leconte)	1, 25, 32
		Epicauta pestifera Werner	1, 32
		Epicauta vittata (F.)	32
		Listroderes costirostris obliquus (Klug)	31, 32
		Locusta migratoria (L.), 0.5%	7
		Locusta migratoria migratorioides (Reiche & Fairmaire), 0.05%	35
Coumestrol: C38	Alfalfa	*Acyrthosiphon pisum* (Harris)	32
		Therioaphis maculata (Buckton)	32
Crotepoxide: C39	*Croton macrostachys* Hochst. ex A. Rich.	*Spodoptera exempta* (Walker)	5
Cucurbitacins	*Citrullus lanatus* (Thunb.) Matsum & Nakai	*Apis mellifera* L.	27, 36
		"Wasps, no sp. specified"	36
Cucurbitacins (?)	*Cucurbita moschata* (Duch.) Duch. ex Poir.	*Epilachna tredecimnotata* (Latreille)	67, 68
		Phaedon cochleariae F.	67, 68
		Phyllotreta undulata Kutschelina	67, 68
Cucurbitacins B: C40, E: C41, and I: C42	*Iberis amara* L.	*Phyllotreta nemorum* (L.); B-40 μg/ml, E-40 μg/ml, I-40 μg/ml	27, 51, 67
Cycloeucalenol	*Swietenia mahagoni* (L.) Jacq.	"Termites, no sp. specified"	36
p-Cymene: C43	*Amorpha fructicosa* L.	*Leptinotarsa decemlineata* (Say)	69, 70
		Locusta migratoria migratoroides (Reiche & Fairmaire)	69, 70
		Pieris brassicae (L.)	69, 70
L-Cystine	—	*Spodoptera littoralis* (Boisduval), 0.125 *M*	24
(Z)-Decalin with an epoxydiacetate	—	*Locusta migratoria* (L.)	71
Demissidine	*Solanum demissum* Lindl.	*Empoasca fabae* (Harris)	27, 31

Compound	Source	Insect	Ref.
Demissine	*Solanum demissum* Lindl., various	"Chrysomelidae (Fam.), no sp. specified	32, 36
		"Pieridae (Fam.), no sp. specified"	32, 36
		Leptinotarsa decemlineata (Say)	31, 32
3-Desacetylsalannin: D1	*Azadirachta indica* A. Juss	*Epilachna varivestis* Mulsant	72
Dhurrin: D3	*Sorghum bicolor* (L.) Moench	*Locusta migratoria* (L.), 0.01 mM HCN	73
1,3-Diacetylvilasinin: D4	*Azadirachta indica* A. Juss.	*Epilachna varivestis* Mulsant	72
2,4-Diaminobutyric acid (DABA)	—	*Anacridium melanorhodon arabafrum* Dirsh	28
		Locusta migratoria migratorioides (Reiche & Fairmaire)	28
2,3-Diaminopropionic acid (DAPA)	—	*Anacridium melanorhodon arabafrum* Dirsh	28
		Locusta migratoria migratorioides (Reiche & Fairmaire)	28
2,3-Dichloro-1,4-naphthoquinone (Dichlone): D5	—	*Periplaneta americana* (L.)	20
		Scolytus multistriatus (Marsham)	20
Dictamine: D6	Various	*Spodoptera litura* (F.), 1000 ppm	27, 47
N,N-Diethyl-*o*-toluamide: D7	—	*Neodiprion sertifer* (Geoffroy), 2%	74
		Neodiprion swainei Middleton, 2%	74
		Neodiprion pratti paradoxicus Ross 2%	74
N,N-Diethyl-*p*-toluamide: D8	—	*Neodiprion sertifer* (Geoffroy), 2%	74
Digitonin: D9	*Digitalis purpurea* L.	*Melanoplus bivittatus* (Say)	27, 31, 32
Dihydrocaryoptin: C15	*Caryopteris divaricata* Maxim.	*Spodoptera litura* (F.), 80 ppm	1, 4, 65
Dihydrocaryoptionol: C16	*Caryopteris divaricata* Maxim.	*Spodoptera litura* (F.), 100 ppm	1, 4, 65
Dihydroclerodin-I: C17	*Caryopteris divaricata* Maxim.	*Spodoptera litura* (F.), 50 ppm	4, 65
2,3-Dihydro-4,6-dimethoxyfuro = [2,3-*b*]quinoline and its 3-hydroxy derivatives: D10	—	*Spodoptera litura* (F.), 300 ppm	75
Dihydroisopimpinellin: D11	—	*Spodoptera litura* (F.), 10—100 ppm	34
Dihydrokokusagine: D12	—	*Spodoptera litura* (F.), 300 ppm	34
(±)-Dihydrolinalool	—	*Reticulitermes lucifugus santonensis* (Feytaud), 0.1 ng/insect	36, 62
Dihydro-α-solanin	*Solanum tuberosum* L.	*Leptinotarsa decemlineata* (Say)	32
Dihydroxanthotoxin: D13	—	*Spodoptera litura* (F.), 100—500 ppm	34
2,4-Dihydroxy-7-methoxy-2*H*-1,4-benzoxazin-3(4*H*)-one (DIMBOA)	Corn	*Ostrinia nubilalis* (Hübner) 0.32 mg/g diet	27, 32, 76
2',6'-Dihydroxy-4'-methoxydihydrochalcone: D14	*Populus deltoides* Bartr. ex Marsh.	*Scolytus multistriatus* (Marsham), 0.002 M	20
5,8-Dihydroxy-1,4-naphthoquinone: D15	—	*Periplaneta americana* (L.)	20
		Scolytus multistriatus (Marsham)	20

TABLE 1 (continued)
Insect Feeding Deterrents

Feeding deterrent	Plant source	Insect scientific name, conc for deterrent activity	Ref.
erythro-9,10-Dihydroxy-1-octadecanol acetate	Aleurites fordii Hemsl.	Anthonomus grandis grandis Boheman, 1%	77
Dimethoxybenzoxazolinone	—	Locusta migratoria migratorioides (Reiche & Fairmaire)	8
4'-(3,3-Dimethyl-1-triazeno) = acetanilide (AC-24055): D16	—	Sitotroga cerealella (Olivier), 0.05%—0.2%	78
		Neodiprion lecontei (Fitch), 2.4 mg/ml	74
		Neodiprion nannulus nannulus Schedl, 2.4 mg/ml	74
		Neodiprion pratti paradoxicus Ross, 2.4 mg/ml	74
		Neodiprion rugifrons Middleton, 2.4 mg/ml	74
		Neodiprion sertifer (Geoffroy), 2.4 mg/ml	74
		Neodiprion swainei Middleton, 2.4 mg/ml	74
Diosgenin: D17	Various	Melanoplus bivittatus (Say), 1.0%	27, 31, 79
Djenkolic acid	Pithecellobium jiringa (Jack) Mansf.	Locusta migratoria migratorioides (Reiche & Fairmaire), 1 — 10%	27, 28
L-Dopa: D18	Mucuna sp.	Spodoptera eridania (Cramer), 5%	32, 80
Ecdysterone: E1		Bombyx mori (L.)	31
		Pieris brassicae (L.), 25—100 ppm	31, 81
Effusin: E2	Plectranthus effusus (Maxim.) Honda	"Lepidoptera, no sp. specified"	82
Elemicin: E3	Various	"No sp. specified"	27, 36
α-Eleostearic acid: E4	Aleurites fordii Hemsl.	Anthonomus grandis grandis Boheman, 1%	77
Emodin: E5	Rhamnus alnifolia L'Her.	Lymantria dispar (L.), 15—450 ppm	83
3-Epicaryoptin: C12	Clerodendrum calamitosum L.	Spodoptera litura (F.), 200 ppm	4, 23, 54, 55, 65
3-Epidihydrocaryoptin: C18	Parabenzoin praecox (Sieb & Zucc.) Nakai	Spodoptera litura (F.), 100 ppm	4, 65
(±)-Epieudesmin: E6	Liriodendron tulipifera L.	Spodoptera litura (F.), 0.05%	4, 36
Epitulipinolide diepoxide		Lymantria dispar (L.)	84
Erlancorin		"No sp. specified"	85
(+)-Eudesmin: E7	Parabenzoin praecox (Sieb. & Zucc.) Nakai	Spodoptera litura (F.), 1%	4, 36
Evoxine: E8	Orixa japonica Thunberg	Spodoptera litura (F.), 500 ppm	47
Farnesol: F1	Various	Locusta migratoria (L.), 0.1%	7, 27
		Lymantria dispar (L.), 3.75 mg/ml	27, 86

Compound	Host plant	Test insect, dose	Ref.
Fenitrothion: F2	—	*Periplaneta americana* (L.), 1 µg/insect	33
Fentin acetate (Brestan or TPTA): F3	—	*Plutella maculipennis* Curtis, 0.05%	87
		Sitotroga cerealella (Olivier), 0.05—0.2%	14, 78
		Stomopteryx subsecivella (Zeller), 0.05%	87
		Spodoptera littoralis (Boisduval)	14
		Spodoptera litura (F.), 0.705%	88
Fentin chloride (Brestanol): F4, see Triphenyltin chloride	—	*Sitotroga cerealella* (Olivier), 0.05%—0.2%	14, 78
		Plutella maculipennis Curtis, 0.05%	87
Fentin hydroxide (TPTH or Duter): F5, see Triphenyltin hydroxide	—	*Stomopteryx subsecivella* (Zeller), 0.05%	87
		Spodoptera littoralis (Boisduval)	14
		Spodoptera litura (F.), 0.075%	14, 88
Ferulic acid: F6	*Sorghum bicolor* (L.) Moench	*Locusta migratoria* (L.), 0.06%	7
		Locusta migratoria migratorioides (Reiche & Fairmaire)	8
2-Formylanthraquinone: F7	Teak, and West Indian mahogany	*Nasutitermes exitiosus* (Hill), 1.0—2.0%	4, 36, 89
Fraxetin: F8	*Fraxinus americana* L.	*Scolytus multistriatus* (Marsham), 0.002 M	20
Friedelin: F9	*Acokanthera oblongifolia* (Hochst.) Codd	*Spodoptera littoralis* (Boisduval), 0.25—0.5%	90
Gentisic acid: G1	*Sorghum bicolor* (L.) Moench	*Locusta migratoria* (L.), 0.01%	7
Geranial: G2	Various	*Locusta migratoria migratorioides* (Reiche & Fairmaire), 0.001%	27, 35
Geraniol: G3	Various	*Locusta migratoria migratorioides* (Reiche & Fairmaire), 0.005%	27, 35, 36
		Lymantria dispar (L.), 3.75 mg/ml	27, 86
		Reticulitermes lucifugus santonensis (Feytaud), 0.1 ng/insect	27, 36, 62
Glaucolide-A: G4	*Vernonia gigantea* Trelease, *Vernonia glauca* Britton	*Diacrisia virginica* (F.), 0.5%	91
		Sibine stimulea (Clemens), 0.5%	91
		Spodoptera eridania (Cramer), 0.5%	91
		Spodoptera frugiperda (J. E. Smith), 0.5%	91
		Spodoptera ornithogalli (Guenée), 0.5%	91
		Trichoplusia ni (Hübner), 0.5%	91
		Manduca sexta (L.)	31
		Boarmia (*Ascotis*) *selenaria* (Dennis and Schiffermuller), 1%	92
		Locusta migratoria migratorioides (Reiche & Fairmaire), 0.05%	35
Glucotropaeolin: G5	—	*Heliothis* spp., 1.2%	93
Gossypol: G6	Cotton	*Spodoptera littoralis* (Boisduval), 1.0%	94

TABLE 1 (continued)
Insect Feeding Deterrents

Feeding deterrent	Plant source	Insect scientific name, conc for deterrent activity	Ref.
Gramine: G7	*Acer negundo* L., *Acer saccharinum* L.	*Scolytus multistriatus* (Marsham), 0.002 M	20
Guazatine triacetate (SN 513)	—	*Epilachna varivestis* Mulsant, 1.92%	95
		Euschistus servus (Say), 1.92%	95
		Heliothis zea (Boddie), 1.92%	95
		Lepinotarsa decemlineata (Say), 1.92%	95
		Manduca sexta (L.), 1.92%	95
		Plathypena scabra (F.), 0.25—0.75%	57, 96, 97
		Plutella xylostella (L.), 1.92%	95
		Pseudoplusia includens (Walker), 0.8 kg/ha	98
		Spodoptera exigua (Hübner), 1.92%	95
		Trichoplusea ni (Hübner), 1.92%	95
Gymnemic acids	*Gymnema sylvestre* (Retz.) Schult.	*Spodoptera eridania* (Cramer), 0.09%	1, 99
Hallactone A	*Podocarpus hallii* T. Kirk	*Musca domestica* L.	32
Hallactone B	*Podocarpus hallii* T. Kirk	*Musca domestica* L.	32
Halostachine: H1		*Locusta migratoria migratorioides* (Reiche & Fairmaire), 0.1%	8, 35
Harrisonin: H2	*Harrisonia abyssinica* Oliv.	*Spodoptera exempta* (Walker), 20 ppm	17, 23, 36, 100
Hecogenin: H3	*Nepeta cataria* L. (catnip)	*Melanoplus bivittatus* (Say), 1.0%	32, 81
L-Histidine	*Acacia* spp.	*Spodoptera littoralis* (Boisduval), 0.125 M	24
Homoarginine		*Anacridium melanorhodon arabafrum* Dirsh, 1.0—5.0%	28
		Locusta migratoria migratorioides (Reiche & Fairmaire), 1.0 — 5.0%	28
Hordenine: H4	*Hordeum* sp.	*Melanoplus bivittatus* (Say), 1.0%	27, 31, 79
Hordenine sulfate		*Laodelphax striatellus* (Fallen), 10—100 ppm	101
		Nephotettix cincticeps (Uhler), 10—100 ppm	101
		Nilaparvata lugens (Stal), 10—100 ppm	101
		Sogatella furcifera (Horvath), 10—100 ppm	101
1-Hydroxyanthraquinone: H5	—	*Reticuliitermes lucifugus santonensis* (Feytaud), 0.1 ng/insect	36, 62

Compound	Source	Insect	Ref.
2-(Hydroxymethyl)anthraquinone: H6	Teak, West Indian mahogany	*Reticulitermes lucifugus santonensis* (Feytaud), 0.1 ng/insect	36, 62
p-Hydroxybenzaldehyde: H7	*Sorghum bicolor* (L.) Moench	*Locusta migratoria* (L.), 0.4%	7
m-Hydroxybenzoic acid: H8	*Sorghum bicolor* (L.) Moench	*Locusta migratoria* (L.), 0.02%	7
p-Hydroxybenzoic acid: H9	*Sorghum bicolor* (L.) Moench	*Locusta migratoria* (L.), 0.2%	7
2-Hydroxy-1,4-naphthoquinone: H10	—	*Periplaneta americana* (L.)	20
		Scolytus multistriatus (Marsham)	20
Indican dioegin	*Nepeta cataria* L. (catnip)	*Melanoplus bivittatus* (Say), 0.65%	32, 79
Inflexin: I1	*Isodon inflexus* Kudo	*Spodoptera exempta* (Walker), 5.4 µg/ml	5, 102
Inokosterone	—	*Pieris brassicae* (L.)	31
Inorganic ions (Na$^+$ & Ca^{++})	—	*Locusta migratoria migratorioides* (Reiche & Fairmaire), 4.0%	35
Isoasaron: I2	*Piper kadsura* (Choisy) Ohwi	*Spodoptera litura* (F.), 0.5%	4
Isobergapten: I3	—	*Spodoptera litura* (F.), 5—100 ppm	34
Isoboldine: I4	*Cocculus trilobus* (Thunb.) DC.	*Calospilos miranda* (Butler), 200 ppm	1, 31, 36, 47
		Spodoptera litura (F.), 200 ppm	1, 31, 36, 47
Isoferulic acid: I5	*Sorghum bicolor* (L.) Moench	*Locusta migratoria* (L.), 0.02%	7
Isolongifolene	—	*Locusta migratoria migratorioides* (Reiche & Fairmaire), 1.0%	35
Isolongifolene ketone A	—	*Locusta migratoria migratorioides* (Reiche & Fairmaire), 0.05%	35
Isolongifolene ketone B	—	*Locusta migratoria migratorioides* (Reiche & Fairmaire), 0.01%	35
Isomedin: I6	*Isodon shikokianus* (Makino) Hara var. *intermedius*	*Spodoptera exempta* (Walker), 4.0 µg/ml	5
Isopimpinellin: I7	*Orixa japonica* Thunberg	*Blatella germanica* (L.), 67 ppm	34
		Neostylopyga rhombifolia (Stoll), 67 ppm	34
		Periplaneta americana (L.), 67 ppm	34
		Spodoptera litura (F.), 5—100 ppm	1, 4, 34, 36, 47
Isothiocyanates	—	*Melanophus sanguinipes* (F.)	31
Jacobine	—	*Locusta migratoria migratorioides* (Reiche & Fairmaire), 0.001%	35
Jaconine	—	*Locusta migratoria migratorioides* (Reiche & Fairmaire), 0.05%	35
Japonin	*Orixa japonica* Thunberg	*Spodoptera litura* (F.), 300 ppm	47

TABLE 1 (continued)
Insect Feeding Deterrents

Feeding deterrent	Plant source	Insect scientific name, conc for deterrent activity	Ref.
Juglone: J1	*Carya ovata* (Mill.) K. Koch	*Acalymma vittatum* (F.), 0.1—0.5%	37
		Periplaneta americana (L.)	20
		Scolytus multistriatus (Marsham), 1 mg disks	20, 31, 32, 36, 103
Kaempferol: K1	*Robinia pseudoacacia* L.	*Scolytus multistriatus* (Marsham), 0.002 *M*	20
(−)-Kaur-16-en-19-oic acid: K2	*Helianthus annuus* L.	*Homoeosoma electellum* (Hulst), 1.0—2.0%	36, 104
		Pectinophora gossypiella (Saunders)	36, 104
13-Keto-8(14)-podo-carpen-18-oic acid: K3	*Pinus banksiana* Lamb.	*Neodiprion rugifrons* Middleton, 0.2—0.5 mg/ml	105, 106
Kokusagine: K4	*Orixa japonica* Thunberg	*Neodiprion swainei* Middleton, 1.0 mg/ml	105, 106
		Spodoptera litura (F.), 100 ppm	1, 4, 34, 36, 47, 75
Lapachol: L1	*Paratecoma peroba* (Record) Kuhlm., *Stereospermum suaveolens* DC., *Tabebuia capitata* (Bur. and Schum.) Sandiv., *Tabebuia flavescens* Benth. and Hook. f. ex Griseb., *Tabebuia ipe* (Mart.) Standley	*Reticulitermes lucifugus santonensis* (Feytaud), 0.1 ng/insect	36, 62
Lapachonone	*Paratecoma peroba* (Record) Kuhlm., *Stereospermum suaveolens* DC., *Tabebuia capitata* (Bur. and Schum.) Sandiv., *Tabebuia flavescens* Benth. and Hook. f. ex Griseb., *Tabebuia ipe* (Mart.) Standley	"Termites, no sp. specified"	36
Leptine I	*Solanum chacoense* Bitter	*Empoasca fabae* (Harris)	31
		Leptinotarsa decemlineata (Say)	31
Leptine III	*Lycopersicon esculentum* Mill., solanaceous plants	*Leptinotarsa decemlineata* (Say)	36
Leptinine I	*Lycopersicon esculentum* Mill., solanaceous plants	*Leptinotarsa decemlineata* (Say)	36
Leptinine II	*Lycopersicon esculentum* Mill., solanaceous plants	*Leptinotarsa decemlineata* (Say)	36

Compound	Plant source	Insect species (concentration)	Ref.
Limonene: L2	*Pinus silvestris* L.	*Dendrolimus pini* (L.), 0.3%	50
Linalool: L3	—	*Locusta migratoria migratorioides* (Reiche & Fairmaire), 0.001%	35
Linamarin: L4	—	*Locusta migratoria migratorioides* (Reiche & Fairmaire), 0.005%	35
Lipiferolide	*Liriodendron tulipifera* L.	*Locusta migratoria migratorioides* (Reiche & Fairmaire), 2.0%	35
		Lymantria dispar (L.)	84
Lobeline: L5	*Lobelia inflata* L.	*Melanoplus bivittatus* (Say), 1.67%	27, 31, 79
Longicyclene: L6	—	*Locusta migratoria migratorioides* (Reiche & Fairmaire), 0.01%	35
Longifolene: L7	—	*Locusta migratoria migratorioides* (Reiche & Fairmaire), 1.0%	35
Lophocereine	*Lophocereus schottii* (Engelman) Britton & Rose (Senita cactus)	*Drosophila pachea* Patterson and Wheeler	32
Lucenine	*Nepeta cataria* L. (catnip)	*Melanoplus bivittatus* (Say)	32
Lupinine: L8		*Melanoplus bivittatus* (Say), 0.37%	31, 79
Lycorine: L9	*Hymenocallis littoralis* (Jacq.) Salisb.	*Schistocerca gregaria* Forsk, 0.05%	107
Magnoline: M1	*Magnolia acuminata* (L.) L.	*Scolytus multistriatus* (Marsham), 0.002 M	20
Malic acid	—	*Locusta migratoria migratorioides* (Reiche & Fairmaire)	35
Meliantriol: M2	*Azadirachta indica* A. Juss., *Melia azedarach* L.	*Schistocerca gregaria* Forsk, 8.0 $\mu g/cm^2$	17, 31, 36, 108
Meliatin	*Melia azedarach* L.	"Beetles, no sp. specified," 0.125 M	36
L-Methionine + 0.125 M sucrose	—	*Spodoptera littoralis* (Boisduval), 0.125 M	24
1-Methoxyanthraquinone: M3	—	*Reticulitermes lucifugus santonensis* (Feytaud), 0.1 ng/insect	36, 62
6-Methoxy-2-benzoxazolinone: M4	*Zea mays* L. subsp. *mays*	*Heliothis zea* (Boddie)	36
		Ostrinia nubilalis (Hübner)	31, 32, 36
7-Methoxyjuglone: M5	Teak, West Indian mahogany	"Termites, no sp. specified"	36
Methyl (Z)-10-acetoxy-8,9-epoxy-2-decen-4,6-diynoate	*Chrysothamnus nauseosus* (Pall.) Britt.	*Leptinotarsa decemlineata* (Say), <35 $\mu g/cm^2$	109
Methyl (Z,Z)-10-acetoxy matriciate: M6	*Chrysothamnus nauseosus* (Pall.) Britt.	*Leptinotarsa decemlineata* (Say), 35—70 $\mu g/cm^2$	109
2-Methylanthraquinone (tectoquinone): M7	*Tectona grandis* L. f. (teak), West Indian mahogany	*Reticulitermes lucifugus santonensis* (Feytaud), 0.1 ng/insect	27, 36, 62
N-Methylflindersine: M8	*Fagara chalybea* (Engl.) Engl., *Xylocarpus granatum* Koen.,	*Spodoptera exempta* (Walker), 1 $\mu g/cm^2$	110, 111

TABLE 1 (continued)
Insect Feeding Deterrents

Feeding deterrent	Plant source	Insect scientific name, conc for deterrent activity	Ref.
Methyl (Z)-10-hydroxy-8,9-epoxy-2-decen-4,6-diynoate: M9	*Chrysothamnus nauseosus* (Pall.) Britt.	*Leptinotarsa decemlineata* (Say), <35 μg/cm²	109
Methyl (Z,Z)-10-hydroxy matricariate: M10	*Chrysothamnus nauseosus* (Pall.) Britt.	*Leptinotarsa decemlineata* (Say), <35 μg/cm²	109
2-Methyl-1-(3-methylbenzoyl) piperidine: M11	—	*Neodiprion lecontei* (Fitch), 2.0%	74
		Neodiprion nannulus nannulus Schedl, 2.0%	74
		Neodiprion pratti paradoxicus Ross, 2.0%	74
		Neodiprion sertifer (Geoffroy), 2.0%	74
		Neodiprion swainei Middleton, 2.0%	74
2-Methyl-1,4-naphthoquinone: M12	—	*Acalymma vittatum* (F.), 0.1—0.5%	37
		Periplaneta americana (L.)	20
		Scolytus multistriatus (Marsham)	20
Morin: M13	*Chlorophora tinctoria* (L.) Gaudich. ex Benth. & Hook. f.	*Anthonomus grandis grandis* Boheman, 0.02 mg/ml	27, 31, 112
Morphine: M14	Various	"Chrysomelidae (Fam.), no sp. specified"	32
		"Pieridae (Fam.), no sp. specified"	32
		Pieris brassicae (L.), 1—5 × 10⁻⁵ M	31, 81
Muzigadial: M15	*Warburgia stuhlmanii* Engl., *Warburgia salutaris* (Bertol f.) Chiov.	*Spodoptera exempta* (Walker), 0.1 ppm	16, 23, 113, 114
Myrcene: M16	Various	*Reticulitermes lucifugus santonensis* (Feytaud), 0.1 ng/insect	27, 36, 62
Myricoside: M17	*Clerodendrum myricoides* (Hochst.) R. Br. ex Vatke	*Spodoptera exempta* (Walker), 10 ppm	115
Myristicin: M18	*Pastinaca sativa* L.	"No sp. specified"	4
Naphthoquinones	*Pinus banksiana* Lamb., synthetics	*Acalymma vittatum* (F.)	36, 37
1,4-Naphthoquinone: N1	—	*Acalymma vittatum* (F.), 0.1—0.5%	36, 37
		Neodiprion rugifrons Middleton, 1 mg/ml	74
		Neodiprion swainei Middleton, 1 mg/ml	74
		Periplaneta americana (L.)	20, 45
		Scolytus multistriatus (Marsham)	20, 45
Nepetalactone: N2	*Nepeta cataria* L. (catnip)	"Formicidae (Fam.), no sp. specified"	32
Nerol: N3	—	*Reticulitermes lucifugus santonensis* (Feytaud), 0.1 ng/insect	36, 62

Compound	Source	Insect, concentration	Ref.
(E,S)-Nerolidol: N4	Melaleuca leucadendron (L.) L.	Lymantria dispar (L.), 3.75 mg/ml	86
Nicandrenones	Nicandra physalodes (L.) Gaertn.	Leptinotarsa decemlineata (Say), 0.01—0.1%	5, 25, 27, 31, 116
Nicotine: N5	Nicotiana tabacum L.	Bombyx mori (L.), 2×10^{-5} M	27, 31, 117
		Locusta migratoria (L.), 0.4%	7, 27
		Locusta migratoria migratorioides (Reiche & Fairmaire), 0.002%	27, 35
		Pieris brassicae (L.)	27, 31
Nimbandiol: N6	Azadirachta indica A. Juss.	Epilachna varivestis Mulsant	72
Norepinephrine: N7	—	Periplaneta americana (L.), 100 µg/insect	33
Nornicotine: N8	Nicotiana tabacum L.	Melanoplus bivittatus (Say)	27, 31
Nornicotine dipicrate	Nepeta cataria L. (catnip)	Melanoplus bivittatus (Say), 1.33%	32, 79
Octopamine: O1		Periplaneta americana (L.), 50 µg/insect	33
Osajin: O2	Maclura pomifera Raf. Schneid.	Chryptotermes brevis (Walker), 0.002—0.2%	36, 60
Oxalic acid	Various	Danaus plexippus (L.), 0.1 M	27, 48
Oxazine	—	Acalymma vittatum (F.), 0.2%	37
		Diabrotica undecimpunctata howardii Barber, 0.2%	37
		Neodiprion lecontei (Fitch), 2.0%	74
		Neodiprion pratti paradoxicus, Ross, 2.0%	74
		Neodiprion sertifer (Geoffroy), 2.0%	74
		Neodiprion swainei Middleton, 2.0%	74
Oxazolidine	—	Acalymma vittatum (F.), 0.05—0.2%	37
		Diabrotica undecimpunctata howardii Barber, 0.2%	37
		Neodiprion lecontei (Fitch), 2.0%	74
		Neodiprion pratti paradoxicus Ross, 2.0%	74
		Neodiprion rugifrons Middleton, 2.0%	74
		Neodiprion sertifer (Geoffroy), 2.0%	74
		Neodiprion swainei Middleton, 2.0%	74
Parthenin: P1	Parthenium hysterophorus L.	Aedes aegypti (L.), 0.1%	118
		Callosobruchus chinensis (Lucas)	118
		Dysdercus koenigii (F.), 0.1%	118
		Periplaneta americana (L.), 0.1%	118
		Phthorimaea operculella (Zeller), 0.1%	118
		Tribolium castaneum (Herbst), 0.1%	118
Pennyroyal oil (85% pulegone): P2	Mentha pulegium L.	Spodoptera frugiperda (J. E. Smith), 5000 ppm	119
Perloline	—	Locusta migratoria migratorioides (Reiche & Fairmaire)	35
Peroxyferolide: P3	Liriodendron tulipifera L.	Lymantria dispar (L.)	120

TABLE 1 (continued)
Insect Feeding Deterrents

Feeding deterrent	Plant source	Insect scientific name, conc for deterrent activity	Ref.
Peucedanin: P4	*Peucedanum officinale* L.	*Spodoptera litura* (F.), 500—1000 ppm	27, 34
Phaseolunatin: P5		"Coleoptera, no sp. specified"	25
(+)-α-Phellandrene: P6	—	*Reticulitermes lucifugus santonensis* (Feytaud), 0.1 ng/ insect	36, 62
(+)-β-Phellandrene: P7	—	*Reticulitermes lucifugus santonensis* (Feytaud), 0.1 ng/ insect	36, 62
Phenacaine: P8	—	*Periplaneta americana* (L.), 50 µg/insect	33
Phenethylamine HCl: P9	—	*Laodelphax striatellus* (Fallen), 100 ppm	101
		Nephotettix cincticeps (Uhler), 100 ppm	101
		Nilaparvata lugens (Stal), 100 ppm	101
		Sogatella furcifera (Horvath), 100 ppm	101
Phenylbenzothiazole: P10	—	*Ostrinia nubilalis* (Hübner), 0.25 mg/g diet	31, 121
2-Phenylethylisothiocyanate: P11	*Brassica rapa* L. (turnip)	*Drosophila melanogaster* Meigen	32
Phenylpropanoids		"No sp. specified"	5
Phloretin: P12	*Malus sylvestris* Mill.	*Scolytus multistriatus* (Marsham), 0.002 M	20
Phlorizin: P13	*Malus* sp.	*Amphorophora agathonica* Hottes, 10^{-5} M	1, 122
		Aphis pomi De Geer, 10^{-5} M	1, 122
		Locusta migratoria migratorioides (Reiche & Fairmaire), 10%	35
		Myzus persicae (Sulzer), 10^{-5} M	1
Phytol: P14	*Clerodendron fragrans* (Vent.) R. Br., *Clerodendrum japonicum* (Thunb.) Sweet	*Spodoptera litura* (F.), 5000 ppm	1, 4
Pilocarpine: P15	*Pilocarpus jaborandi* Holmes	*Bombyx mori* (L.) 2 × 10^{-7} M	27, 31, 117
		Pieris brassicae (L.), 1—5 × 10^{-5} M	27, 31, 79
Pilocereine: P16	*Lophocereus schottii* (Engelman) Britton & Rose (Senita cactus)	*Drosophila pachea* Patterson and Wheeler	32
Pimpinellin	—	*Spodoptera litura* (F.), 5—100 ppm	34
α-Pinene: P17	*Picea abies* (L.) Karst.,	*Dendrolimus pini* (L.), 0.3%/0.5 cm³	36, 50
	Pinus silvestris L.	*Locusta migratoria migratorioides* (Reiche & Fairmaire), 0.01%	35
(+)-β-Pinene: P18	—	"Beetles, no sp. specified"	36

Compound	Plant source	Insect (concentration)	Ref.
Pipecolic acid: P19	*Acacia* spp.	*Anacridium melanorhodon arabafrum* Dirsh, 1.0—5.0%	28
		Locusta migratoria (L.), 3.0%	7
		Locusta migratoria migratorioides (Reiche & Fairmaire), 0.1 M	28
Piperenone: P20	*Piper kadsura* (Choisy) Ohwi	*Spodoptera litura* (F.), 0.05—0.005%	4, 36
Piperidine: P21	—	*Danaus plexippus* (L.), 0.1 M	48
Plumbagin: P22	*Plumbago auriculata* Lam.	*Spodoptera exempta* (Walker), 10 ppm	27, 123
		Spodoptera littoralis (Boisduval), 20 ppm	27, 123
Polygodial: P23	*Warburgia stuhlmannii* Engl., *Warburgia salutaris* (Bertol f.) Chiov.	*Spodoptera exempta* (Walker), >0.1 ppm	17, 23, 63, 113, 124
		Spodoptera littoralis (Boisduval), 0.1 ppm	124
Pomiferin	*Maclura pomifera* Raf. Schmeid.	*Chryptotermes brevis* (Walker), 0.5—1.0%	36, 60
Ponasterone: P24	—	*Pieris brassicae* (L.)	31
Potassium nitrate	—	*Sitona cylindricollis* Fahraeus	1
Propranolol: P25	—	*Periplaneta americana* (L.), 100 µg/insect	33
Protocatechuic acid: P26	*Sorghum bicolor* (L.) Moench	*Locusta migratoria* (L.), 0.02%	7
o-Protocatechuic acid	*Sorghum bicolor* (L.) Moench	*Locusta migratoria* (L.), 0.02%	7
Psoralen: P27	—	*Spodoptera litura* (F.), 500—1000 ppm	34
(+)-Pulegone: P2		*Reticulitermes lucifugus santonensis* (Feytaud), 0.1 ng/insect	36, 62
Quercitin: Q1	*Quercus macrocarpa* Michx.	*Bombyx mori* (L.), 5 × 10⁻⁴ M	31, 117
Quinine: Q2	*Cinchona officinalis* L.	*Scolytus multistriatus* (Marsham), 0.002 M	20
		Bombyx mori (L.), 5 × 10⁻⁴ M	27, 31, 117
		Locusta migratoria migratorioides (Reiche & Fairmaire), 0.01%	27, 35
Quinine HCl	—	*Pieris brassicae* (L.), 2 × 10⁻⁶ M	27, 31, 79
Rutin: R1	Various	*Danaus plexippus* (L.), 0.001 M	48
		Anthonomus grandis grandis Boheman, 0.03 mg/ml	27, 31, 112
		Bombyx mori (L.), 5 × 10⁻⁴ M	27, 31, 117
		Locusta migratoria migratorioides (Reiche & Fairmaire), 0.01%	27, 35
Salannin: D2	*Azadirachta indica* A. Juss.	*Acalymma vittatum* (F.)	3, 37, 38
		Aonidiella aurantii (Maskell)	3
		Diabrotica undecimpunctata howardii Barber	38
		Earias insulana (Boisduval), 0.01%	39
		Locusta spp.	3
		Musca domestica L., 0.1%	125
		Popillia japonica Newman, 0.25%	3
		Schistocerca gregaria Forsk	3
		Spodoptera littoralis (Boisduval), 0.005%	39

TABLE 1 (continued)
Insect Feeding Deterrents

Feeding deterrent	Plant source	Insect scientific name, conc for deterrent activity	Ref.
Salannol: S1	*Azadirachta indica* A. Juss	*Epilachna varivestis* Mulsant	72
Salicin: S2	Various	*Bombyx mori* (L.), 2×10^{-7} *M*	27, 31, 117
		Locusta migratoria migratorioides (Reiche & Fairmaire), 1.0%	27, 35
		Manduca sexta (L.)	27, 31
Salicylic acid: S3	*Sorghum bicolor* (L.) Moench	*Locusta migratoria* (L.), 0.01%	7
Salts (anions and cations)		*Culiseta inornata* (Williston)	126
Santonin: S4	*Nepeta cataria* L. (catnip)	*Melanoplus bivittatus* (Say), 0.54%	32, 79
Saponins	*Nepeta cataria* L. (catnip)	*Danaus plexippus* (L.)	48
		Melanoplus bivittatus (Say), 1.0%	32, 79
Sativan: S5	*Lotus pedunculatus* Cav.	*Costelytra zealandica* (White)	127
Schkuhrin-I: S6	*Schkuhria pinnata* (Lam.) Ktze.	*Epilachna varivestis* Mulsant	128
		Spodoptera exempta (Walker)	128
Schkuhrin-II: S7	*Schkuhria pinnata* (Lam.) Ktze.	*Epilachna varivestis* Mulsant	128
		Spodoptera exempta (Walker)	128
Scillaren (mixture of A: S8, and B)	*Urginea maritima* (L.) Baker	*Schistocerca gregaria* Forsk	27, 31
Scopolamine: S9	*Datura metel* L., various	*Bombyx mori* (L.), 2.3×10^{-3} *M*	31, 117,
		"Chrysomelidae (Fam.), no sp. specified"	32
		"Pieridae (Fam.), no sp. specified"	32
Senecionine: S10	*Senecio* spp.	*Locusta migratoria migratorioides* (Reiche & Fairmaire), 0.001%	27, 35
Seneciphylline	—	*Locusta migratoria migratorioides* (Reiche & Fairmaire), 0.001%	35
—			
L-Serine	—	*Spodoptera littoralis* (Boisduval), 0.125 *M*	24
Shiromodiol diacetate: S11	*Parabenzoin trilobum* (Sieb. & Zucc.) Nakai	*Calospilos miranda* (Butler), 0.13%	1, 32, 47, 63, 129
		Spodoptera litura (F.), 0.5%	1, 32, 47, 63, 129
Shiromodiol monoacetate: S12	*Parabenzoin trilobum* (Sieb. & Zucc.) Nakai	*Calospilos miranda* (Butler), 0.13%	1, 32, 47, 63, 129
		Spodoptera litura (F.), 0.5%	1, 32, 47, 63, 129

Compound	Plant source	Insect (test)	Ref.
Shiromool: S13	—	"No sp. specified"	5, 63
Sinigrin: S14	Cruciferae (Fam.)	*Locusta migratoria* (L.), 0.5%	7
		Locusta migratoria migratorioides (Reiche & Fairmaire), 0.05%	35
		Myzus persicae (Sulzer), 0.1%	31, 130
		Papillo polyxenes asterius Stoll, 0.1%	32, 131
Skimmianine: S15	Various	*Spodoptera litura* (F.), 1000 ppm	27, 47
Sodium nitrate	—	*Sitona cylindricollis* Fahraeus	1
Solacaulin	*Solanum chacoense* Bitter, wild potato	*Leptinotarsa decemlineata* (Say)	32
Soladulcin	*Solanum dulcamara* L.	*Leptinotarsa decemlineata* (Say)	32
Solanidine: S16	*Solanum* spp.	*Empoasca fabae* (Harris)	27, 31
Solanine: S17	*Solanum punae* Juz.,	*Empoasca fabae* (Harris)	31
	Solanum schreiteri Bukasov, *Nepeta cataria* L. (catnip)	*Leptinotarsa decemlineata* (Say)	32
Sparteine: S18	Various	*Melanoplus bivittatus* (Say), 1.89%	32, 79
		"Chrysomelidae (Fam.), no sp. specified"	32
		"Pieridae (Fam.), no sp. specified"	32
Strophanthidin: S19, and strophanthidol glycosides	—	*Phaedon cochleariae* F.	51
		Phyllotreta tetrastigma (Cornolli)	51
		Phyllotreta undulata Kutschelina	51
Strychnine: S20	*Strychnos* spp.	*Bombyx mori* (L.), 10^{-7} M	27, 31, 117
Syringic acid: S21	*Sorghum bicolor* (L.) Moench	*Pieris brassicae* (L.), 10^{-5} M	27, 31, 79
Tannic acid neutralized	—	*Locusta migratoria* (L.), 0.01%	7
		Locusta migratoria migratorioides (Reiche & Fairmaire), 1.2%	35
Tannic acid pH 2	—	*Locusta migratoria migratorioides* (Reiche & Fairmaire), 0.8%	35
Terpinen-4-ol: T1	*Amorpha fructicosa* L.	*Leptinotarsa decemlineata* (Say)	69, 70
		Locusta migratoria migratoroides (Reiche & Fairmaire)	69, 70
α-Terpineol: T2	*Amorpha fructicosa* L., *Pinus silvestris* L.	*Pieris brassicae* (L.)	69, 70
		Dendrolimus pini (L.), 0.3%/0.5 cm³	50
		Leptinotarsa decemlineata (Say)	69, 70
		Locusta migratoria migratoroides (Reiche & Fairmaire)	69, 70
Terpineol acetate: T3	*Pinus silvestris* L.	*Pieris brassicae* (L.)	69, 70
		Dendrolimus pini (L.), 0.3%/0.5 cm³	50

TABLE 1 (continued)
Insect Feeding Deterrents

Feeding deterrent	Plant source	Insect scientific name, conc for deterrent activity	Ref.
Terpinolene: T4	—	*Reticulitermes lucifugus santonensis* (Feytaud), 0.1 ng/insect	36, 62
Tetraphenylphosphonium chloride (Ph₄PCl)	—	*Tribolium castaneum* (Herbst), 500—2000 μmol/kg	132
Tetraphenylstibonium chloride (Ph₄SbCl)		*Tribolium confusum* Jacquelin duVal, 500—2000 μmol/kg	132
		Tribolium castaneum (Herbst), 500—2000 μmol/kg	132
	—	*Tribolium confusum* Jacquelin duVal, 500—2000 μmol/kg	132
Thiabendazole (TBZ): T5	—	*Myzus persicae* (Sulzer), 0.01%	44
Tomatidine: T6	—	*Empoasca fabae* (Harris)	31
Tomatine: T7	*Nepeta cataria* L., *Solanum polyadenium* Greenman, various	"Chrysomelidae (Fam.), no sp. specified"	31, 36
		"Pieridae (Fam.), no sp. specified"	31, 36
		Empoasca fabae (Harris)	31
		Leptinotarsa decemlineata (Say)	31, 32
		Locusta migratoria (L.), 3.0%	7
		Locusta migratoria migratorioides (Reiche & Fairmaire), 0.15%	8, 35
		Manduca sexta (L.)	31
		Melanoplus bivittatus (Say), 2.24%	32, 79
		Epilachna varivestis Mulsant	18
Toonacilin: A3	*Cedrela toona* Roxb. ex Willd. var. *australis*	*Spodoptera frugiperda* (J. E. Smith), 500 ppm	119
Toxol	—	*Spodoptera frugiperda* (J. E. Smith), 500 ppm	119
Toxyl angelate	*Isocoma wrightii* (Gray) Rydb.	*Homoeosoma electellum* (Hulst), 0.5—2.0%	36, 104
Trachyloban-19-oic acid: T8	*Helianthus annuus* L.	*Pectinophora gossypiella* (Saunders)	36, 104
		Danaus plexippus (L.)	48
Trichlorophenoxyacetic acid: T9	—	*Danaus plexippus* (L.)	48
Trichlorophenoxyethanol: T10	—	*Epilachna implicata* Mulsant, 0.2%	14, 133
Tricyclohexyltin hydroxide (Plictran)	—	*Euproctis fraterna* (Moore), 0.2%	14, 133
		Pericallia ricini (F.), 0.2%	14, 133
		Spodoptera litura (F.), 0.2%	14, 133
		Sitotroga cerealella (Olivier), 0.2%	78

Compound	Source	Insect (concentration)	Ref.
Triphenyllead chloride (Ph₃PbCl)	—	Tribolium castaneum (Herbst), 500 µmol/kg	134
Triphenylgermanium chloride (Ph₃GeCl)	—	Tribolium castaneum (Herbst), 2000 µmol/kg	134
Triphenylmethyl chloride (Ph₃CCl₃)	—	Tribolium castaneum (Herbst), >2000 µmol/kg	134
Triphenylsilanol (Ph₃SiOH)	—	Tribolium castaneum (Herbst), 2000 µmol/kg	134
Triphenylstilbine (Ph₃Sb)	—	Tribolium castaneum (Herbst), 500—2000 µmol/kg	132
		Tribolium confusum Jacquelin duVal, 500—2000 µmol/kg	132
Triphenyltin chloride (Ph₃SnCl)	—	Diacrisia obliqua (Walker), 0.09%	135
		Tribolium castaneum (Herbst), 500 µmol/kg	134
Triphenyltin hydroxide (Ph₃SnOH)	—	Neodiprion rugifrons Middleton, 8.6 mg/ml	74
		Neodiprion swainei Middleton, 8.6 mg/ml	74
Tulirinol: T11	Liriodendron tulipifera L.	Spodoptera littoralis (Boisduval), 0.25—0.5%	90
Tyramine HCl: T12	—	Lymantria dispar (L.), 50—250 µg/ml	84
		Laodelphax striatellus (Fallen), 10—1000 ppm	101
		Nephotettix cincticeps (Uhler), 10—1000 ppm	101
		Nilaparvata lugens (Stal), 10—1000 ppm	101
		Sogatella furcifera (Horvath), 10—1000 ppm	101
L-Tyrosine	—	Spodoptera littoralis (Boisduval), 0.125 M	24
Ugandensidial: U1	Warburgia stuhlmannii Engl., Warburgia salutaris (Bertol f.) Chiov.	Spodoptera exempta (Walker), >0.1 ppm	17, 23, 63, 113, 124
		Spodoptera littoralis (Boisduval), 0.1 ppm	124
Unedoside: U2	Arbutus unedo L., Canthium euryoides Bullock ex Hutch. & Dalz.	"No sp. specified"	5
Ursolic acid: U3	—	Locusta migratoria migratorioides (Reiche & Fairmaire), 0.1%	35
Uzarigenin: U4, and its glycoside	Gomphocarpus sp.	Phaedon cochleariae F.	27, 51
Vanilic acid: V1	Sorghum bicolor (L.) Moench	Locusta migratoria (L.), 0.02%	7
Veratrine (mixture of alkaloids)	Veratrum spp.	Melanoplus bivittatus (Say), 1.0%	27, 31, 79
(3R)-(−)-Vestitol: V2	Lotus pedunculatus Cav.	Costelytra zealandica (White), 8.5—100 µg/g medium	127
Vulpinic acid: V3	Letharia vulpina (L.) Vin. (wolf lichen)	Spodoptera ornithogalli (Guenee), 0.6%	2
Warburganal: W1	Warburgia stuhlmannii Engl.	Spodoptera exempta (Walker), 0.1 ppm	16, 23, 113, 114, 124
	Warburgia salutaris (Bertol f.) Chiov.	Spodoptera littoralis (Boisduval), 0.1 ppm	17, 136, 124
Withaferin A: W2	Withania somnifera (L.) Dun.	Locusta migratoria migratorioides (Reiche & Fairmaire), 0.01%	27, 35

TABLE 1 (continued)
Insect Feeding Deterrents

Feeding deterrent	Plant source	Insect scientific name, conc for deterrent activity	Ref.
Xanthotoxin: X1	*Orixa japonica* Thunberg	*Blattella germanica* (L.), 67 ppm	34
		Neostylopyga rhombifolia (Stoll), 67 ppm	34
		Periplaneta americana (L.), 67 ppm	34
		Spodoptera litura (F.), 500—1000 ppm	34, 47
		Spodoptera exempta (Walker)	17, 23, 36, 137
Xylomollin: X2	*Xylocarpus molluccensis* (Lam.) M. Roemer		
Zanthophylline: Z1	*Zanthoxylum monophyllum* (Lam.) P. Wilson	*Hemileuca oliviae* Cockerell, 200—500 ppm	110
		Hypera postica (Gyllenhal), 200—500 ppm	110
		Melanoplus sanguinipes (F.), 200—500 ppm	110

ᵃ Alphanumeric codes refer to selected chemical structures in Figure 1.

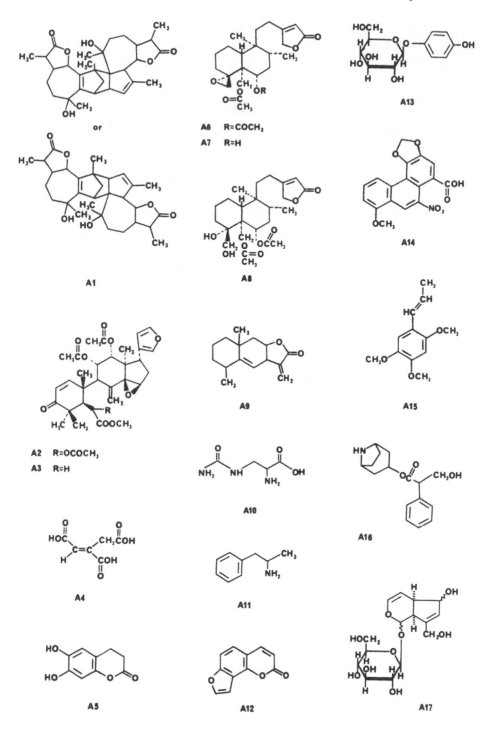

FIGURE 1. Structural formulas of selected feeding deterrents in alphanumeric order as they appear in Table 1. (A3, C10-12, C14-18, C29, and D2 are out of order in Table 1 because of the names of their parent compounds.) The codes also appear in Tables 2 and 3.

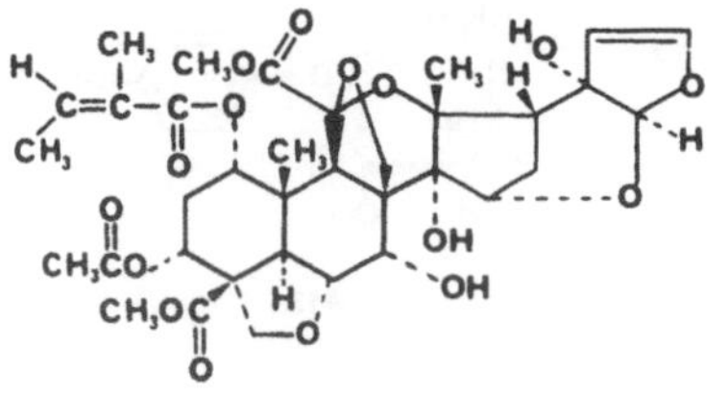

FIGURE 1.   Continued.

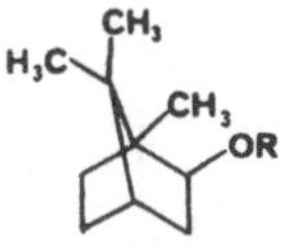

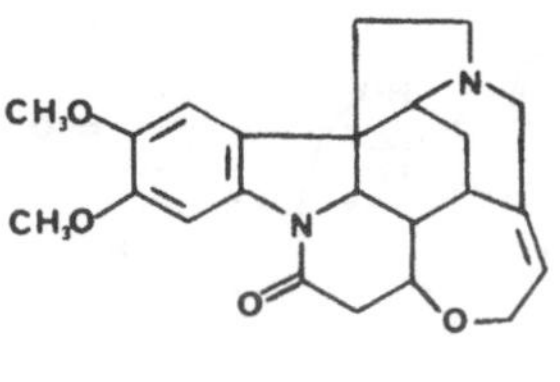

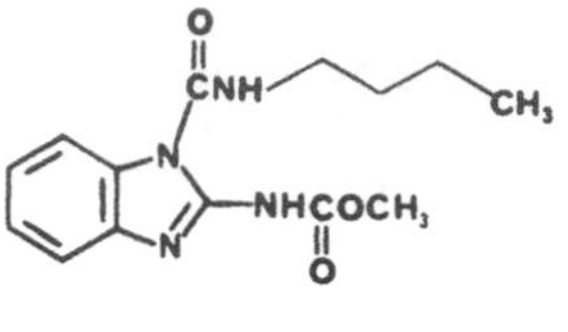

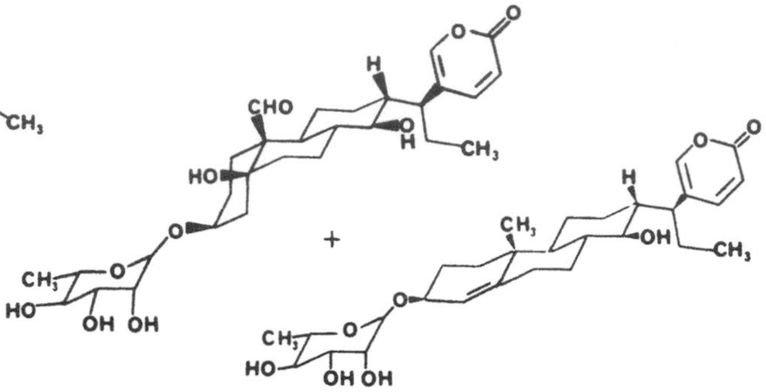

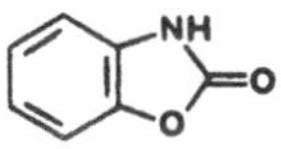

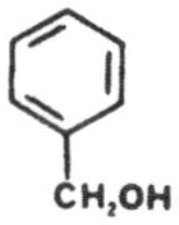

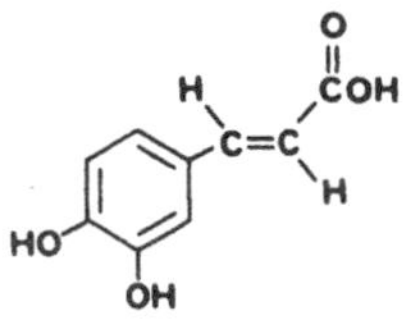

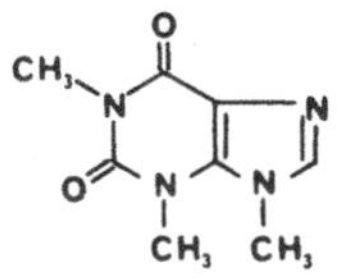

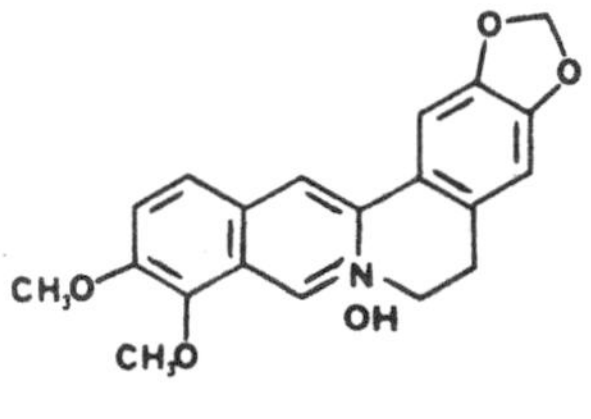

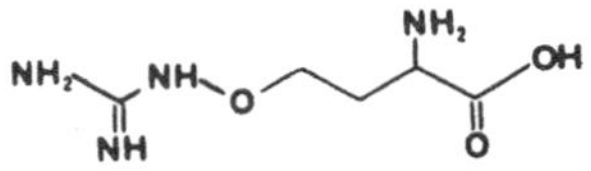

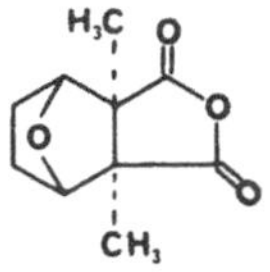

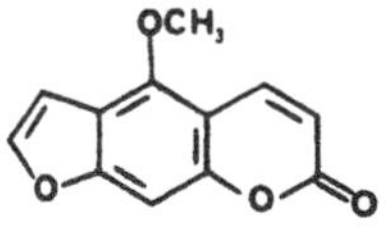

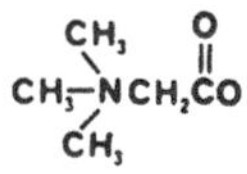

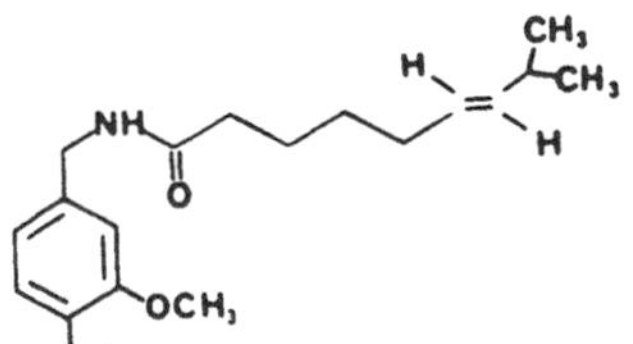

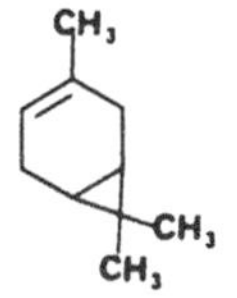

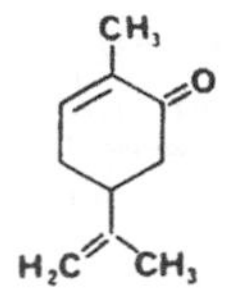

C7

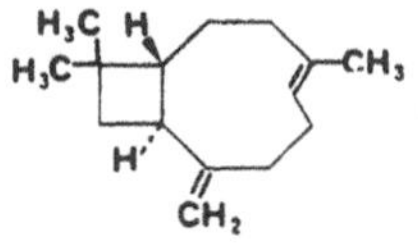

C8

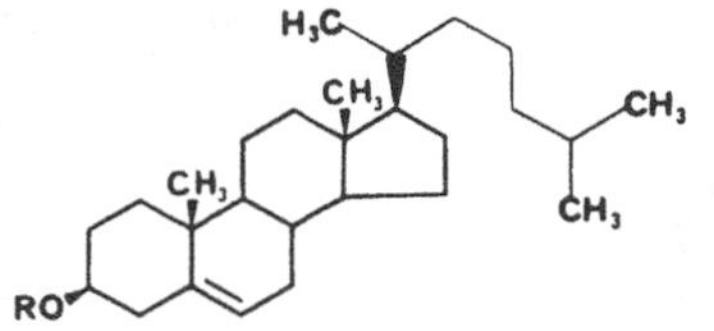

C20 R=COCH$_3$
C21 R=C$_{14}$H$_{27}$O
C22 R=C$_{18}$H$_{33}$O

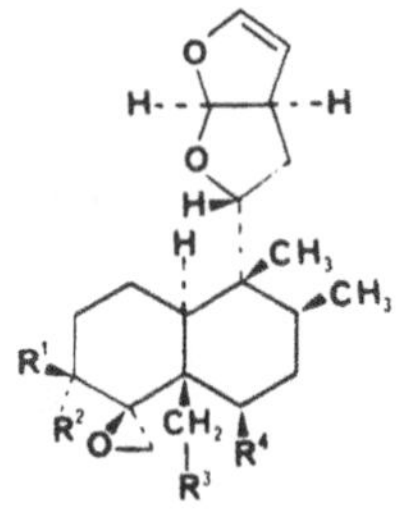

C9   R^1=R^3=R^4=OCOCH$_3$,R^2=H
C10 R^1=OH,R^2=H,R^3=R^4=OCOCH$_3$
C11 R^1=R^2=H,R^3=R^4=OCOCH$_3$
C12 R^1=H,R^2=R^3=R^4=OCOCH$_3$

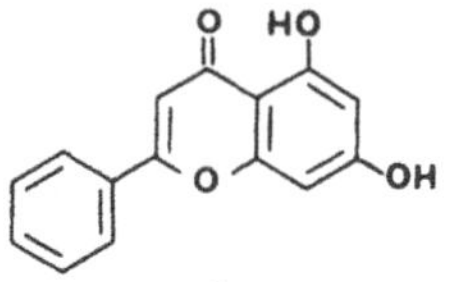

C23

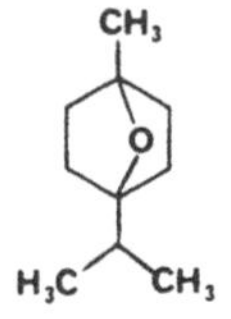

C24

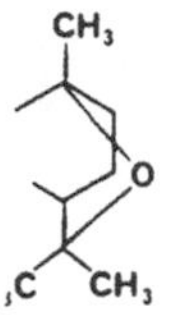

C25

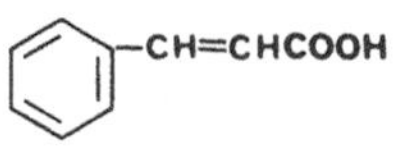

C26

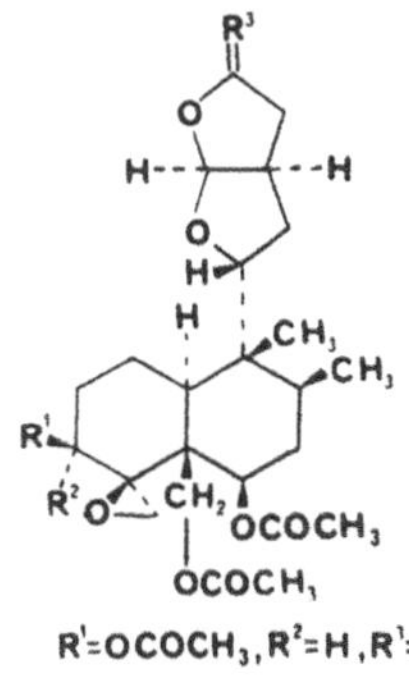

C13 R^1=OCOCH$_3$,R^2=H,R^3=H,OH
C14 R^1=R^2=H,R^3=H,OH
C15 R^1=OCOCH$_3$,R^2=H,R^3=H$_2$
C16 R^1=OH,R^2=H,R^3=H$_2$
C17 R^1=R^2=H,R^3=H$_2$
C18 R^1=H,R^2=OCOCH$_3$,R^3=H

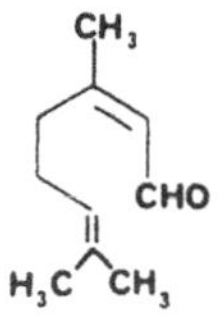

C27

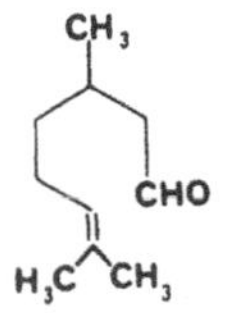

C28

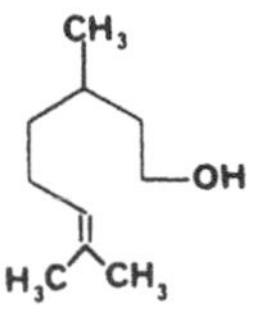

C29

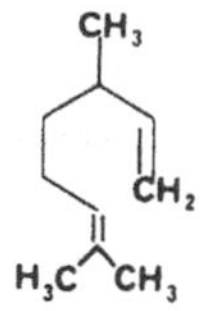

C19

C30

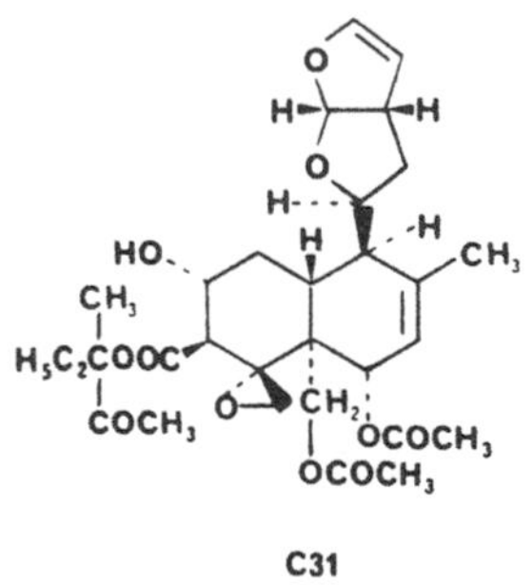

C31

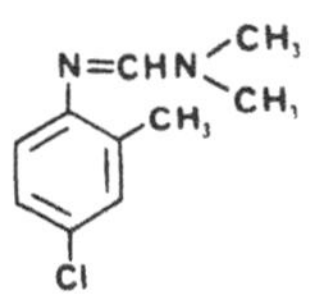

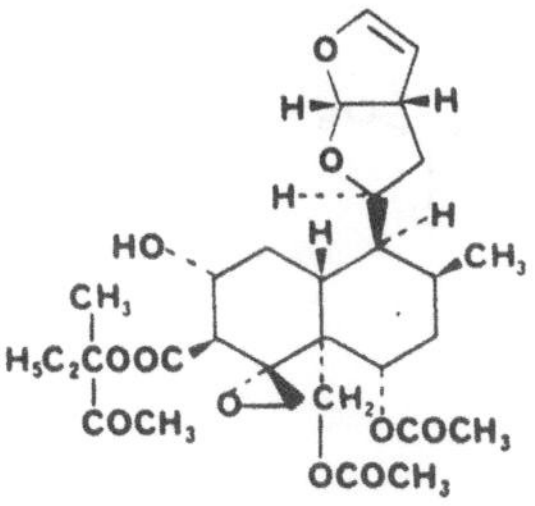

C32

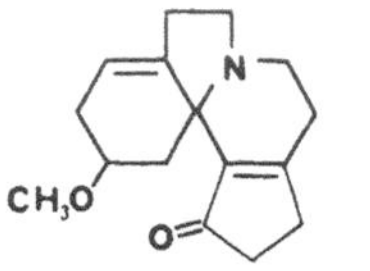

C33

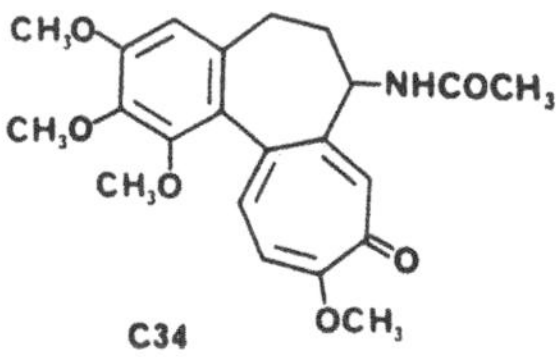

C34

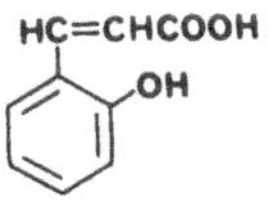

C35

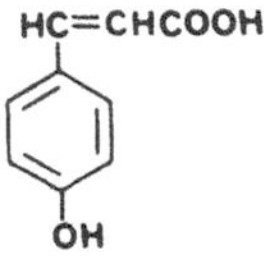

C36

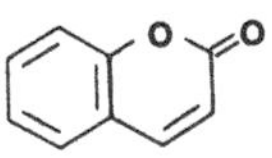

C37

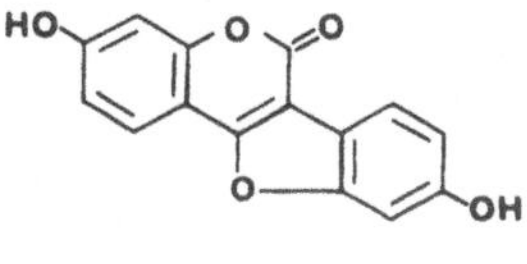

C38

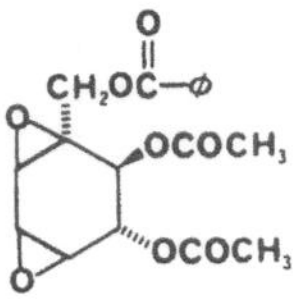

C39

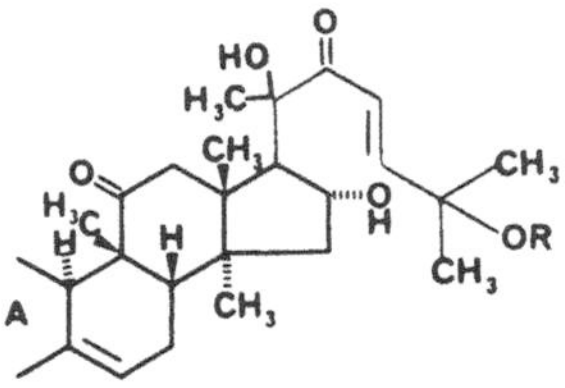

C40
R=COCH₃

C41
R=COCH₃

C42
R=H

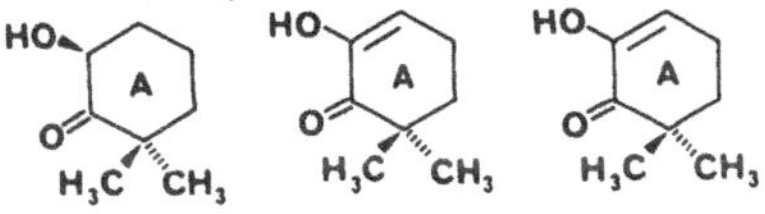

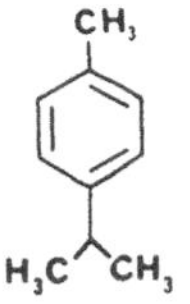

C43

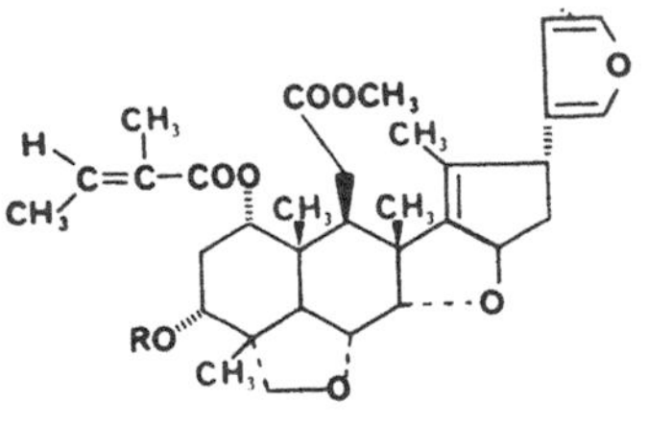

D1   R=H

D2   R=COCH₃

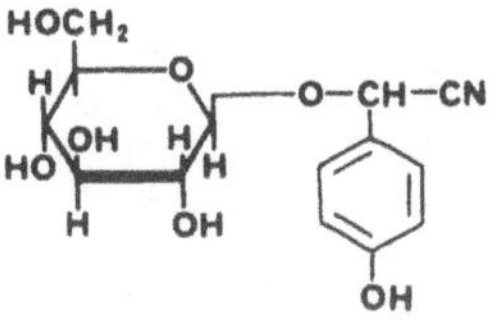

D3

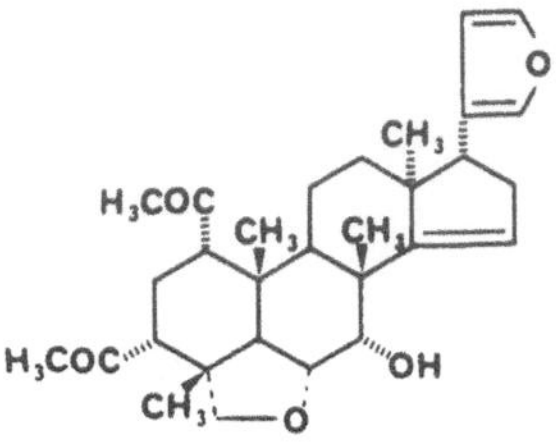

D4

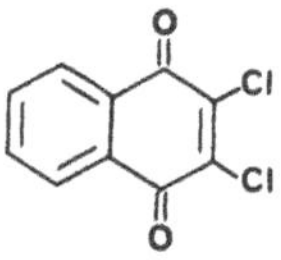

D5

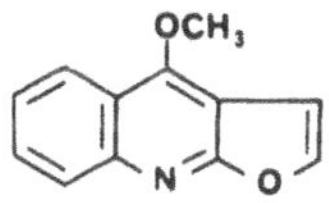

D6

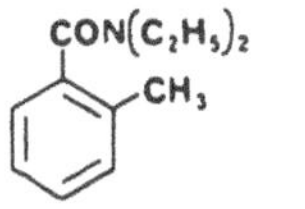

D7

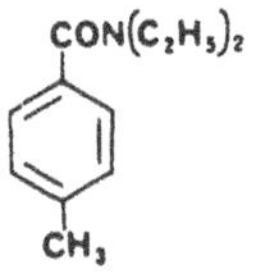

D8

FIGURE 1.   Continued.

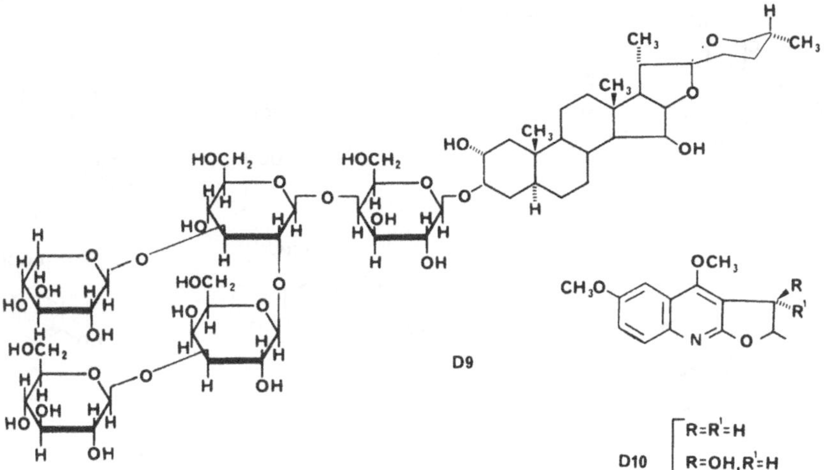

D9

D10

$$\begin{bmatrix} R=R'=H \\ R=OH, R'=H \\ R=H, R'=OH \end{bmatrix}$$

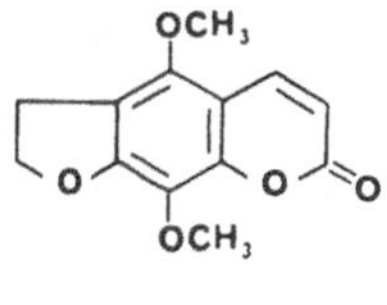

D11

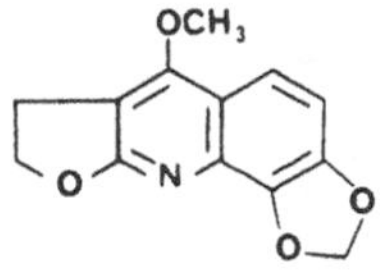

D12

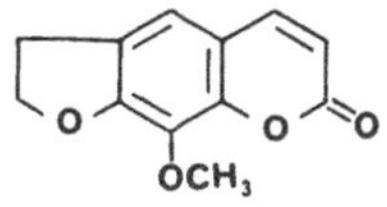

D13

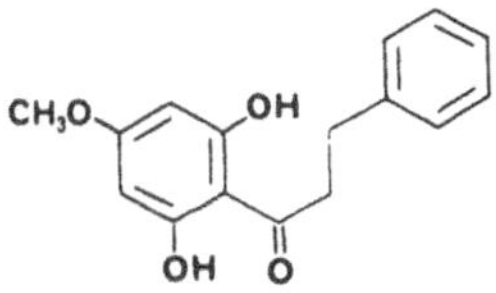

D14

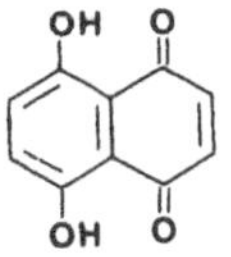

D15

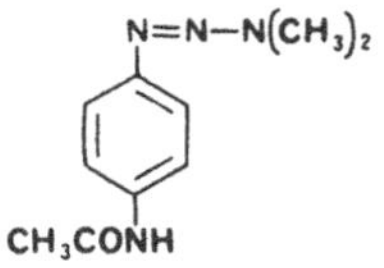

D16

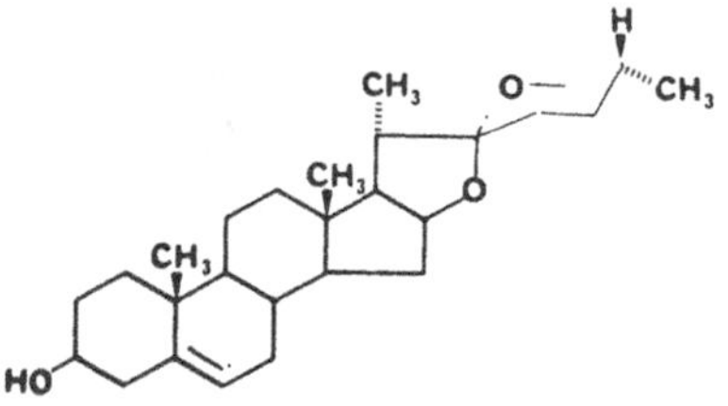

D17

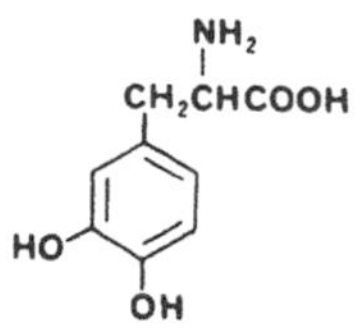

D18

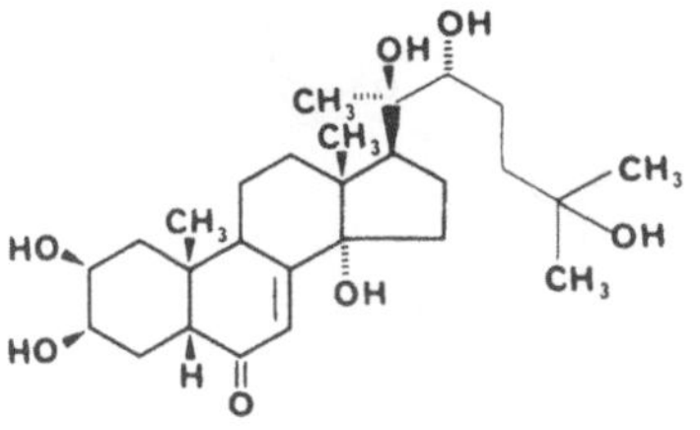

FIGURE 1. Continued.

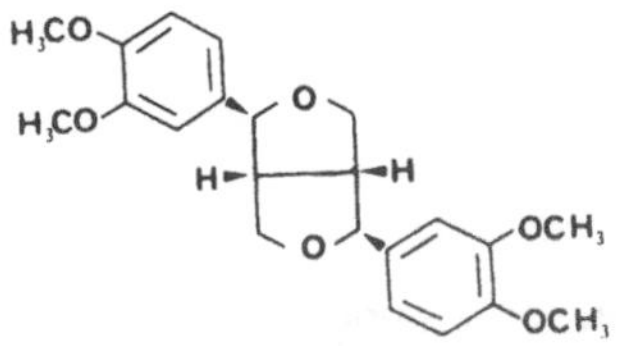

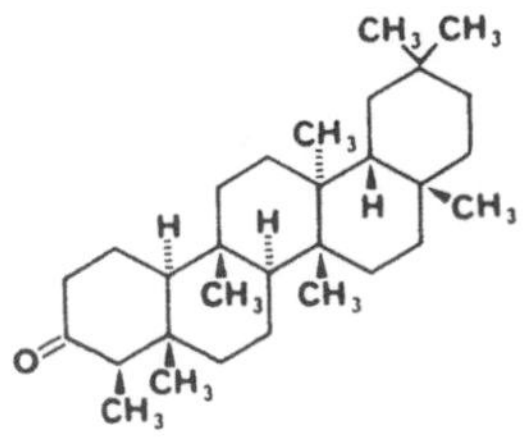

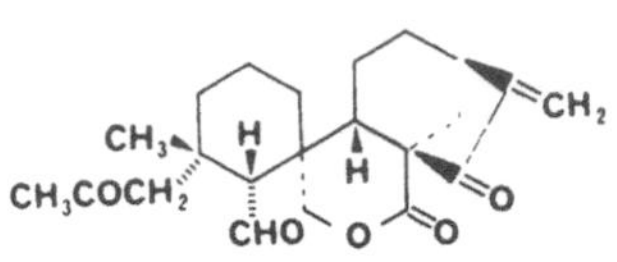

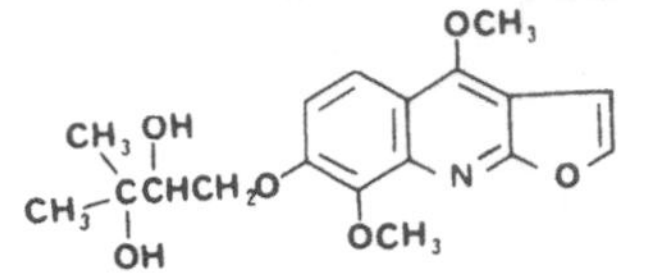

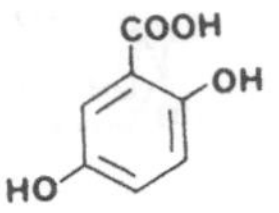

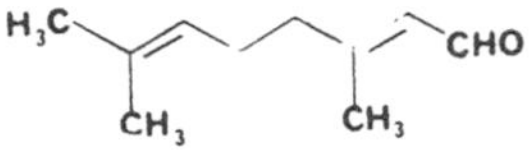

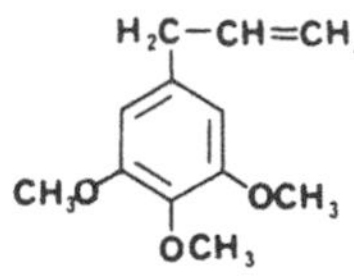

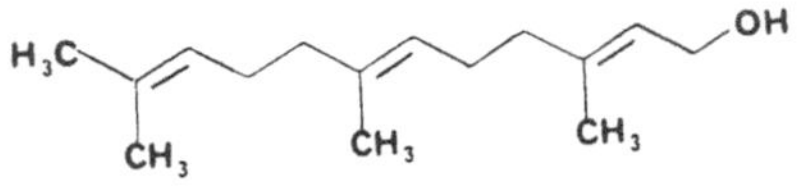

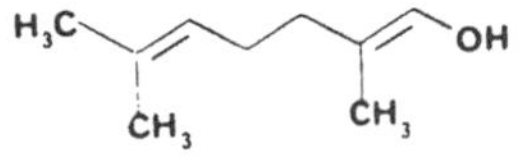

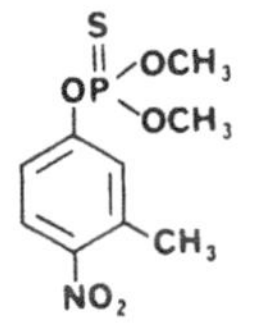

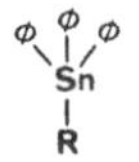

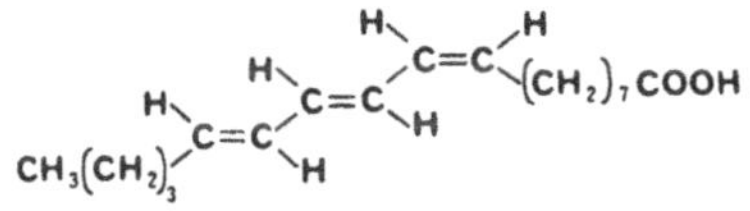

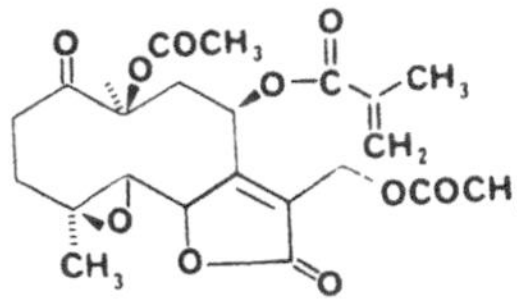

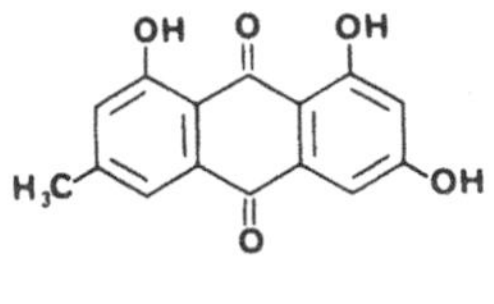

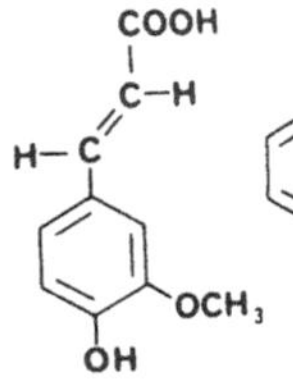

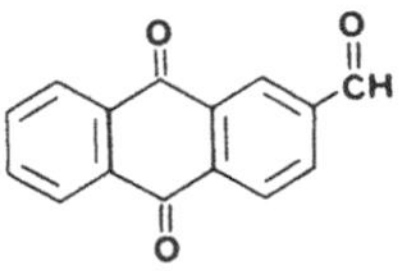

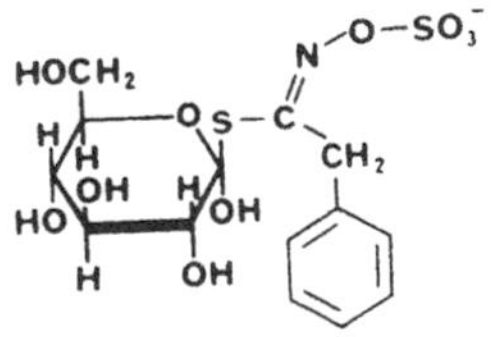

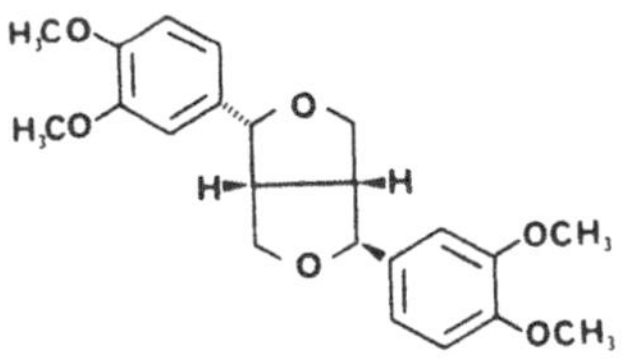

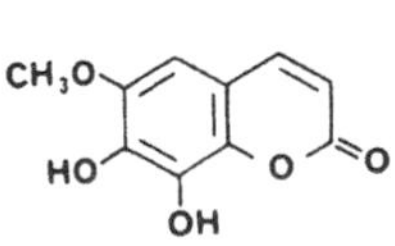

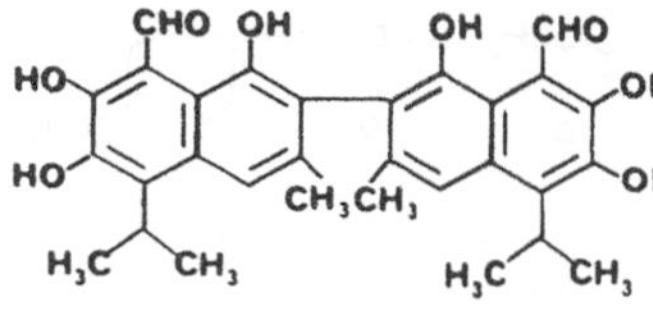

G6

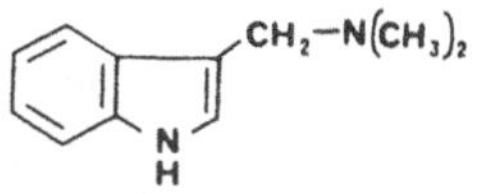

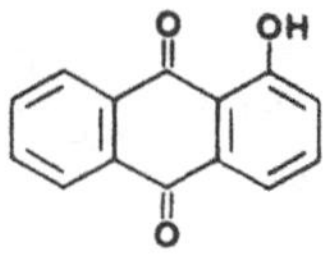

H5

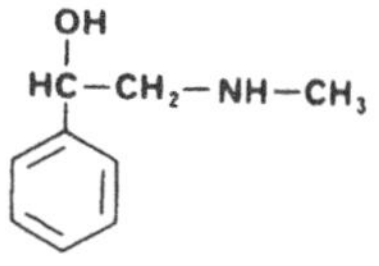

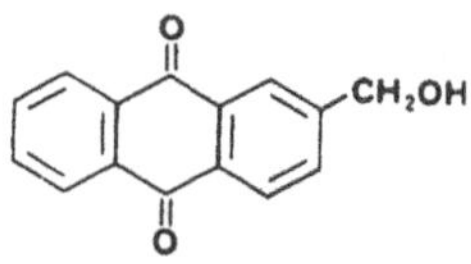

I3

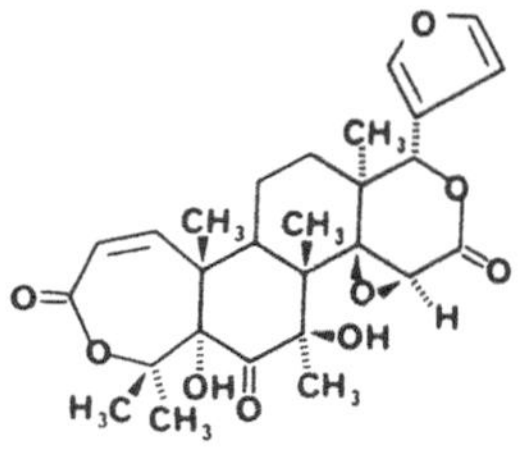

G7

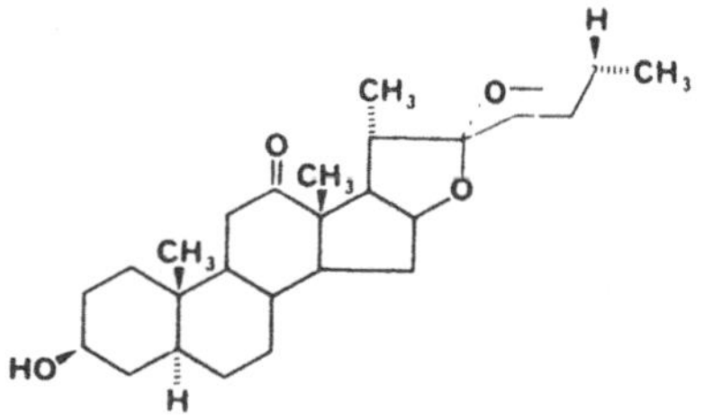

H6

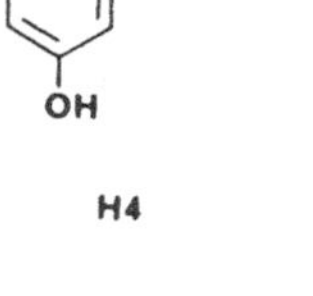

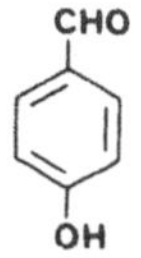

I4

H1

H7

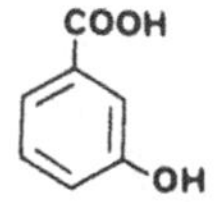

H8

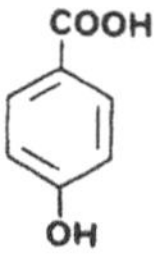

I5

H2

H9

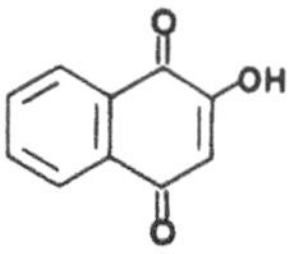

H10

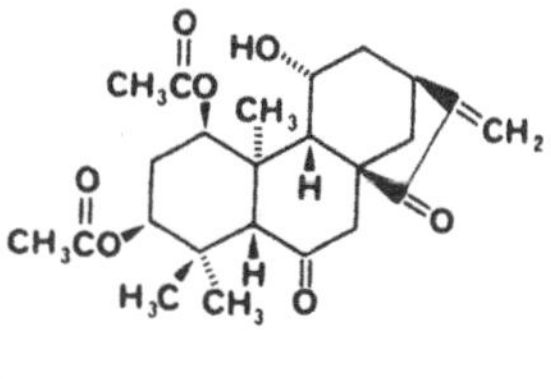

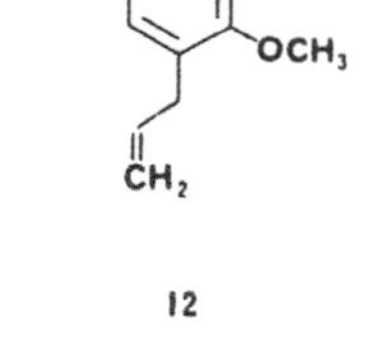

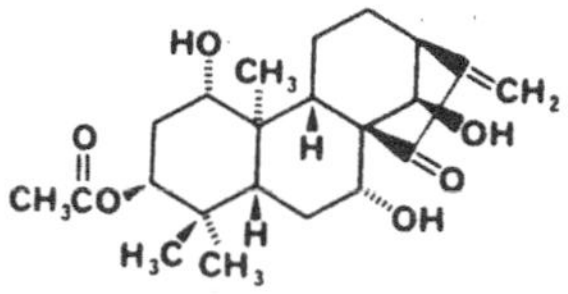

I6

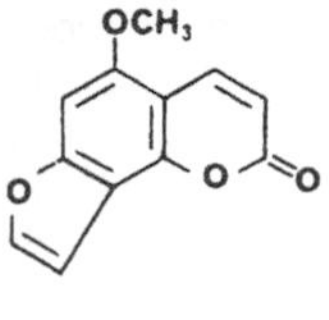

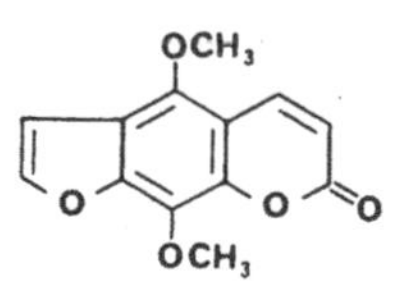

H3

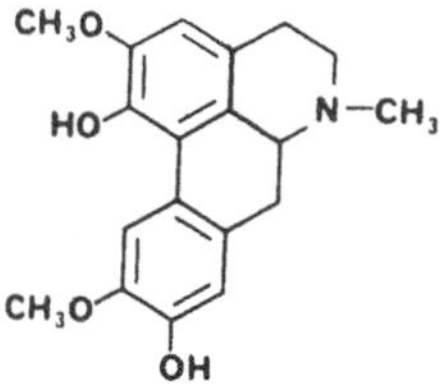

I1

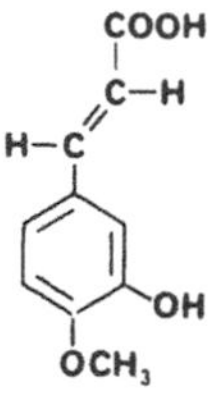

I7

H4

I2

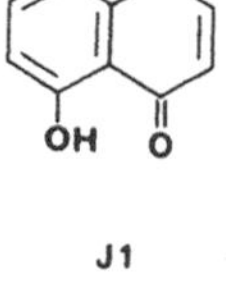

J1

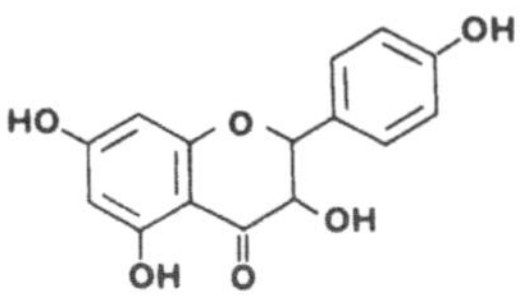

FIGURE 1. Continued.

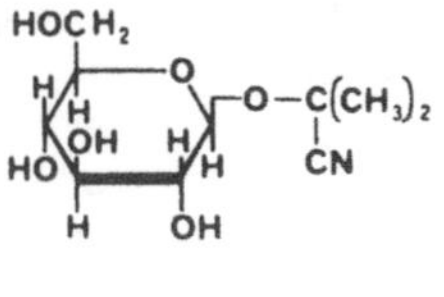

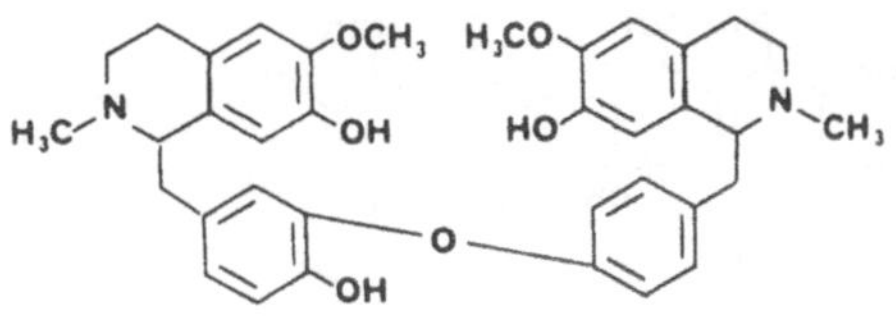

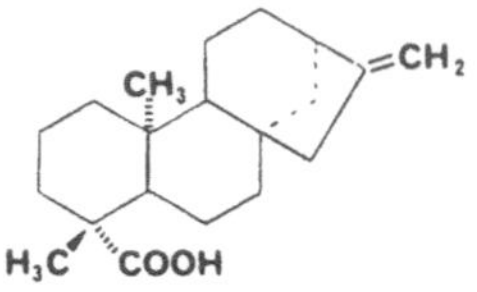

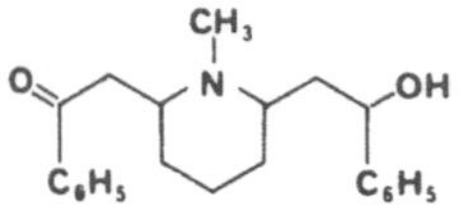

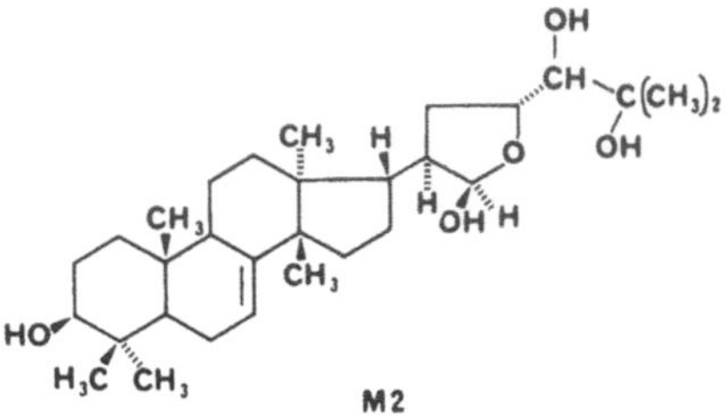

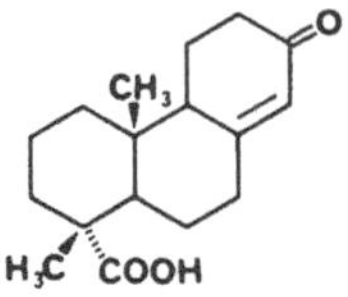

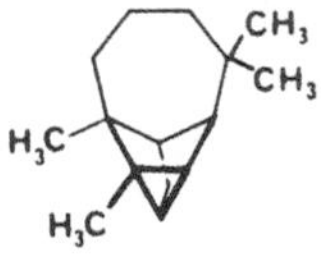

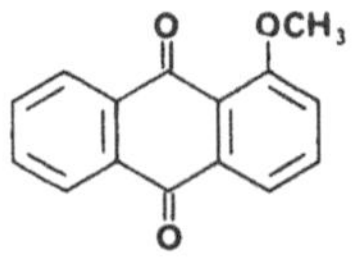

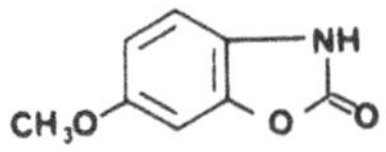

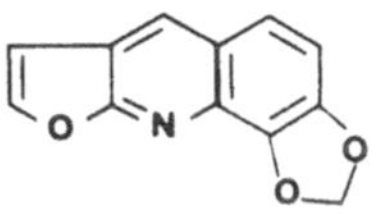

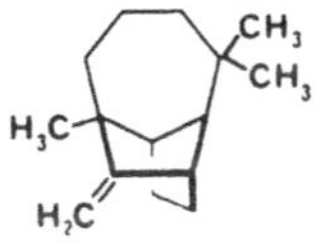

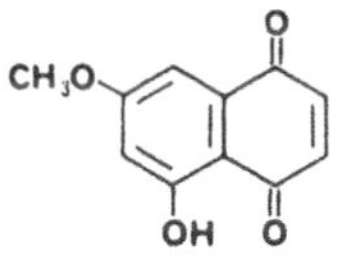

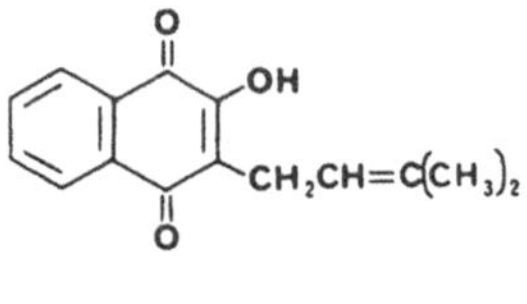

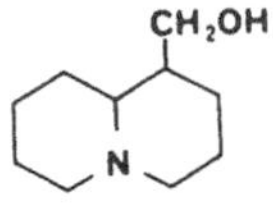

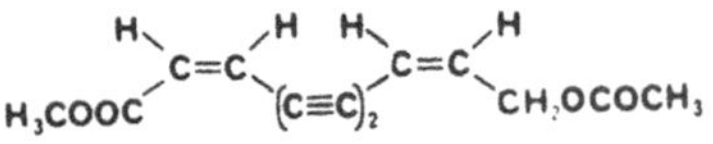

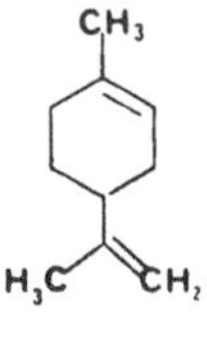

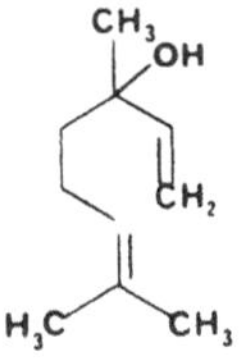

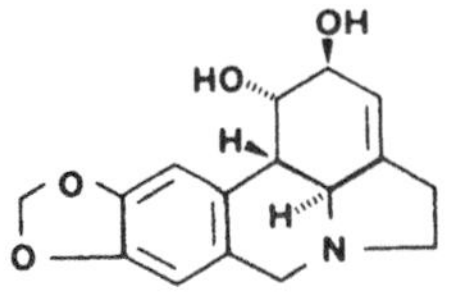

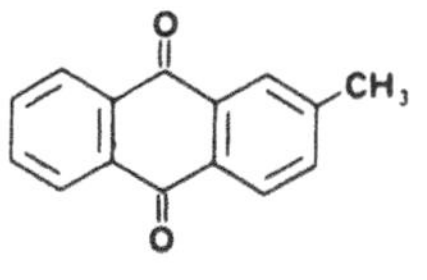

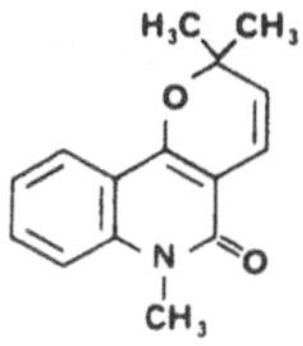

M8

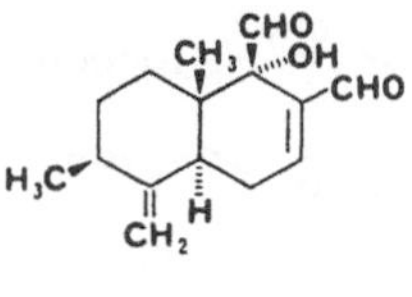

M15

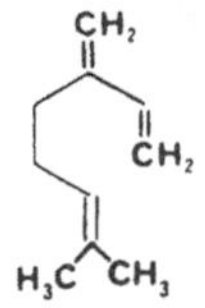

M16

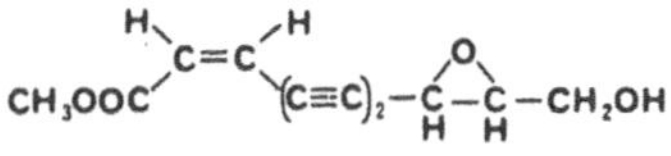

M9

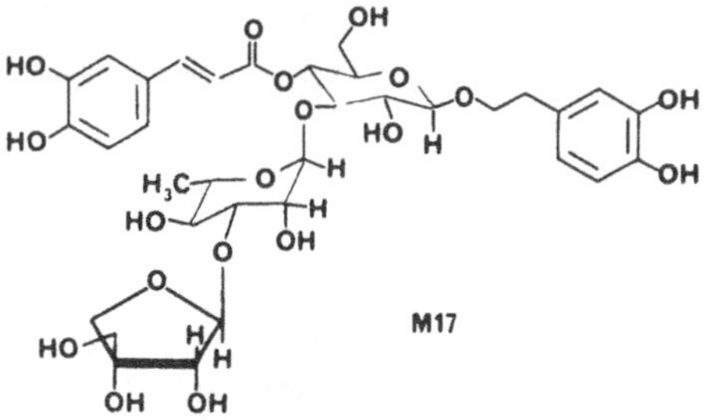

M17

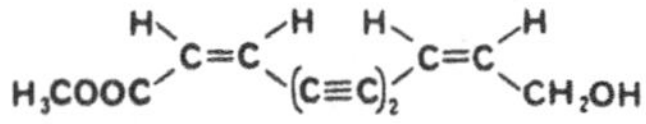

M10

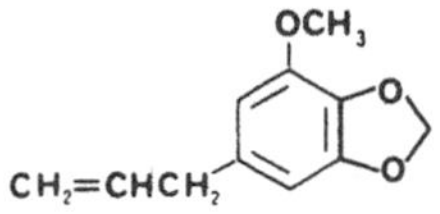

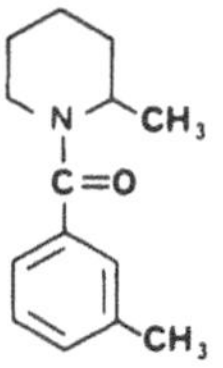

M11

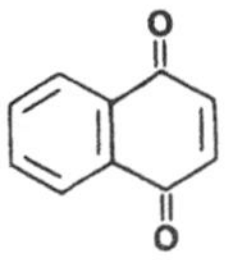

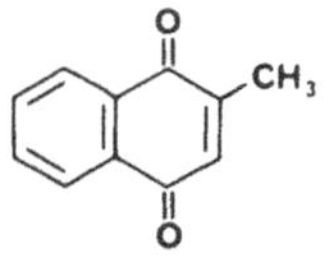

M12

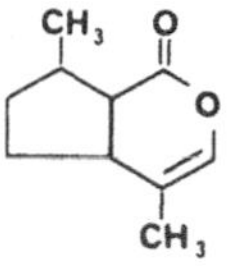

M18

N1

M13

N2

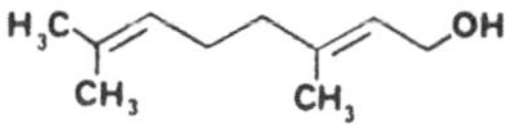

N3

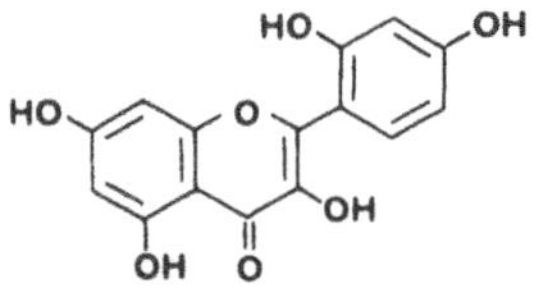

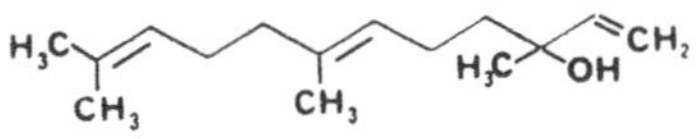

N4

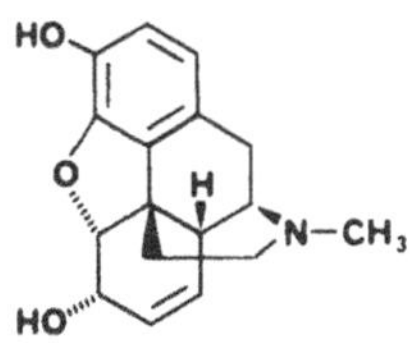

M14

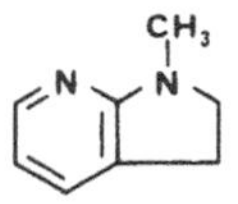

N5

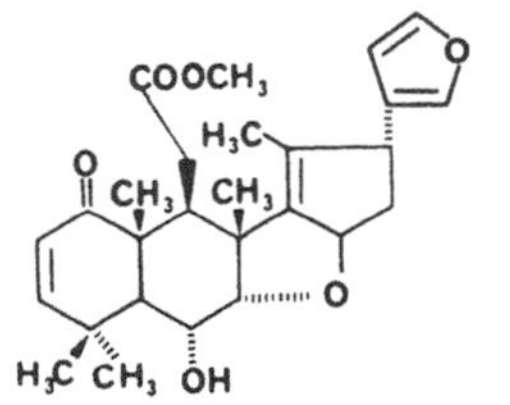

N6

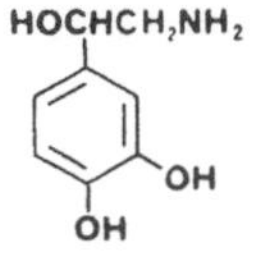

FIGURE 1. Continued.

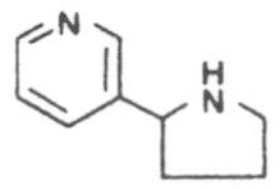

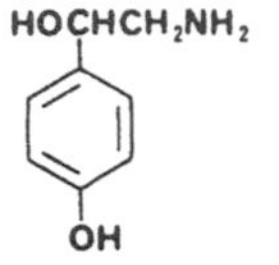

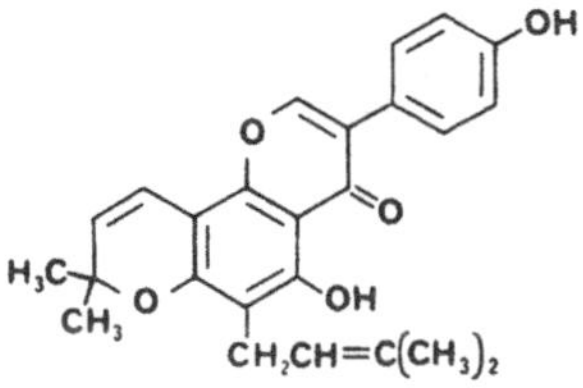

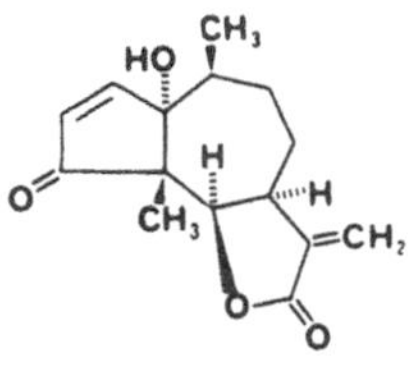

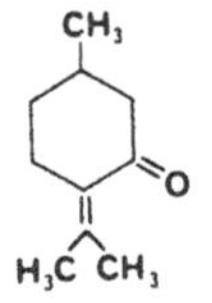

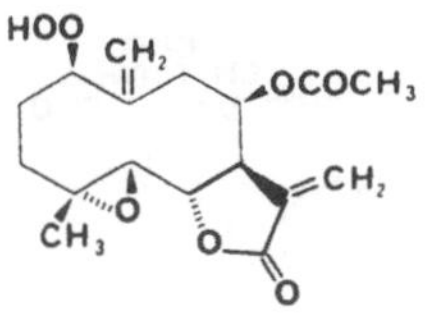

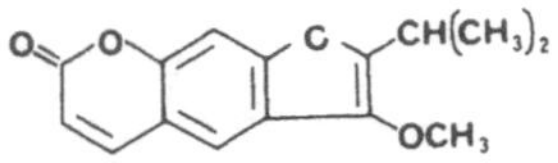

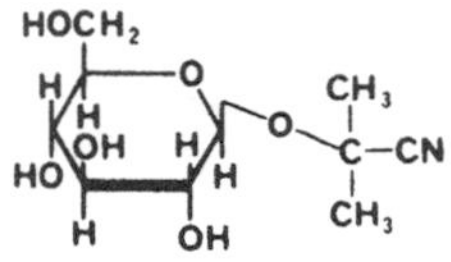

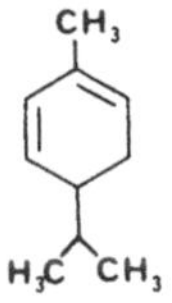

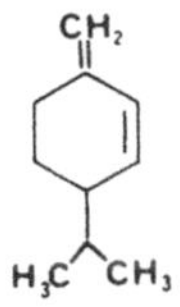

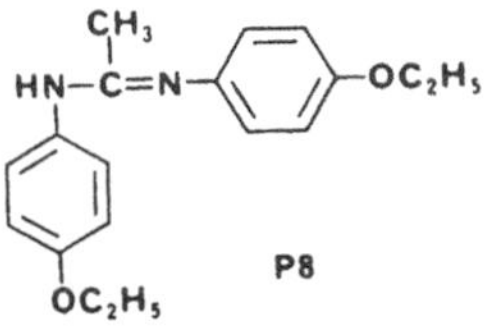

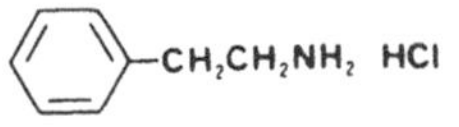

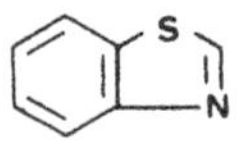

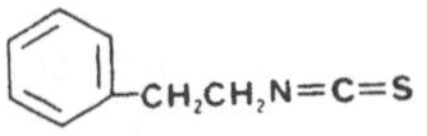

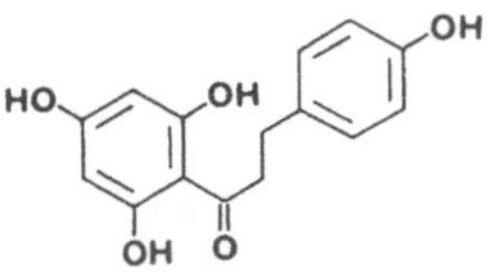

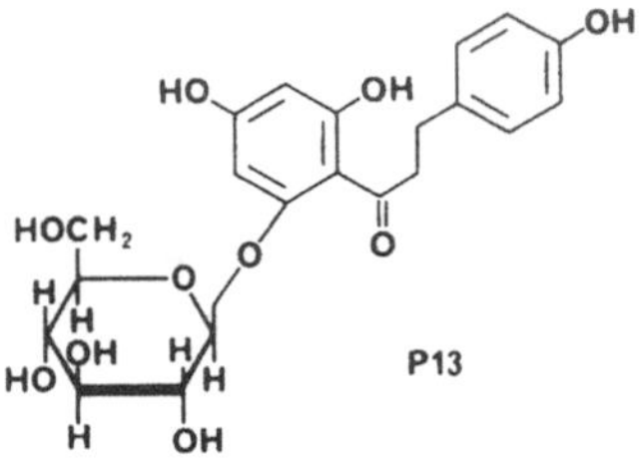

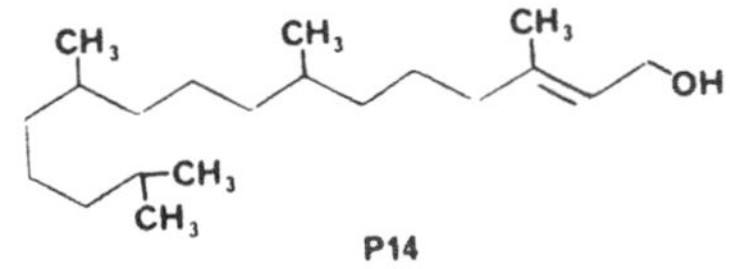

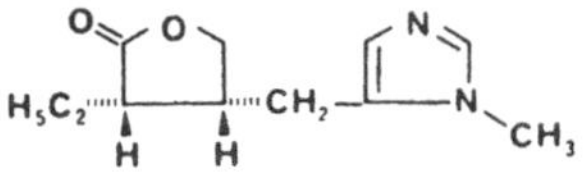

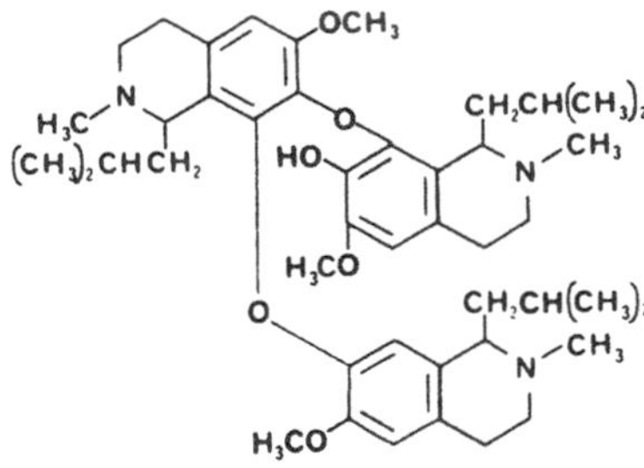

17

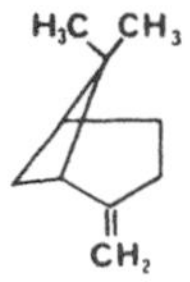

P18

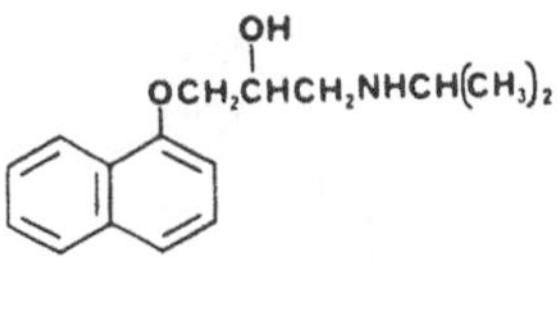

P25

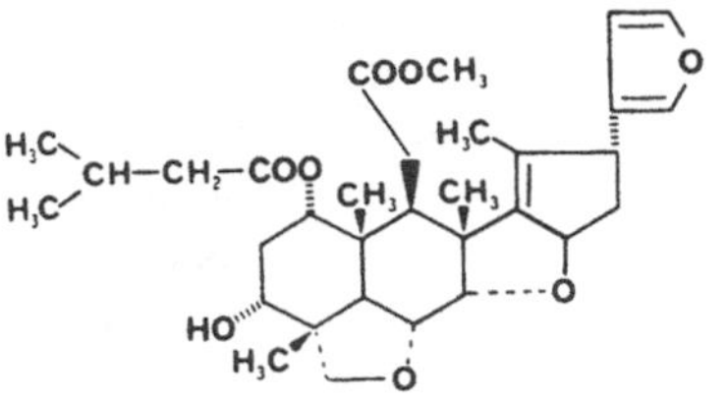

S1

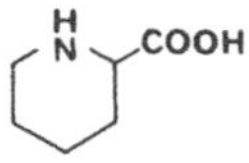

P19

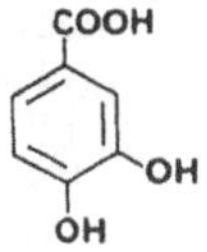

P26

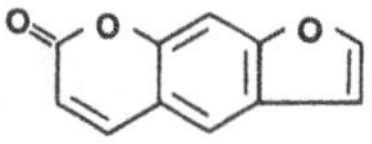

P27

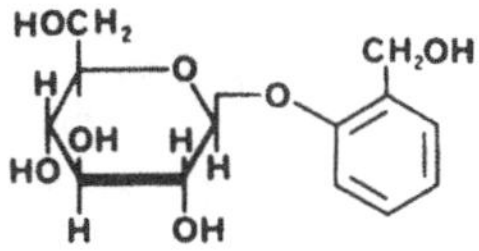

S2

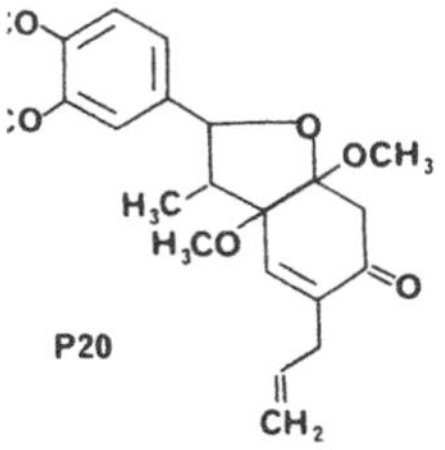

P20

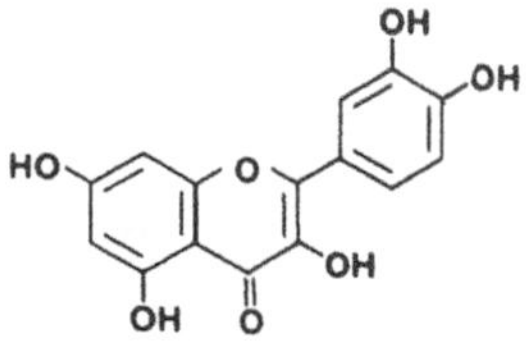

Q1

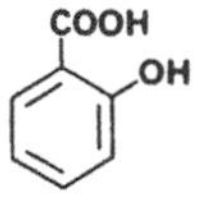

S3

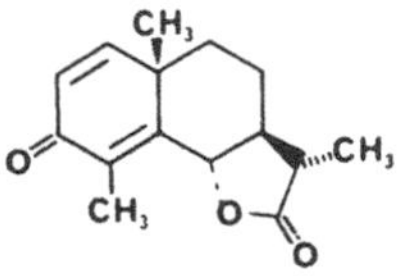

S4

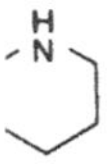

P21

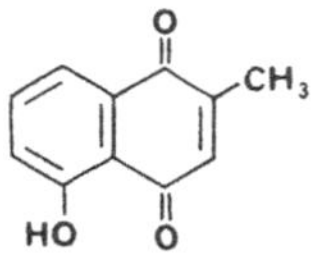

P22

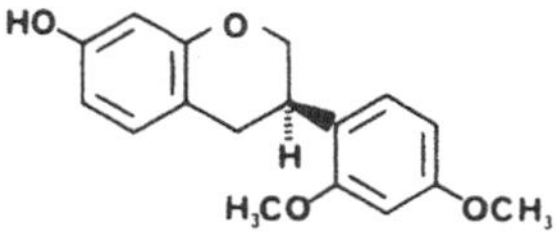

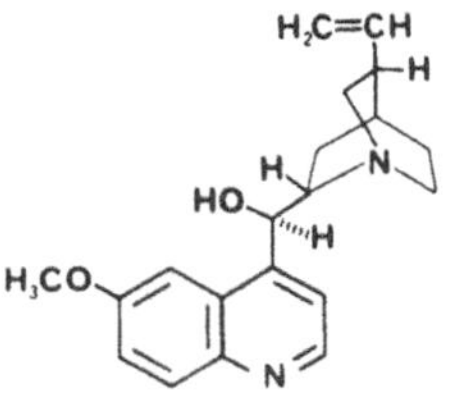

Q2

S5

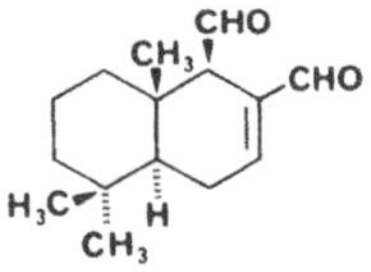

P23

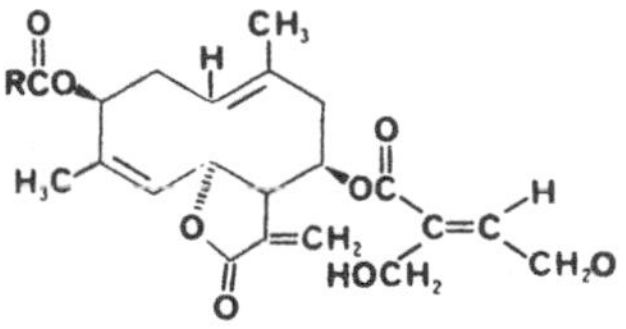

S6 R=CH₃
S7 R=CH(OH)CH(CH₃)₂

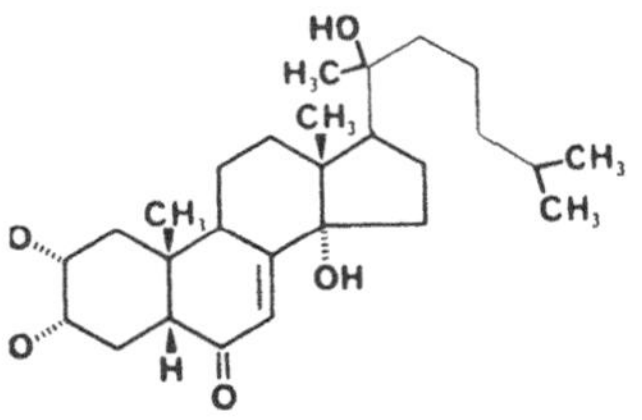

P24

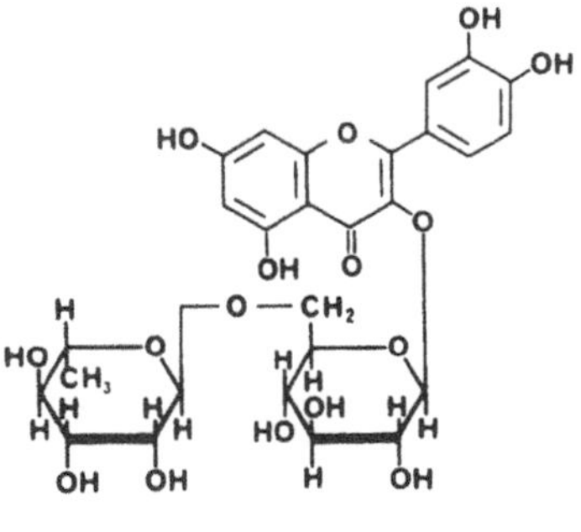

R1

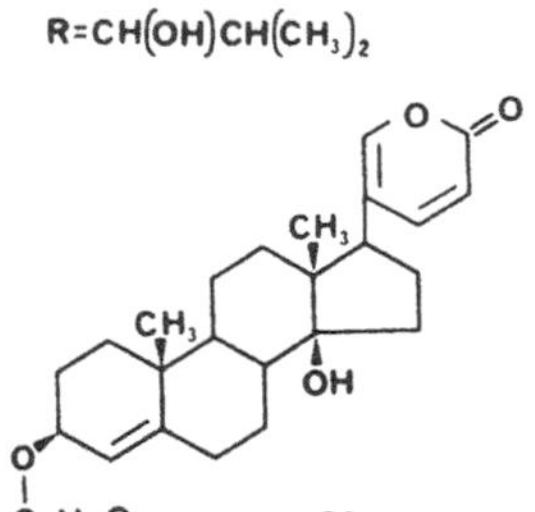

S8

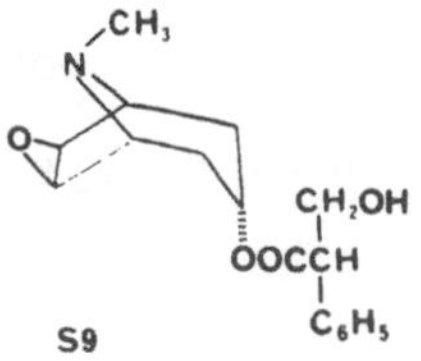

S9

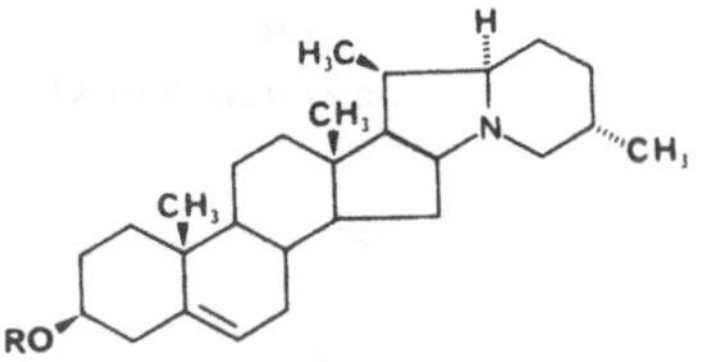

S16 R=H

S17 R=

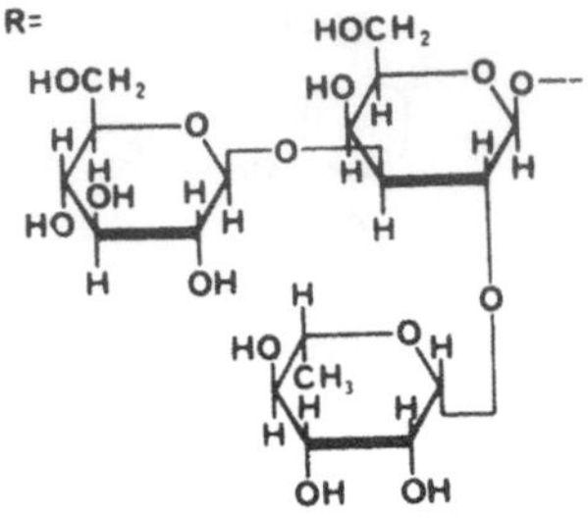

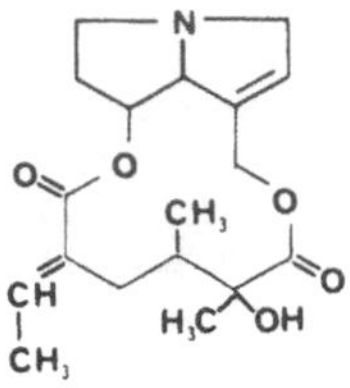

S10

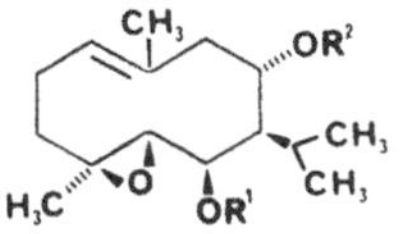

S11 R^1=R^2=COCH$_3$

S12 R^1=H, R^2=COCH$_3$

S13 R^1=R^2=H

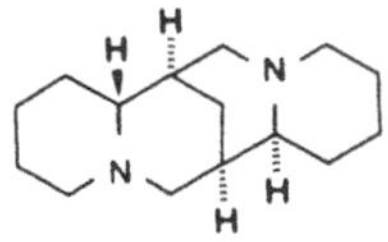

S18

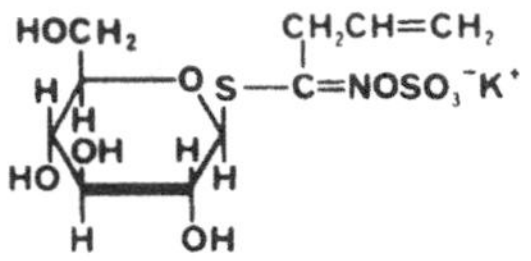

S14

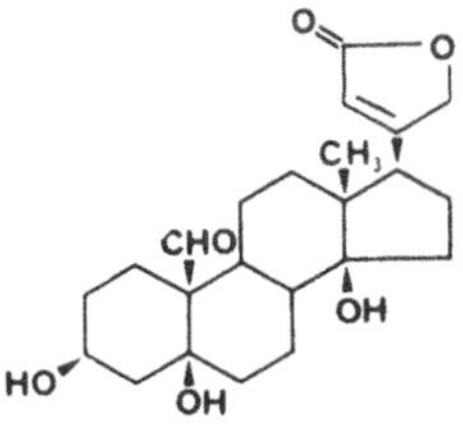

S19

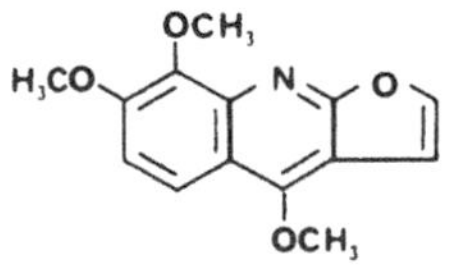

S15

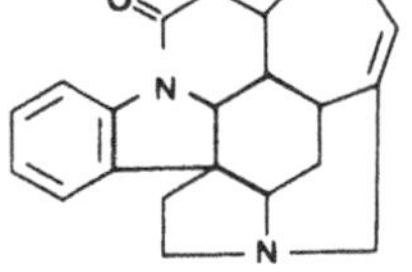

S20

FIGURE 1. Continued.

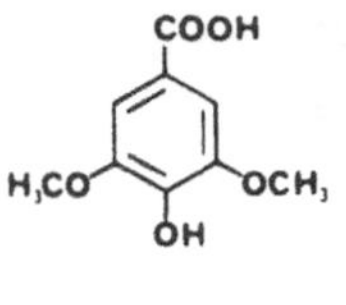

S21

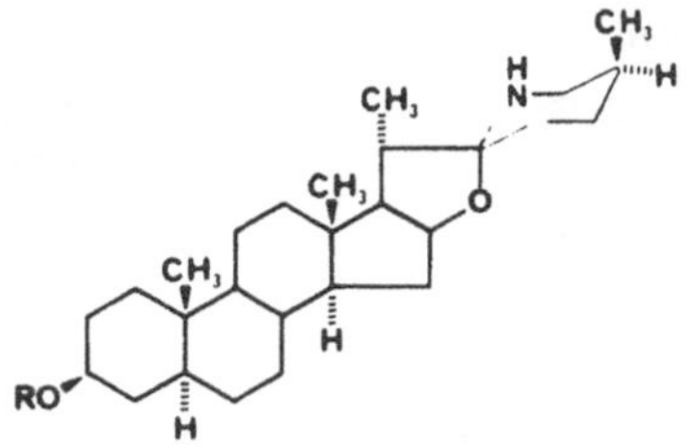

T6   R=H

T7   R=

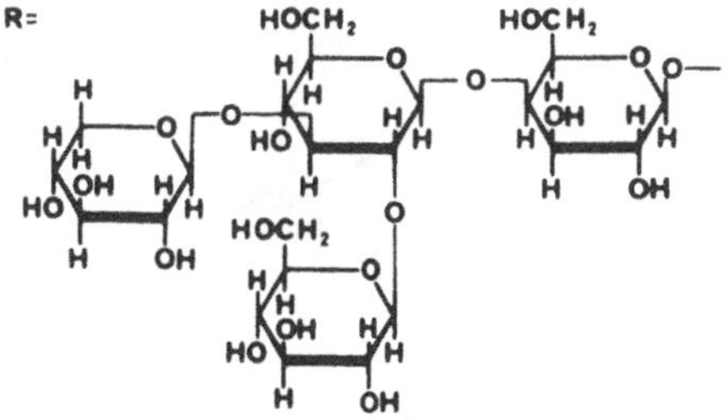

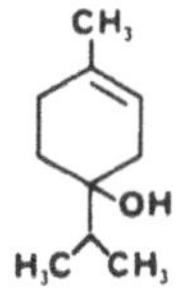

T1

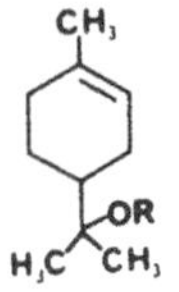

T2   R=H

T3   R=COCH₃

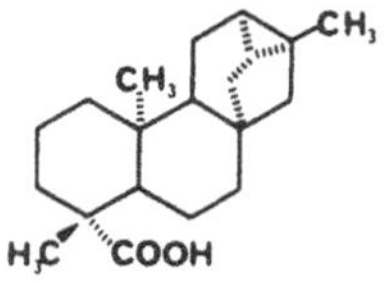

T8

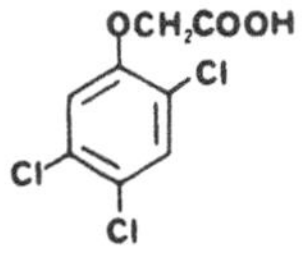

T9

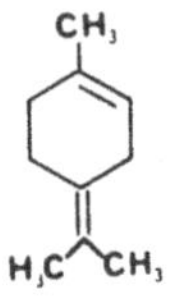

T4

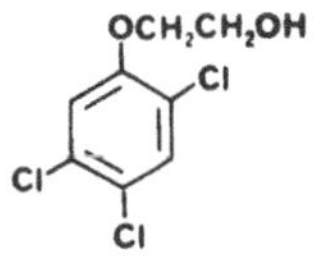

T10

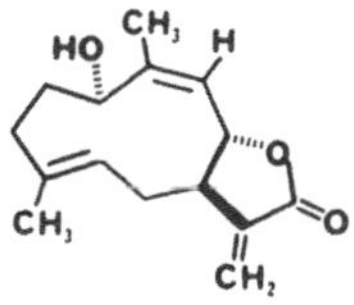

T11

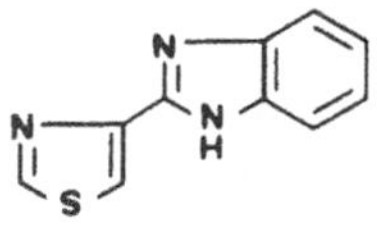

T5

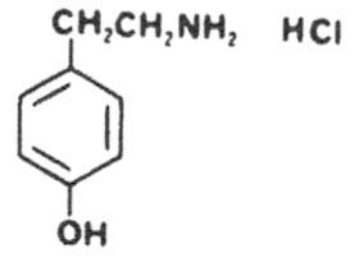

T12

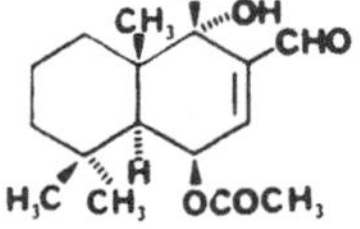

U1

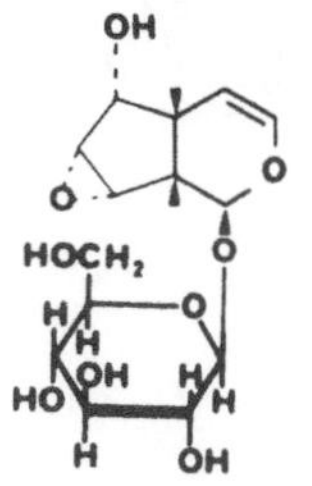

FIGURE 1.   Continued.

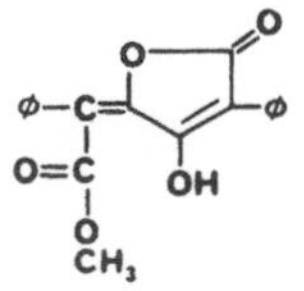

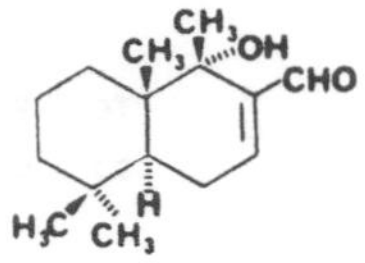

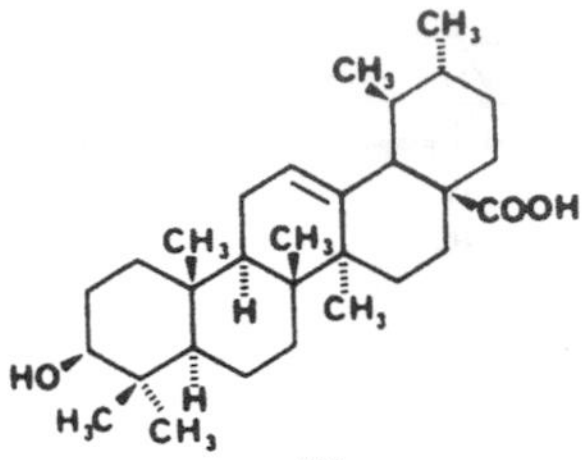

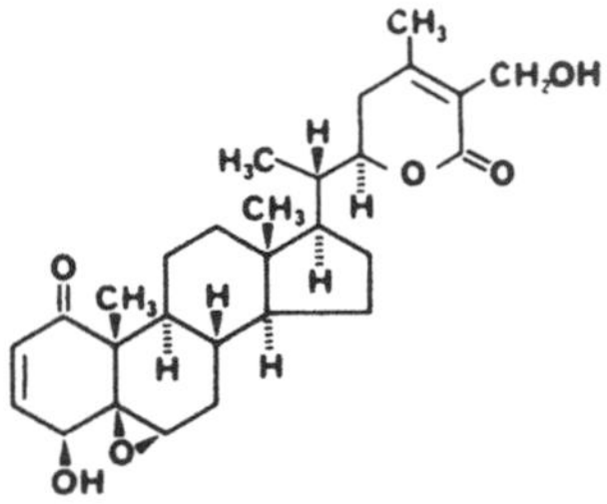

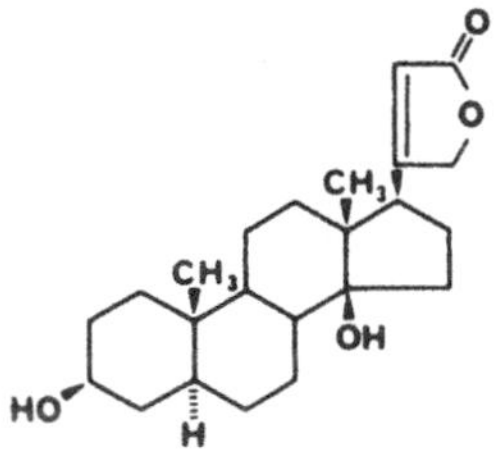

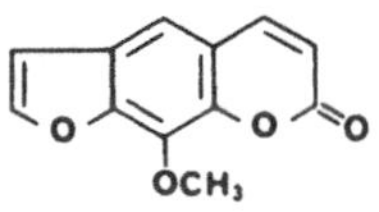

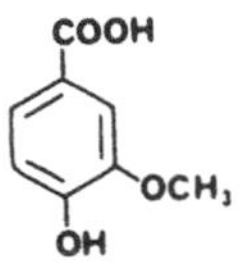

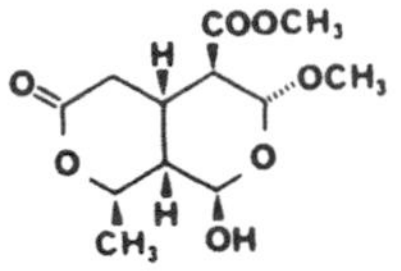

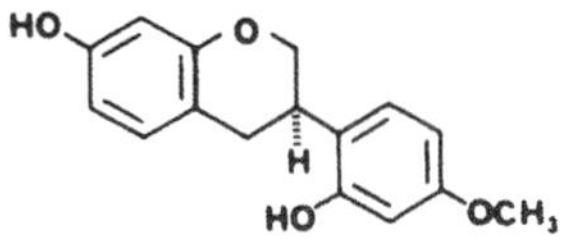

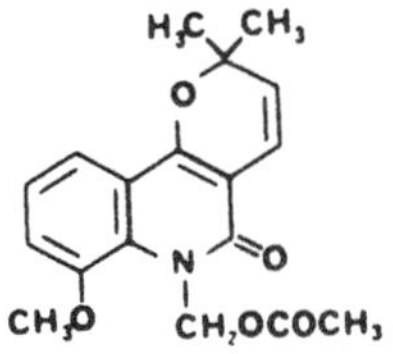

TABLE 2
Order, Scientific Name, and Feeding Deterrent Cross Reference for Table 1

Insect		
Scientific name	**Common name**	**Feeding deterrent**

COLEOPTERA

Scientific name	Common name	Feeding deterrent
"No sp. specified"	"Beetles"[a]	Meliatin
		α-Pinene: P17[b]
		(+)-β-Pinene: P18
"No sp. specified"	"Epilachninae"	Phaseolunatin: P5
"No sp. specified"	"Weevils"	Benomyl: B1
Acalymma vittatum (F.)	Striped cucumber beetle	Azadirachtin: A
		Juglone: J1
		2-Methyl-1,4-naphthoquinone: M12
		1,4-Naphthoquinone: N1
		Naphthoquinones
		Oxazine
		Oxazolidine
		Salannin: D2
Anthonomus grandis grandis Boheman	Boll weevil	*erythro*-9,10-Dihydroxy-1-octadecanol acetate
		α-Eleostearic acid: E4
		Morin: M13
		Rutin: R1
Callosobruchus chinensis (Lucas)	—	Cocculolidine: C33
		Parthenin: P1
"Chrysomelidae (Fam.), no sp. specified"	"Leaf beetles"	Atropine: A16
		Capsaicin: C5
		Demissine
		Morphine: M14
		Scopolamine: S9
		Sparteine: S18
		Tomatine: T7
Costelytra zealandica (White)	—	Sativan: S5
		(3*R*)-(-)-Vestitol: V2
Diabrotica undecim-punctata howardii Barber	Spotted cucumber beetle	Azadirachtin: A
		Oxazine
		Oxazolidine
		Salannin: D2
Epicauta fabricii (Leconte)	Ashgray blister beetle	(Z)-*o*-Coumaric acid glucoside Coumarin: C37
Epicauta pestifera Werner	Margined blister beetle	(Z)-*o*-Coumaric acid glucoside Coumarin: C37
Epicauta spp.	Blister beetle	Cantharidin: C4
Epicauta vittata (F.)	Striped blister beetle	(Z)-*o*-Coumaric acid glucoside Coumarin: C37
Epilachna implicata Mulsant	—	Tricyclohexyltin hydroxide (Plictran)
Epilachna tredecimno-tata (Latreille)	—	Cucurbitacins (?)
Epilachna varivestis Mulsant	Mexican bean beetle	6-Acetoxytoonacilin: A2
		1,3-Diacetylvilasinin: D4
		3-Desacetylsalannin: D1
		Guazatine triacetate (SN 513)
		Nimbandiol: N6
		Salannol: S1
		Schkuhrin-I: S6
		Schkuhrin-II: S7
		Toonacilin: A3
Hypera postica (Gyllenhal)	Alfalfa weevil	Zanthophylline: Z1

TABLE 2 (continued)
Order, Scientific Name, and Feeding Deterrent Cross Reference for Table 1

Insect		Feeding deterrent
Scientific name	**Common name**	
Leptinotarsa decemlineata (Say)	Colorado potato beetle, "tomato beetle"	Azadirachtin: A
		N,3-Bis(4-chlorophenyl)-4,5-dihydro-1*H*-pyrazole-1-carboxamide (TH 6041)
		p-Cymene: C43
		Demissine
		Dihydro-α-solanin
		Guazatine triacetate (SN 513)
		Leptine I
		Leptine III
		Leptinine I
		Leptinine II
		Methyl (Z)-10-acetoxy-8,9-epoxy-2-decen-4,6-diynoate
		Methyl (Z,Z)-10-acetoxymatricariate: M6
		Methyl (Z)-10-hydroxy-8,9-epoxy-2-decen-4,6-diynoate: M9
		Methyl (Z,Z)-10-hydroxymatricariate: M10
		Nicandrenones
		Solacaulin
		Soladulcin
		Solanine: S17
		Terpinen-4-ol: T1
		α-Terpineol: T2
		Tomatine: T7
Listroderes costirostris obliquus (Klug)	Vegetable weevil	Coumarin: C37
Phaedon cochleariae F.	—	Bufadienolides: B11
		Cardenolides
		Cucurbitacins (?)
		Strophanthidin: S19, and strophanthidol glycosides
		Uzarigenin: U4, and its glycoside
Phyllotreta tetrastigma (Comolli)	—	Strophanthidin: S19, and strophanthidol glycosides
Phyllotreta nemorum (L.)	—	Cucurbitacins B: C40, E: C41, and I: C42
Phyllotreta undulata Kutschelina	—	Cucurbitacins (?)
		Strophanthidin: S19, and strophanthidol glycosides
Popillia japonica Newman	Japanese beetle	Azadirachtin: A18
		Salannin: D2
Scolytus multistriatus (Marsham)	Smaller European elm bark beetle	Aesculetin: A5
		p-Benzoquinone: B2
		o-Coumaric acid: C35
		p-Coumaric acid: C36
		2,3-Dichloro-1,4-naphthoquinone (Dichlone): D5
		2',6'-Dihydroxy-4'-methoxydihydrochalcone: D14
		5,8-Dihydroxy-1,4-naphthoquinone: D15
		Fraxetin: F8
		Gramine: G7
		2-Hydroxy-1,4-naphthoquinone: H10
		Juglone: J1
		Kaempferol: K1
		Magnoline: M1
		2-Methyl-1,4-naphthoquinone: M12
		1,4-Naphthoquinone: N1
		Phloretin: P12
		Quercitin: Q1

TABLE 2 (continued)
Order, Scientific Name, and Feeding Deterrent Cross Reference for Table 1

Insect		
Scientific name	**Common name**	**Feeding deterrent**
Sitona cylindricollis Fahraeus	Sweetclover weevil	Ammonium nitrate
		Potassium nitrate
		Sodium nitrate
Tribolium castaneum (Herbst)	Red flour beetle	Parthenin: P1
		Tetraphenylphosphonium chloride (Ph_4PCl)
		Tetraphenylstibonium chloride (Ph_4SbCl)
		Triphenylgermanium chloride (Ph_3GeCl)
		Triphenyllead chloride (Ph_3PbCl)
		Triphenylmethyl chloride (Ph_3CCl_3)
		Triphenylsilanol (Ph_3SiOH)
		Triphenylstilbine (Ph_3Sb)
		Triphenyltin chloride (Ph_3SnCl): F4
Tribolium confusum Jacquelin duVal	Confused flour beetle	Alantolactone: A9
		Tetraphenylphosphonium chloride (Ph_4PCl)
		Tetraphenylstibonium chloride (Ph_4SbCl)
		Triphenylstilbine (Ph_3Sb)
	DIPTERA	
"No sp. specified"	"Mosquito"	Asarone: A15
		Cardol
		1,4-Cineole: C24
		1,8-Cineole: C25
Aedes aegypti (L.)	Yellow-fever mosquito	Parthenin: P1
Culiseta inornata (Williston)	—	Salts (anions and cations)
Drosophila pachea Patterson and Wheeler	—	Lophocereine
		Pilocereine: P16
Drosophila melanogaster Meigen	Vinegar fly	2-Phenylethylisothiocyanate: P11
Musca domestica L.	House fly	Asarone: A15
		Cardol
		1,4-Cineole: C24
		1,8-Cineole: C25
		Hallactone A
		Hallactone B
		Salannin: D2
	HETEROPTERA	
Dysdercus koenigii (F.)	—	Parthenin: P1
Euschistus servus (Say)	Brown stink bug	Guazatine triacetate (SN 513)
	HOMOPTERA	
Acyrthosiphon pisum (Harris)	Pea aphid	Coumestrol: A38
Amphorophora agathonica Hottes	—	Phlorizin: P13
Aonidiella aurantii (Maskell)	California red scale	Salannin: D2
Aphis pomi De Geer	Apple aphid	Phlorizin: P13

TABLE 2 (continued)
Order, Scientific Name, and Feeding Deterrent Cross Reference for Table 1

Insect		Feeding deterrent
Scientific name	**Common name**	
Empoasca fabae (Harris)	Potato leafhopper	Demissidine
		Leptine I
		Solanidine: S16
		Solanine: S17
		Tomatidine: T6
		Tomatine: T6
Hyadaphis erysimi (Kaltenback)	Turnip aphid	Chlordimeform: C19
Laodelphax striatellus (Fallen)	"Smaller brown planthopper"	Chlordimeform: C19
		Hordenine sulfate
		Phenethylamine HCl: P9
		Tyramine HCl: T12
Myzus persicae (Sulzer)	Green peach aphid	Benomyl: B1
		Chlordimeform: C19
		Phlorizin: P13
		Sinigrin: S14
		Thiabendazole (TBZ): T5
Nephotettix cincticeps (Uhler)	"Green rice leafhopper"	Chlordimeform: C19
		Cocculolidine: C33
		Hordenine sulfate
		Phenethylamine HCl: P9
		Tyramine HCl: T12
Nilaparvata lugens (Stal)	"Brown planthopper"	(E)-Aconitic acid: A4
		Chlordimeform: C19
		Hordenine sulfate
		Phenethylamine HCl: P9
		Tyramine HCl: T12
Schizaphis graminum (Rondani)	Greenbug	Benzyl alcohol: B4
Sogatella furcifera (Horvath)		Hordenine sulfate
		Phenethylamine HCl: P9
		Tyramine HCl: T12
Therioaphis maculata (Buckton)	Spotted alfalfa aphid	Coumestrol: C38

HYMENOPTERA

"No sp. specified"	"Wasps"	Cucurbitacins
Apis mellifera L.	Honey bee	Cucurbitacins
"Formicidae (Fam.), no sp specified"	—	Nepetalactone: N2
Neodiprion lecontei (Fitch)	Redheaded pine sawfly	4'-(3,3-Dimethyl-1-triazeno)acetanilide (AC-24055): D16
		2-Methyl-1-(3-methylbenzoyl)piperidine: M11
		Oxazine
		Oxazolidine
Neodiprion nannulus nannulus Schedl	Red pine sawfly	4'-(3,3-Dimethyl-1-triazeno)acetanilide (AC-24055): D16
		2-Methyl-1-(3-methylbenzoyl)piperidine: M11
Neodiprion pratti paradoxicus Ross	Jack pine sawfly	N,N-Diethyl-p-toluamide: D8
		4'-(3,3-Dimethyl-1-triazeno)acetanilide (AC-24055): D16
		2-Methyl-1-(3-methylbenzoyl)piperidine: M11
		Oxazine
		Oxazolidine

TABLE 2 (continued)
Order, Scientific Name, and Feeding Deterrent Cross Reference for Table 1

Insect		Feeding deterrent
Scientific name	Common name	
Neodiprion rugifrons Middleton	Redheaded jack pine sawfly	4'-(3,3-Dimethyl-1-triazeno)acetanilide (AC-24055): D16 13-Keto-8(14)-podocarpen-18-oic acid: K3 1,4-Naphthoquinone: N1 Oxazolidine Triphenyltin hydroxide (Ph₃SnOH): F5
Neodiprion sertifer (Geoffroy)	European pine sawfly	N,N-Diethyl-o-toluamide: D7 N,N-Diethyl-p-toluamide: D8 4'-(3,3-Dimethyl-1-triazeno)acetanilide (AC-24055): D16 2-Methyl-1-(3-methylbenzoyl)piperidine: M11 Oxazine Oxazolidine
Neodiprion swainei Middleton	Swaine jack pine sawfly	N,N-Diethyl-o-toluamide: D7 4'-(3,3-Dimethyl-1-triazeno)acetanilide (AC-24055): D16 13-Keto-8(14)-podocarpen-18-oic acid: K3 2-Methyl-1-(3-methylbenzoyl)piperidine: M11 1,4-Naphthoquinone: N1 Oxazine Oxazolidine Triphenyltin hydroxide (Ph₃SnOH): F5

ISOPTERA

"No sp. specified"	"Termites"	Cycloeucalenol Lapachonone 7-Methoxyjuglone: M5
Chryptotermes brevis (Walker)	—	Chrysin: C23 Osajin: 02 Pomiferin
Nasutitermes exitiosus (Hill),	—	2-Formylanthraquinone: F6
Reticulitermes lucifugus santonensis (Feytaud)	—	(+)-Citronellene: C30 (−)-Citronellol: C29 (±)-Dihydrolinalool Geraniol: G3 1-Hydroxyanthraquinone: H5 2-(Hydroxymethyl)anthraquinone: H6 Lapachol: L1 1-Methoxyanthraquinone: M3 2-Methylanthraquinone (tectoquinone): M7 Myrcene: M16 Nerol: N3 (+)-α-Phellandrene: P6 (+)-β-Phellandrene: P7 (+)-Pulegone: P2 Terpinolene: T4
Reticulitermes sp.	"Termites"	Azadirachtin: A

LEPIDOPTERA

"No sp. specified"	"Moths"	Asarone: A15 Effusin: E2
Boarmia (Ascotis) selenaria (Dennis and Schiffermuller)	"Giant looper"	Gossypol: G6

TABLE 2 (continued)
Order, Scientific Name, and Feeding Deterrent Cross Reference for Table 1

Insect		Feeding deterrent
Scientific name	**Common name**	
Bombyx mori" (L.)	Silkworm	Berberine: B5
		Brucine: B10
		Chlordimeform HCl (Fundal™)
		Ecdysterone: E1
		Nicotine: N5
		Pilocarpine: P15
		Quercitin: Q1
		Quinine: Q2
		Rutin: R1
		Salicin: S2
		Scopolamine: S9
		Strychnine: S20
Calospilos miranda (Butler)	—	Chlordimeform: C19
		Clerodendrin-A: C31
		Clerodendrin-B: C32
		Isoboldine: I4
		Shiromodiol diacetate: S11
		Shiromodiol monoacetate: S12
Ceramica picta (Harris)	Zebra caterpillar	N,3-Bis(4-chlorophenyl)-4,5-dihydro-1H-pyrazole-1-carboxamide (TH 6041)
Danaus plexippus (L.)	Monarch butterfly	Betaine: B7
		Brucine: B10
		Caffeine: C2
		Oxalic acid
		Piperidine: P21
		Quinine HCl
		Saponins
		Trichlorophenoxyacetic acid: T9
		Trichorophenoxyethanol: T10
Dendrolimus pini (L.)	"Pine moth"	Borneol: B8
		Borneol acetate: B9
		Δ^3-Carene: C6
		Limonene: L2
		α-Pinene: P17
		α-Terpineol: T2
		Terpineol acetate: T3
Diacrisia obliqua (Walker)	—	Triphenyltin chloride (Ph$_3$SnCl): F4
Diacrisia virginica (F.)	Yellow woolybear	Glaucolide-A: G4
Diatraea grandiosella (Dyar)	Southwestern corn borer	Cholesterol acetate: C20
		Cholesterol myristate: C21
		Cholesterol oleate: C22
Earias insulana (Boisduval)	"Spiny bollworm"	Azadirachtin: A18
		Salannin: D2
Euproctis fraterna (Moore)	—	Tricyclohexyltin hydroxide (Plictran)
Euproctis subflava (Bremer)	"Oriental tussock moth"	Clerodendrin-A: C31
		Clerodendrin-B: C32
Galleria mellonella (L.)	Greater wax moth	Azadirachtin: A30
Heliothis armigera (Hübner)	—	Amitraz (Mitac®) and *Bacillus thuringiensis* var *alesti* Berliner, HD-1 strain (Dipel®)
		Bacillus thuringiensis var *alesti* Berliner HD-1 strain (Dipel®)
		Chlordimeform HCl (Fundal™) and *Bacillus thuringiensis* var *alesti* Berliner HD-1, strain (Dipel®)
		Chlordimeform HCl (Fundal™)

TABLE 2 (continued)
Order, Scientific Name, and Feeding Deterrent Cross Reference for Table 1

Insect		Feeding deterrent
Scientific name	**Common name**	
Heliothis spp.	—	Gossypol: G6
Heliothis virescens (F.)	Tobacco budworm,	Azadirachtin: A30
Heliothis zea (Boddie)	Corn earworm	Azadirachtin: A30
		2-Benzoxazolinone: B3
		Guazatine triacetate (SN 513)
		6-Methoxy-2-benzoxazolinone: M4
Hemileuca oliviae Cockerell	Range caterpillar	Zanthophylline: Z1
Homoeosoma electellum (Hulst)	Sunflower moth	(−)-Kaur-16-en-19-oic acid: K2 Trachyloban-19-oic acid: T8
Hypsipyla grandella (Zeller)	—	Azadirachtin: A30
Lymantria dispar (L.)	Gypsy moth	Emodine: E5
		Epitulipinolide diepoxide
		Farnesol: F1
		Geraniol: G3
		Lipiferolide
		(*E,S*-Nerolidol: N4
		Peroxyferolide: P3
		Tulirinol: T11
Manduca sexta (L.)	Tobacco hornworm	Caffeine: C2
		L-Canavanine: C3
		Glucotropaeolin: G5
		Guazatine triacetate (SN 513)
		Salicin: S2
		Tomatine: T7
Ostrinia nubilalis (Hübner)	European corn borer	2-Benzoxazolinone: B3
		Clerodendrin-A: C31
		Clerodendrin-B: C32
		2,4-Dihydroxy-7-methoxy-2*H*-1,4-benzoxazin-3-(4*H*)-one (DIMBOA)
		6-Methoxy-2-benzoxazolinone: M4
		Phenylbenzothiazole: P10
Papillo polyxenes asterius Stoll	Black swallowtail	Sinigrin: S14
Pectinophora gossypiella (Saunders)	Pink bollworm	(−)-Kaur-16-en-19-oic acid: K2 Trachyloban-19-oic acid: T8
Pericallia ricini (F.)		Tricyclohexyltin hydroxide (Plictran)
Phthorimaea operculella (Zeller)	Potato tuberworm	Parthenin: P1
"Pieridae (Fam.), no sp. specified"	"White and sulfur butterfiles"	Atropine: A16
		Capsaicin: C5
"Pieridae (Fam.), no sp. specified"	—	Demissine
		Morphine: M14
		Scopolamine: S9
		Sparteine: S18
		Tomatine: T7
Pieris brassicae (L.)	"Cabbage moth," "cabbage white"	Atropine: A16
		Berberine: B5
		Brucine: B10
		p-Cymene: C43
		Ecdysterone: E1
		Inokosterone
		Morphine: M14
		Nicotine: N5

TABLE 2 (continued)
Order, Scientific Name, and Feeding Deterrent Cross Reference for Table 1

Insect		
Scientific name	**Common name**	**Feeding deterrent**
		Pilocarpine: P15
		Ponasterone: P24
		Quinine: Q2
		Strychnine: S20
		Terpinen-4-ol: T1
		α-Terpineol: T2
Plathypena scabra (F.)	Green cloverworm	Chlordimeform HCl (Fundal™)
		Guazatine triacetate SN 513
Plutella maculipennis Curtis	—	Fentin acetate (Brestan or TPTA): F3
		Fentin hydroxide (TPTH or Du-ter): F5
Plutella xylostella (L.)	Diamondback moth	Azadirachtin: A30
		Guazatine triacetate (SN 513)
Pryeria sinica Moore	—	Chlordimeform: C19
Pseudoplusia includens (Walker)	Soybean looper	Guazatine triacetate (SN 513)
Sibine stimulea (Clemens)	Saddleback caterpillar	Glaucolide-A: G4
Sitotroga cerealella (Olivier)	Angoumois grain moth, "paddy moth"	4'-(3,3-Dimethyl-1-triazeno)acetanilide (AC-24055): D16
		Fentin acetate (Brestan or TPTA): F3
		Fentin chloride (Brestanol): F4
		Tricyclohexyltin hydroxide (Plictran)
Spodoptera eridania (Cramer)	Southern armyworm	L-Dopa: D18
		Glaucolide-A: G4
		Gymnemic acids
Spodoptera exempta (Walker)	Nutgrass armyworm, "African armyworm"	Ajugarin I: A6, II: A7, III: A8
		Crotepoxide: C39
		Harrisonin: H2
		Inflexin: I1
		Isomedin: I6
		N-Methylflindersine: M8
		Muzigadial: M15
		Myricoside: M17
		Plumbagin: P22
		Polygodial: P23
		Schkuhrin-I: S6
		Schkuhrin-II: S7
		Ugandensidial: U1
		Warburganal: W1
		Xylomollin: X2
Spodoptera exigua (Hübner)	Beet armyworm	Guazatine triacetate (SN 513)
Spodoptera frugiperda (J. E. Smith)	Fall armyworm	Azadirachtin: A30
		Glaucolide-A: G4
		Pennyroyal oil (85% pulegone): P2
		Toxol
		Toxyl angelate
Spodoptera littoralis (Boisduval)	"African armyworm," "Egyptian cotton leafworm"	Ajugarin I: A6, II: A7, III: A8
		L-Alanine
		DL-α-Aminobutyric acid
		Azadirachtin: A30
		L-Cystine
		Fentin acetate (Brestan or TPTA): F3
		Fentin hydroxide (TPTH or Du-ter): F5
		Friedelin: F9
		Gossypol: G6

TABLE 2 (continued)
Order, Scientific Name, and Feeding Deterrent Cross Reference for Table 1

Insect		Feeding deterrent
Scientific name	**Common name**	
		L-Histidine
		L-Methionine + 0.125 M sucrose
		Plumbagin: P22
		Polygodial: P23
		Salannin: D2
		L-Serine
		Triphenyltin hydroxide (Ph₃SnOH): F5
		L-Tyrosine
		Ugandensidial: U1
		Warburganal: W1
Spodoptera litura (F.)	"Tobacco cutworm," "tobacco caterpillar," "a fruit piercing moth"	Absinthin: A1
		Angelicin: A12
		Bergapten: B6
		Caryoptin: C9
		Caryoptin hemiacetal: C13
		Caryoptionol: C10
		Chlordimeform: C19
		Clerodendrin A: C31
		Clerodendrin B: C32
		Clerodin: C11
		Clerodin hemiacetal: C14
		Dictamnine: D6
		Dihydrocaryoptin: C15
		Dihydrocaryoptionol: C16
		Dihydroclerodin-I: C17
		2,3-Dihydro-4,6-dimethoxyfuro[2,3-*b*]quinoline and its 3-hydroxy derivatives: D10
		Dihydroisopimpinellin
		Dihydrokokusagine: D11
		Dihydroxanthotoxin: D13
		3-Epicaryopterin: C12
		3-Epidihydrocaryoptin: C18
		(±)-Epieudesmin: E6
		(+)-Eudesmin: E7
		Evoxine: E8
		Fentin acetate: F3
		Fentin hydroxide (TPTH or Du-ter): F5
		Isoasaron: I2
		Isobergapten: I3
		Isoboldine: I4
		Isopimpinellin: I7
		Japonin
		Kokusagine: K4
		Peucedanin: P4
		Phytol: P14
		Pimpinellin
		Piperenone: P20
		Psoralen: P27
		Shiromodiol diacetate: S11
		Shiromodiol monoacetate: S12
		Skimmianine: S15
		Tricyclohexyltin hydroxide (Plictran): F5
		Xanthotoxin: X1
Spodoptera ornithogalli (Guenée)	Yellowstriped armyworm	Glaucolide-A: G4
		Vulpinic acid: V3

TABLE 2 (continued)
Order, Scientific Name, and Feeding Deterrent Cross Reference for Table 1

Insect		Feeding deterrent
Scientific name	**Common name**	
Stomopteryx subseci-vella (Zeller)	"Peanut leaf folder"	Fentin acetate (Brestan or TPTA): F3 Fentin hydroxide (TPTH or Du-ter): F5
Trichoplusia ni (Hübner)	Cabbage looper	Ascorbic acid Glaucolide-A: G4 Guazatine triacetate (SN 513)

ORTHOPTERA

"No sp. specified"	"Grasshoppers"	Alanine + serine + aminobutyric acid
Anacridium melanorho-don arabafrum Dirsh	"Tree locust"	Albizziin: A10 2-Amino-3-acetylaminopropionic acid (ADAP) 2-Amino-4-(oxalylamino)butyric acid (ODAB) 2-Amino-3-(oxalylamino)propionic acid (ODAP) 2,4-Diaminobutyric acid (DABA) 2,3-Diaminopropionic acid (DAPA) Homoarginine Pipecolic acid: P19
Blattella germanica (L.)	German cockroach	Bergapten: B6 Isopimpinellin: I7 Xanthotoxin: X1
Chrotoicetes terminifera (Walker)	—	Amino acids (nonprotein)
Locusta spp.	—	Salannin: D2
Locusta migratoria (L.)	"Migratory locust"	Albizziin: A10 Arbutin: A13 Azadirachtin: A30 Caffeic acid: C1 1,8-Cineole: C25 *o*-Coumaric acid: C35 *p*-Coumaric acid: C36 Coumarin: C37 (Z)-Decalin with an epoxydiacetate Dhurrin: D3 Farnesol: F1 Ferulic acid: F6 Gentisic acid: G1 *p*-Hydroxybenzaldehyde: H7 *m*-Hydroxybenzoic acid: H8 *p*-Hydroxybenzoic acid: H9 Isoferulic acid: I5 Nicotine: N5 Pipecolic acid: P19 Protocatechuic acid: P26 *o*-Protocatechuic acid Salicylic acid: S3 Sinigrin: S14 Syringic acid: S21 Tomatine: T7 Vanillic acid: V1
Locusta migratoria migratorioides (Reiche & Fairmaire)	—	Albizziin: A10 Amino acids (non-protein) 2-Amino-4-(oxalylamino)butyric acid (ODAB) 2-Amino-3-(oxalylamino)propionic acid (ODAP) Arbutin: A13 Aristolochic acid: A14

TABLE 2 (continued)
Order, Scientific Name, and Feeding Deterrent Cross Reference for Table 1

Insect		Feeding deterrent
Scientific name	Common name	

Scientific name	Common name	Feeding deterrent
		Aucubin: A17
		Azadirachtin: A30
		Carvone: C7
		Caryophyllene: C8
		1,8-Cineole: C25
		Cinnamic acid: C26
		Citral: C27
		Citronellal: C28
		Colchicine: C34
		Condensed tannin
		Coumarin: C37
		p-Cymene: C43
		2,3-Diaminopropionic acid (DAPA)
		2,4-Diaminobutyric acid (DABA)
		Dimethoxybenzoxazolinone
		Djenkolic acid
		Ferulic acid: F6
		Geranial. G2
		Geraniol: G3
		Gossypol: G6
		Halostachine: H1
		Homoarginine
		Inorganic ions (Na^+ & Ca^{++})
		Isolongifolene
		Isolongifolene ketone A
		Isolongifolene ketone B
		Jacobine
		Jaconine
		Limonene: L2
		Linalool: L3
		Linamarin: L4
		Longicyclene: L6
		Longifolene: L7
		Malic acid
		Nicotine: N5
		Perloline
		Phlorizin: P13
		α-Pinene: P17
		Pipecolic acid: P19
		Quinine: Q2
		Rutin: R1
		Salicin: S2
		Senecionine: S10
		Seneciphylline
		Sinigrin: S14
		Tannic acid neutralized
		Tannic acid pH 2
		Terpinen-4-ol: T1
		α-Terpineol: T2
		Tomatine: T7
		Ursolic acid: U3
		Withaferin A: W2
Melanoplus bivittatus (Say)	Twostriped grasshopper	Atropine: A16
		Digitonin: D9
		Diosgenin: D17

TABLE 2 (continued)
Order, Scientific Name, and Feeding Deterrent Cross Reference for Table 1

Insect		
Scientific name	Common name	Feeding deterrent
		Hecogenin: H3
		Hordenine: H4
		Indican dioegin
		Lobeline: L5
		Lucenine
		Lupinine: L8
		Nornicotine: N8
		Nornicotine dipicrate
		Santonin: S4
		Saponins
		Solanine: S17
		Tomatine: T7
		Veratrine (mixture of alkaloids)
Melanoplus sanguinipes (F.)	Migratory grasshopper	Isothiocyanates
		Zanthophylline: Z1
Neostylopyga rhombifolia (Stoll)	Harlequin cockroach	Bergapten: B6
		Isopimpinellin: I7
		Xanthotoxin: X1
Periplaneta americana (L.)	American cockroach	Amphetamine: A11
		Bergapten: B6
		Chlorodimeform HCl (Fundal™)
		2,3-Dichloro-1,4-naphthoquinone (Dichlone): D5
		5,8-Dihydroxy-1,4-naphthoquinone: D15
		Fenitrothion: F2
		2-Hydroxy-1,4-naphthoquinone: H10
		Isopimpinellin: I7
		Juglone: J1
		2-Methyl-1,4-naphthoquinone: M12
		1,4-Naphthoquinone: Ni
		Norepinephrine: N7
		Octopamine: O1
		Parthenin: P1
		Phenacaine: P8
		Propranolol: P25
		Xanthotoxin: X1
Schistocerca americana gregaria (Dirsh)	—	Amino acids (nonprotein)
Schistocerca gregaria Forsk	"Desert locust"	Azadirachtin: A30
		Lycorine: L9
		Meliantriol: M2
		Salannin: D2
		Scillaren (mixture of A: S8, and B)

SIPHONAPTERA

"No sp. specified"	"Fleas"	Asarone: A15

[a]   Common names of insects in "quotes" do not appear in *Common Names of Insects and Related Organisms* (*1978 Revision*), Special Publication 78-1, Sutherland, D. W. S., Chairman, Committee on Common Names of Insects, Entomological Society of America, March 1978.

[b]   Alphanumeric codes refer to selected chemical structures in Figure 1.

TABLE 3
Physical Constants and Sources of Spectral Data of Selected Feeding Deterrents

Name and Formula — CRC Atlas No. (138), [Merck Index No.] (139)	Mol. wt.	mp°C	IR	UV	NMR	MS	X-ray	Structure references
Absinthin: A1ᵃ $C_{30}H_{40}O_6$ — [M7]	496.62	179—180d	IR	—	—	—	—	139, 140, 141, 142
6-Acetoxytoonacilin: A2 $C_{31}H_{38}O_9$	554.61	215	IR	—	$^1H, ^{13}C$	MS	X-ray	18
(E)-Aconitic acid: A4 $C_6H_6O_6$ — p1024, [M112]	174.11	198—199d	IR	UV	—	—	—	138, 139
Aesculetin: A5 $C_9H_6O_4$ — c335	178.14	276	—	UV	—	—	—	138
Ajugarin I: A6 $C_{24}H_{34}O_7$	434.51	155—157	IR	UV	$^1H, ^{13}C$	MS	—	22
Ajugarin II: A7 $C_{26}H_{36}O_8$	476.55	188—189	—	—	—	—	—	22
Ajugarin III: A8 $C_{24}H_{36}O_6$	420.53	243—245	—	—	$^1H,^{13}C$	MS	—	22
Alantolactone: A9 $C_{15}H_{20}O_2$ — h5, [M196]	232.31	76, 78—79	IR	UV	$^1H,^{13}C$	MS	—	138, 139, 143
Albizziin: A10 $C_4H_9N_3O_3$ — [M199]	147.14	218—220d	—	—	^1H	—	—	139, 144
Amphetamine: A11 $C_9H_{13}N$ — a513, [M616]	135.20	—	—	—	—	MS	—	138, 139
Angelicin: A12 $C_{11}H_6O_3$	186.16	—	—	—	—	—	—	34
Arbutin: A13 $C_{12}H_{16}O_7$ — a755, [M800]	272.25	199.5—200	IR	UV	^1H	—	—	138, 139
Aristolochic acid: A14 $C_{17}H_{11}NO_7$ — [M811]	341.29	281—286d	—	UV	—	—	—	139
Asarone: A15 $C_{12}H_{16}O_3$ — b578, [M847]	208.25	62—63, 67	—	—	^1H	—	—	138, 139
Atropine: A16 $C_{17}H_{23}NO_3$ — a787, [M892]	289.36	118—119, 114—116	IR	UV	^1H	—	—	138, 139
Aucubin: A17 $C_{15}H_{22}O_9$ — [M896]	346.33	181	IR	UV	^1H	—	—	139, 145, 146
Azadirachtin: A30 $C_{35}H_{44}O_{16}$	720.70	154—158	IR	UV	$^1H, ^{13}C$	MS	—	217—221
Benomyl: B1 $C_{14}H_{18}N_4O_3$ — [M1047]	290.32	—	—	—	—	—	—	139
p-Benzoquinone: B2 $C_6H_4O_2$ — b1216	108.09	115.7	IR	UV	^1H	MS	—	138
2-Benzoxazolinone: B3 $C_7H_5NO_2$	135.12	137—139	IR	UV	NMR	—	—	148
Benzyl alcohol: B4 C_7H_8O — t309, [M1138]	108.13	−15.3	IR	UV	^1H	MS	—	138, 139
Berberine: B5 $C_{20}H_{19}NO_5$ — b1270, [M1177]	353.36	145	IR	UV	^1H	—	—	138, 139, 149
Bergapten: B6 $C_{12}H_8O_4$	216.18	188—191	—	—	—	—	—	150
Betaine: B7 $C_5H_{11}O_2$ — b1272, [M1207]	117.15	293d, 310d	IR	—	^1H	—	—	138, 139
Borneol: B8 $C_{10}H_{18}O$ — b1411, [M1350]	154.24	208	—	—	—	MS	—	138, 139
Borneol acetate: B9 $C_{18}H_{20}O_2$ — b1414, [M1351]	196.28	29	—	—	—	MS	—	138, 139
Brucine: B10 $C_{23}H_{26}N_2O_4$ — b1424, [M1453]	394.45	178	IR	UV	^1H	—	—	138, 139
Bufadienolides: B11								
Desglucohellebrin $C_{30}H_{42}O_9$	546.64	—	—	—	—	—	—	51
Proscillaridin A $C_{30}H_{42}O_8$ — [M7662]	530.64	219—222	—	—	—	—	—	139

TABLE 3 (continued)
Physical Constants and Sources of Spectral Data of Selected Feeding Deterrents

Name and Formula — CRC Atlas No. (138), [Merck Index No.] (139)	Mol. wt.	mp°C	IR	UV	NMR	MS	X-ray	Structure references
Caffeic acid: C1 $C_9H_8O_4$ — C239, [M1622]	180.15	225d	IR	UV	—	—	—	138, 139
Caffeine: C2 $C_8H_{10}N_4O_2$ — C2, [M1623]	194.19	238	IR	UV	^{1}H	MS	—	138, 139
L-Canavanine: C3 $C_5H_{12}N_4O_3$ — [M1741]	176.18	184	—	—	—	—	—	139, 151, 152
Cantharidin: C4 $C_{10}H_{12}O_4$ — c36, [M1752]	196.20	218	IR	UV	—	—	—	138, 139
Capsaicin: C5 $C_{18}H_{27}NO_3$ — c37, [M1766]	305.40	65	IR	UV	^{1}H	MS	—	138, 139
Δ^3-Carene: C6 $C_{10}H_{16}$ — c98, c99; [M1838]	136.23	168—169	IR	UV	^{1}H	MS	—	138, 139
Carvone: C7 $C_{10}H_{14}O$ — c111, [M1867]	150.21	—	IR	UV	—	MS	—	138, 139
Caryophyllene: C8 $C_{15}H_{24}$ — c116, [M1868]	204.34	—	—	—	—	MS	—	54, 153
Caryoptin: C9 $C_{26}H_{36}O_9$	492.55	176—177	IR	—	^{1}H	MS	—	153
Caryoptin hemiacetal: C13 $C_{26}H_{38}O_{10}$	510.56	188—189	—	—	—	—	—	153
Caryoptionol: C10 $C_{24}H_{34}O_8$	450.51	—	—	—	—	—	—	139
Chlordimeform: C19 $C_{10}H_{13}N_2Cl$ — [M2055]	196.67	35	—	—	—	—	—	139
Cholesterol acetate: C20 $C_{29}H_{48}O_2$ — c168, [M2192]	428.67	115—116	IR	UV	—	MS	—	138, 139
Cholesterol myristate: C21 $C_{41}H_{72}O_2$	596.99	—	—	—	—	—	—	59
Cholesterol oleate: C22 $C_{45}H_{78}O_2$	651.07	—	—	—	—	—	—	59
Chrysin: C23 $C_{15}H_{10}O_4$ — f19, [M2254]	254.23	275, 285	—	UV	—	—	—	138, 139
1,4-Cineole: C24 $C_{10}H_{18}O$ — c192	154.24	1	IR	—	—	—	—	138
1,8-Cineole: C25 $C_{10}H_{18}O$ — c193, [M2280]	154.24	1.5	IR	—	^{1}H	—	—	138, 139
Cinnamic acid: C26 $C_9H_8O_2$ — c201-4, [M2288]	148.15	—	IR	UV	^{1}H	MS	—	138, 139
Citral: C27 $C_{10}H_{16}O$ — c273-4, [M2303]	152.23	—	IR	UV	^{1}H	MS	—	138, 139
Citronellal: C28 $C_{10}H_{18}O$ — c279-281, [M2310]	154.24	—	IR	UV	^{1}H	MS	—	138, 139
(+)-Citronellene: C30 $C_{10}H_{18}$	138.24	—	—	—	—	—	—	154
(−)-Citronellol: C29 $C_{10}H_{20}O$ — c284, [M2311]	156.26	—	—	UV	^{1}H	MS	—	138, 139
Clerodendrin-A: C31 $C_{31}H_{42}O_{12}$	606.65	164—165	IR	—	^{1}H	MS	X-ray	155, 156, 157
Clerodendrin-B: C32 $C_{31}H_{44}O_{12}$	608.67	—	—	UV	—	—	—	1
Clerodin: C11 $C_{24}H_{34}O_7$	434.51	164—165	IR	UV	^{1}H	—	X-ray	54, 158, 159, 160
Clerodin hemiacetal: C14 $C_{26}H_{36}O_8$	476.55	179—181	—	UV	^{1}H	—	—	1, 4, 65, 153
Cocculolidine: C33 $C_{15}H_{19}NO_3$	261.31	144—146	IR	UV	^{1}H	MS	—	161, 162
Colchicine: C34 $C_{22}H_{25}NO_6$ — c298, [M2436]	399.43	155—157, 142—150	IR	UV	^{1}H	—	—	138, 139
o-Coumaric acid: C35 $C_9H_8O_3$ — c248	164.15	217d	IR	UV	—	—	—	138

Compound	MW	mp	IR	UV	NMR	MS	X-ray	Ref.
p-Coumaric acid: C36 C9H8O3 — c250, [M2545]	164.15	215d, 210—213	IR	—	—	—	—	138, 139
Coumarin: C37 C9H6O2 — c330, [M2547]	146.14	71, 68—70	IR	UV	1H	MS	—	138, 139
Coumestrol: C38 C15H8O5 — c359, [M2549]	268.21	385d	—	UV	—	—	—	138, 139, 163
Crotepoxide: C39 C18H18O8	362.32	150—151	IR	UV	1H	—	X-ray	164
Cucurbitacins B: C40 C32H46O8 — [M2610]	558.69	184—186	—	—	—	—	—	139, 165, 166
Cucurbitacins E: C41 C32H44O8 — [M2611]	556.67	235—236d, 233—235	—	—	—	—	—	139, 165, 166
Cucurbitacins I: C42 C30H42O7 — [M2611]	514.64	148—149d	IR	UV	1H	—	—	139, 166
p-Cymene: C43 C10H14 — b532, [M2765]	134.21	-67.94	—	—	1H	MS	—	138, 139
3-Desacetylsalannin: D1 C32H42O8	538.72	200d	—	—	NMR	MS	X-ray	72
Dhurrin: D3 C14H17NO7 — [M2915]	311.28	—	—	UV	—	—	—	139
1,3-Diacetylvilasinin: D4 C30H40O7	512.62	—	—	—	—	—	—	72
2,3-Dichloro-1,4-naphthoquinone (Diclone): D5 C10H4Cl2O2 — [M3016]	227.06	193	—	—	—	—	—	139
Dictamnine: D6 C12H9NO2 — d83, [M3065]	199.2	133—134, 133	—	UV	—	—	—	138, 139
N,N-Diethyl-o-toluamide: D7 C12H17NO	191.26	—	—	—	—	—	—	74
N,N-Diethyl-p-toluamide: D8 C12H17NO	191.26	—	—	—	—	—	—	74
Digitonin: D9 C56H92O29 — [M3137]	1229.30	235—240, 244—285	—	—	—	—	—	139, 167
Dihydrocaryoptin: C15 C26H38O9	494.56	198.5—199.5	—	—	1H	MS	—	153
Dihydrocaryoptionol: C16 C24H36O8	452.53	—	IR	—	—	—	—	153
Dihydroclerodin-I: C17 C24H36O7	436.53	169—170	IR	UV	1H	MS	—	153, 160
2,3-Dihydro-4,6-dimethoxyfuro[2,3-b]quinoline C14H15NO and its 3-hydroxy derivatives C14H15NO2: D10	213.27	133—134	IR	UV	1H	—	—	75
	229.2 Z	255—256	—	UV	—	MS	—	75
	E	210—212	—	UV	—	MS	—	75
Dihydroisopimpinellin: D11 C13H12O5	248.23	162—163	IR	—	1H	MS	—	34
Dihydrokokusagine: D12 C13H11NO4	245.23	169	—	UV	1H	MS	—	34, 168
Diosgenin: D17 C27H42O3 — [M3295]	414.61	204—207	—	—	—	—	—	139, 171
L-Dopa: D18 C9H11NO4 — a395, [M5310]	197.19	285.5d, 276—278d, 284—286	IR	—	—	—	—	138, 139
Ecdysterone: E1 C27H44O7 — [M3459]	480.62	—	IR	UV	1H, 13C	—	—	31
Effusin: E2 C22H28O6	388.44	206—209	IR	—	—	—	—	82
Elemicin: E3 C12H16O3	208.25	—	—	—	—	—	—	154

TABLE 3 (continued)
Physical Constants and Sources of Spectral Data of Selected Feeding Deterrents

Name and Formula — CRC Atlas No. (138), [Merck Index No.] (139)	Mol. wt.	mp°C	IR	UV	NMR	MS	X-ray	Structure references
α-Eleostearic acid: E4 $C_{18}H_{30}O_2$ — o30	278.42	49	IR	UV	^{1}H	—	—	138
Emodin: E5 $C_{15}H_{10}O_5$ — [M3512]	270.23	256—257	—	UV	—	—	—	139
3-Epicaryoptin: C12 $C_{26}H_{36}O_9$	492.55	171—172	IR	—	^{1}H	MS	—	54, 55, 172
3-Epidihydrocaryoptin: C18 $C_{26}H_{38}O_9$	494.56	161—162	—	—	^{1}H	MS	—	4, 65, 172
(±)-Epieudesmin: E6 $C_{22}H_{26}O_6$	386.43	133—134	—	—	—	MS	—	4, 173
(+)-Eudesmin: E7 $C_{22}H_{26}O_6$	386.43	107—108	—	—	—	MS	—	4, 173
Evoxine: E8 $C_{18}H_{21}NO_6$	347.36	154—157	IR	UV	^{1}H	MS	—	47, 174
Farnesol: F1 $C_{15}H_{26}O$ — f4, [M3878]	222.36	—	IR	UV	—	MS	—	138, 139
Fenitrothion: F2 $C_9H_{12}NO_5PS$ — [M3905]	277.25	—	—	UV	—	—	—	139
Fentin acetate (Brestan or TPTA): F3 $C_{20}H_{18}O_2Sn$ — [M9412]	409.07	122—124	—	—	—	—	—	139
Fentin chloride (Brestanol or Triphenyltin chloride): F4 $C_{18}H_{15}SnCl$	385.48	—	—	—	—	—	—	14
Fentin hydroxide (TPTH, Du-ter, or Triphenyltin hydroxide): F5 $C_{18}H_{16}OSn$ — [M9412]	367.03	122—123.5	—	—	—	—	—	139
Ferulic acid: F6 $C_{10}H_{10}O_4$ — c252, [M3986]	194.18	171, 174	IR	UV	—	—	—	138, 139
2-Formylanthraquinone: F7 $C_{15}H_8O_3$	236.21	—	—	—	—	—	—	4, 36
Fraxetin: F8 $C_{10}H_8O_5$ — [M4116]	208.16	228	—	—	—	—	—	139
Friedelin: F9 $C_{30}H_{50}O$	426.70	263—263.5	IR	—	—	—	—	175, 176, 177
Gentisis acid: G1 B925, [M4230]	—	205, 199—200	IR	UV	^{1}H	—	—	138, 139
Geranial: G2 $C_{10}H_{16}O$ — c273	152.23	—	IR	UV	^{1}H	MS	—	138
Geraniol: G3 $C_{10}H_{18}O$ — g7, [M4235]	154.24	—	IR	UV	^{1}H	MS	—	138, 139
Glaucolide-A: G4 $C_{23}H_{28}O_8$	432.45	—	—	—	—	—	—	91
Glucotropaeolin: G5 $C_{14}H_{18}O_9NS_2$	—	—	—	—	—	—	—	31
Gossypol: G6 $C_{30}H_{30}O_8$ — [M4377]	518.54	184, 199, 214	IR	UV	—	—	—	139
Gramine: G7 $C_{11}H_{14}N_2$ — g105, [M4381]	174.24	138—139	IR	UV	—	—	—	138, 139
Halostachine: H1 $C_9H_{13}NO$ — [M4454]	151.20	43—45	—	—	—	—	—	139
Harrisonin: H2 $C_{27}H_{32}O_{10}$	516.53	155—156	IR	UV	^{1}H, ^{13}C	MS	—	100
Hecogenin: H3 $C_{27}H_{42}O_4$ — h3, [M4476]	430.61	265—268, 264—266	IR	UV	—	—	—	138, 139

Compound	M.W.	m.p.	IR	UV	NMR	MS	X-ray	Ref.
Hordenine: H4 C$_{10}$H$_{15}$NO — [M4625]	165.23	117—118	—	—	—	—	—	139
1-Hydroxyanthraquinone: H5 C$_{14}$H$_8$O$_3$ — a714	224.20	194—5	IR	UV	—	MS	—	138
2-(Hydroxymethyl)anthraquinone: H6 C$_{15}$H$_{10}$O$_3$	238.23	—	—	UV	—	—	—	62
p-Hydroxybenzaldehyde: H7 C$_7$H$_6$O$_2$ — b78, [M4710]	122.12	116, 117	IR	UV	^{1}H	MS	—	138, 139
m-Hydroxybenzoic acid: H8 C$_7$H$_6$O$_3$ — b1008	138.13	201.5—203	IR	UV	^{1}H	MS	—	138
p-Hydroxybenzoic acid: H9 C$_7$H$_6$O$_3$ — b1014, [M4711]	138.13	214.5—215.5, 213—214	IR	UV	^{1}H	MS	—	138, 139
2-Hydroxy-1,4-naphthoquinone: H10 C$_{10}$H$_6$O$_3$; n238	174.15	192d	IR	UV	—	—	—	138
Inflexin: I1 C$_{24}$H$_{32}$O$_7$	432.50	203—205	IR	UV	^{1}H, ^{13}C	MS	—	102
Isoasaron: I2 C$_{12}$H$_{16}$O$_3$	208.25	23—23.5	—	—	—	MS	—	4, 173
Isobergapten: I3 C$_{12}$H$_8$O$_4$ — i96	216.18	222—223	IR	UV	—	—	—	138
Isoboldine: I4 C$_{19}$H$_{21}$NO$_4$	327.37	127	IR	UV	—	MS	—	178
Isoferulic acid: I5 C$_{10}$H$_{10}$O$_4$ — C251	194.18	228	IR	UV	^{1}H	—	—	138
Isomedin: I6 C$_{20}$H$_{32}$O$_5$	352.46	—	—	—	—	—	—	5
Isopimpinellin: I7 C$_{13}$H$_{10}$O$_5$ — i139	246.21	151	—	UV	—	—	—	139
Juglone: J1 C$_{10}$H$_6$O$_3$ — n239, [M5119]	174.15	154, 155	IR	UV	—	—	—	138, 139
Kaempferol: K1 C$_{15}$H$_{10}$O$_6$ — f26, [M5126]	286.23	276—278	IR	UV	—	MS	—	138, 139
(−)-Kaur-16-en-19-oic acid: K2 C$_{20}$H$_{30}$O$_2$	302.44	—	IR	—	^{1}H	MS	—	179
13-Keto-8(14)-podo-carpen-18-oic acid: K3 C$_{17}$H$_{24}$O$_3$	396.46	—	IR	UV	^{1}H	MS	—	105, 106
Kokusagine: K4 C$_{12}$H$_7$NO$_3$	213.18	202—203	IR	UV	^{1}H	MS	—	47
Lapachol: L1 C$_{15}$H$_{24}$O$_3$ — 16, [M5211]	242.26	139.5—140.5, 140	—	UV	—	MS	—	138, 139
Limonene: L2 C$_{10}$H$_{16}$ — 119—21, [M5333]	136.23	—	IR	UV	^{1}H	—	—	138, 139
Linalool: L3 C$_{10}$H$_{18}$O — 122, [M5335]	154.24	—	IR	UV	^{1}H	—	—	138, 139
Linamarin: L4 C$_{10}$H$_{17}$NO$_6$ — [M5337]	247.24	142—143	—	—	—	—	—	139
Lobeline: L5 C$_{22}$H$_{27}$NO$_2$ — 128, [M5397]	337.47	130—131	IR	UV	—	—	—	138, 139
Longicyclene: L6 C$_{15}$H$_{24}$	204.34	—	IR	UV	NMR	MS	X-ray	180
Longifolene: L7 C$_{15}$H$_{24}$ — 129	204.34	—	IR	UV	NMR	MS	—	138, 180
Lupinine: L8 C$_{10}$H$_{19}$NO — 136, [M5425]	169.27	70, 68.5—69.2	—	—	—	MS	—	138, 139
Lycorine: L9 C$_{16}$H$_{17}$NO$_4$ — 139, [M5444]	287.30	280, 275—280	—	UV	—	—	—	138, 139
Magnoline: M1 C$_{36}$H$_{40}$N$_2$O$_6$ — [M5525]	596.70	178—179	—	—	^{1}H	—	—	139
Meliantriol: M2 C$_{30}$H$_{50}$O$_5$	490.70	176—178	—	—	—	MS	—	108
1-Methoxyanthraquinone: M3 C$_{15}$H$_{10}$O$_2$	218.25	—	—	—	—	—	—	36
6-Methoxy-2-benzoxazolinone: M4 C$_8$H$_7$NO$_3$	165.14	154—155	IR	UV	—	—	—	181, 182, 183
7-Methoxyjuglone: M5 C$_{11}$H$_8$O$_4$	204.17	—	—	—	—	—	—	36
Methyl (Z,Z)-10-acetoxy matricariate: M6 C$_{13}$H$_{12}$O$_4$	232.23	—	IR	UV	^{1}H, ^{13}C	MS	—	109

TABLE 3 (continued)
Physical Constants and Sources of Spectral Data of Selected Feeding Deterrents

Name and Formula — CRC Atlas No. (138), [Merck Index No.] (139)	Mol. wt.	mp°C	IR	UV	NMR	MS	X-ray	Structure references
2-Methylanthraquinone (tectoquinone): M7 $C_{15}H_{10}O_2$ — [M5894]	222.23	177	—	—	—	—	—	139
N-Methylflindersine: M8 $C_{15}H_{15}NO_2$ — [M4013]	241.28	84	—	UV	—	—	—	139, 184, 185
Methyl (Z)-10-hydroxy-8,9-epoxy-2-decen-4,6-diynoate: M9 $C_{11}H_{10}O_4$	206.19	53.0—54	IR	UV	^1H, ^{13}C	MS	—	109
Methyl (Z,Z)-10-hydroxymatricariate: M10 $C_{11}H_{10}O_3$	190.19	38.5—40.5	IR	UV	^1H, ^{13}C	MS	—	109
2-Methyl-1-(3-methylbenzoyl)piperidine: M11 $C_{14}H_{19}NO$	217.30	—	—	—	—	—	—	74
2-Methyl-1,4-naphthoquinone: M12 $C_{11}H_8O_2$ — N244	172.17	107	IR	UV	^1H	—	—	138
Morin: M13 $C_{15}H_{10}O_7$ — f22, [M6101]	302.23	303—304, 285—290	IR	UV	—	—	—	138, 139
Morphine: M14 $C_{17}H_{19}NO_3$ — m217, [M6108]	285.33	254—256.4d, 254d	IR	UV	—	MS	—	138, 139
Muzigadial: M15 $C_{15}H_{22}O_3$	250.33	122—124	IR	UV	^1H, ^{13}C	MS	—	15, 113
Myrcene: M16 $C_{10}H_{16}$ — m236, [M6157]	136.23	—	—	UV	—	MS	—	138, 139
Myricoside: M17 $C_{33}H_{42}O_{19}$	742.67	165—167	IR	UV	^1H, ^{13}C	MS	—	115
Myristicin: M18 $C_{11}H_{12}O_3$ — m237	192.21	—	IR	—	—	MS	—	138, 186
1,4-Naphthoquinones: N1 $C_{10}H_6O_2$ — n225, [M6222]	158.15	128.5, 126	IR	UV	^1H	MS	—	138, 139
Nepetalactone: N2 $C_{10}H_{14}O_2$ — [M6293]	166.21	—	—	—	—	—	—	139, 187
Nerol: N3 $C_{10}H_{18}O$ — [M6299]	154.24	—	—	UV	—	—	—	139
(E,S)-Nerolidol: N4 $C_{15}H_{26}O$ — n257, [M6300]	222.36	—	IR	—	^1H, ^{13}C	MS	—	86, 138, 139
Nicotine: N5 $C_{10}H_{14}N_2$ — n260-2, [M6342]	162.23	—	IR	UV	—	MS	—	138, 139
Nimbandiol: N6 $C_{27}H_{34}O_6$	454.54	—	—	—	NMR	MS	X-ray	72
Norepinephrine: N7 $C_8H_{11}NO_3$ — n342, [M6504]	169.18	216.5—218	—	UV	^1H, ^{13}C	—	—	138, 139
Normicotine: N8 $C_9H_{12}N_2$ — n345, [M6521]	148.20	—	IR	UV	^1H, ^{13}C	—	—	138, 139
Octopamine: O1 $C_8H_{11}NO_2$ — [M6568]	153.18	>250d	—	UV	—	—	—	139
Osajin: O2 $C_{25}H_{24}O_5$ — [M6723]	404.44	193	—	UV	—	—	—	139
Parthenin: P1 $C_{15}H_{18}O_4$ — [M6849]	262.31	163—166	IR	UV	^1H	MS	—	139, 188
Pennyroyal oil (85% pulegone): P2 $C_{10}H_{16}O$ — p1107, [M7725]	152.23	—	IR	UV	^1H	MS	—	138, 139
Peroxyferolide: P3 $C_{17}H_{22}O_7$	338.35	190—300d	IR	UV	^1H, ^{13}C	MS	—	120

Compound	MW	mp	IR	UV	NMR	MS	X-ray	Ref.
Peucedanin: P4 $C_{15}H_{14}O_4$ — p271, [M6976]	258.26	87 & 97, 109	—	UV	—	—	—	138, 139
Phaseolunatin: P5 $C_{10}H_{17}NO_6$	247.24	—	—	—	—	—	—	154
(+)-α-Phellandrene: P6 $C_{10}H_{16}$ — p272, [M6983]	136.23	—	IR	UV	—	MS	—	138, 139
(+)-β-Phellandrene: P7 $C_{10}H_{16}$ — p273, [M6984]	136.23	—	—	UV	—	MS	—	138, 139
Phenacaine: P8 $C_{18}H_{22}N_2O_2$ — [M6985]	310.38	—	—	—	—	—	—	139
Phenethylamine HCl: P9 $C_8H_{11}N.HCl$ — [M7016]	156.63	217	—	—	—	—	—	139
Phenylbenzothiazole: P10 C_7H_5NS — b1244, [M1118]	135.18	2	IR	UV	—	MS	—	138, 139
2-Phenylethylisothiocyanate: P11 C_9H_9NS	162.23	—	IR	—	—	—	—	154
Phloretin: P12 $C_{15}H_{14}O_5$ — p431, [M7136]	274.26	262—264d, 262d	—	—	—	—	—	138, 139, 189
Phlorizin: P13 $C_{21}H_{24}O_{10}$ — [M7137]	436.40	110	—	—	—	—	—	139
Phytol: P14 $C_{20}H_{40}O$ — p527, [M7194]	296.52	—	IR	UV	¹H	MS	—	138, 139
Pilocarpine: P15 $C_{11}H_{16}N_2O_2$ — p533 [M7224]	208.25	34	IR	UV	—	—	—	138, 139
Pilocereine: P16 $C_{45}H_{65}N_3O_6$ — [M7227]	744.04	176.5—177	—	—	—	—	—	139, 190
α-Pinene: P17 $C_{10}H_{16}$ — p540, [M7242]	136.23	—	IR	UV	¹H	MS	—	138, 139
(+)-β-Pinene: P18 $C_{10}H_{16}$ — p541	136.23	—	IR	UV	¹H	MS	—	138
Pipecolic acid: P19 $C_6H_{11}NO_2$ — [M7251]	129.16	264	—	—	—	—	—	139
Piperenone: P20 $C_{22}H_{28}O_6$	388.44	86—88	IR	UV	¹H	MS	X-ray	173, 191, 192, 193
Piperidine: P21 $C_5H_{11}N$ — p556, [M7261]	85.15	—	IR	UV	—	MS	—	138, 139
Plumbagin: P22 $C_{11}H_8O_3$ — [M7316]	188.17	78—79	IR	UV	¹H, ¹³C	MS	—	123, 139
Polygodial: P23 $C_{15}H_{22}O_2$	234.33	57	IR	UV	¹H	MS	—	194
Ponasterone: P24 $C_{27}H_{44}O_6$	464.62	96	—	—	—	—	—	195
Propranolol: P25 $C_{16}H_{21}NO_2$ — [M7628]	259.34	—	—	—	—	—	—	139
Protocatechuic acid: P26 $C_7H_6O_4$ — b927, [M7679]	154.12	200—202d, 200d	—	UV	—	MS	—	138, 139
Psoralen: P27 $C_{11}H_6O_3$ — [M7713]	186.16	163—164, 169—179	—	—	—	—	—	139
Quercitin: Q1 $C_{15}H_{10}O_7$ — f23, [M7821]	302.23	316—317, 314d	IR	UV	—	MS	—	138, 139, 196
Quinine: Q2 $C_{20}H_{24}N_2O_2$ — [M7853]	324.41	177	—	—	—	—	—	139
Rutin: R1 $C_{27}H_{30}O_{16}$ — [M8062]	610.51	214—215d	—	—	—	—	—	139
Salannin: D2 $C_{34}H_{44}O_9$	596.69	167—170	IR	UV	¹H	MS	—	125, 197, 198, 199
Salannol: S1 $C_{33}H_{44}O_8$	568.68	—	—	—	NMR	MS	X-ray	72
Salicin: S2 $C_{13}H_{18}O_7$ — t273, [M8085]	286.27	204.7—208.7, 199—202	—	UV	—	—	—	138, 139
Salicylic acid: S3 $C_7H_6O_3$ — b989, [M8093]	138.12	159, 157—159	IR	UV	—	MS	—	138, 139

TABLE 3 (continued)
Physical Constants and Sources of Spectral Data of Selected Feeding Deterrents

Name and Formula — CRC Atlas No. (138), [Merck Index No.] (139)	Mol. wt.	mp°C	IR	UV	NMR	MS	X-ray	Structure references
Santonin: S4 $C_{15}H_{18}O_3$ — s13, [M8118]	246.29	174—176, 170—173	—	UV	—	MS	—	138, 139
Sativan: S5 $C_{17}H_{18}O_4$	286.31	125—127	—	UV	^{1}H, ^{13}C	MS	—	127, 200
Schkuhrin-I: S6 $C_{22}H_{28}O_8$	420.44	59—61	IR	UV	^{1}H	MS	—	133
Schkuhrin-II: S7 $C_{25}H_{34}O_9$	478.52	65—66	IR	UV	^{1}H	MS	—	133
Scillaren: S8 $C_{36}H_{52}O_{13}$ — [M8151]	692.78	270	IR	—	—	—	—	139
Scopolamine: S9 $C_{17}H_{21}NO_4$ — h33, [M8158]	303.35	—	IR	—	—	—	—	138, 139
Senecionine: S10 $C_{18}H_{25}NO_5$ — [M8197]	335.39	236d	IR	UV	^{1}H	MS	—	139, 201
Shiromodiol diacetate: S11 $C_{19}H_{30}O_5$	338.43	112	IR	—	^{1}H	—	—	129, 202
Shiromodiol monoacetate: S12 $C_{17}H_{28}O_4$	296.39	80	IR	—	—	—	—	129, 202
Shiromool: S13 $C_{15}H_{26}O_2$	238.36	72—73	—	—	^{1}H	MS	—	202
Sinigrin: S14 $C_{10}H_{16}KNO_9S_2$ — [M8289]	397.48	179	—	—	—	—	—	139
Skimmianine: S15 $C_{14}H_{13}NO_4$ — f1, [M8298]	259.25	177, 178	IR	UV	—	—	—	138, 139
Solanidine: S16 $C_{27}H_{43}NO$ — s54, [M8478]	397.62	218—219	IR	—	—	—	—	138, 139
Solanine: S17 $C_{45}H_{73}NO_{15}$ — [M8479]	868.04	255d	—	UV	—	—	—	139
Sparteine: S18 $C_{15}H_{26}N_2$ — s59, [M8510]	234.37	30—31	—	UV	—	MS	—	138, 139
Strophanthidin: S19 $C_{23}H_{32}O_6$ — s139, [M8645]	404.49	235, 230	—	UV	—	—	—	138, 139
Strychnine: S20 $C_{21}H_{22}N_2O_2$ — s141, [M8649]	334.40	286—288	IR	UV	—	—	—	138, 139
Syringic acid: S21 $C_9H_{10}O_5$	198.17	—	—	—	—	—	—	7
Terpinen-4-ol: T1 $C_{10}H_{18}O$	154.24	—	—	—	—	MS	—	150
α-Terpineol: T2 $C_{10}H_{18}O$ — t14, [M8886]	154.24	—	IR	UV	^{1}H	MS	—	138, 139
Terpineol acetate: T3 $C_{12}H_{20}O_2$ — t15	196.28	—	IR	—	^{1}H	MS	—	138
Terpinolene: T4 $C_{10}H_{16}$ — t17	136.23	—	IR	—	—	MS	—	138
Thiabendazole (TBZ): T5 $C_{10}H_7N_3S$ — [M9017]	201.26	—	—	UV	—	—	—	139
Tomatidine: T6 $C_{27}H_{45}NO_2$ — t407, [M9245]	415.64	210—211, 202—206	—	—	—	—	—	138, 139
Tomatine: T7 $C_{50}H_{83}NO_{21}$ — [M9247]	1034.22	263—268	—	—	^{1}H, ^{13}C	—	—	139, 203
Toonacilin: A3 $C_{31}H_{38}O_9$	554.61	151—152	IR	UV	^{1}H	MS	—	18
Trachyloban-19-oic acid: T8 $C_{20}H_{30}O_2$	302.44	98—100	IR	—	—	MS	—	179
Trichlorophenoxyacetic acid: T9 $C_8H_5Cl_3O_3$ — [M9324]	255.49	153	—	—	—	—	—	139

Compound	MW	mp	IR	UV	NMR	MS	X-ray	Ref.
Trichlorophenoxyethanol: T10 $C_8H_7Cl_3O_2$	241.51	—	—	—	—	—	—	48
Tulirinol: T11 $C_{17}H_{22}O_5$	306.35	204—206	IR	—	$^1H, ^{13}C$	MS	X-ray	84
Tyramine HCl: T12 $C_8H_{11}NO.HCl$ — t458, [M9489]	173.64	—	IR	UV	^1H	—	—	138, 139
Ugandensidial: U1 $C_{17}H_{24}O_5$	308.36	137—140	IR	UV	—	MS	—	204
Unedoside: U2 $C_{14}H_{20}O_9$	332.30	232—234	IR	UV	^1H	—	—	205
Ursolic acid: U3 $C_{30}H_{48}O_3$ — [M9552]	456.68	285—288	—	—	—	—	—	139
Uzarigenin: U4 $C_{23}H_{34}O_4$ — [M9561]	374.50	240—256	—	—	—	—	—	139
Vanillic acid: V1 $C_8H_8O_4$ — b1024, [m9590]	168.14	213—215, 210	IR	UV	^1H	—	—	138, 139
(3R)-(−)-Vestitol: V2 $C_{16}H_{16}O_4$	272.29	144—145	—	UV	^1H	MS	—	127
Vulpinic acid: V3 $C_{19}H_{14}O_5$	322.30	—	—	—	—	MS	—	154
Warburganal: W1 $C_{15}H_{22}O_3$	250.33	134—136	IR	UV	$^1H, ^{13}C$	MS	—	113, 124
Withaferin A: W2 $C_{28}H_{38}O_6$ — [M9712]	470.58	252—253, 243—245	IR	UV	^1H	MS	X-ray	139, 206, 207
Xanthotoxin: X1 $C_{12}H_8O_4$	216.18	145—146	—	—	$^1H, ^{13}C$	—	—	150
Xylomollin: X2 $C_{12}H_{18}O_7$	274.26	138—139	IR	—	$^1H, ^{13}C$	MS	—	137
Zanthophylline: Z1 $C_{18}H_{19}NO_5$	329.34	138—139	IR	UV	$^1H, ^{13}C$	MS	—	110, 208

a Alphanumeric codes refer to selected chemical structures in Figure 1.

In looking at the structures of these very active feeding deterrents, one can see some similarities. The complexity of the A and B rings in azadirachtin (A18) are similar to those of warburganal (W1), muzigadial (M15), ugandensidial (U1), and polygodial (P23); pilocarpine (P15) and strychnine (S20) are heterocyclics containing nitrogen; and the aromatic nature of aristolochic acid (A14) is slightly similar to salicin (S2) with its aromatic ring.

Other less active deterrents in Figure 1 that are steroidal in structure like azadirachtin include the toonacilins (A2 and A3), the cholesterols (C20—22), the cucurbitacins (C40—42), 1,3-diacetylvilasinin (D4), digitonin (D9), diosgenin (D17), ecdysterone (E1), friedelin (F9), harrisonin (H2), hecogenin (H3), meliantriol (M2), nimbandiol (N6), ponasterone (P24), salannol (S1), solanidine (S16), solanine (S17), strophanthidin (S19), tomatidine (T6), tomatine (T7), ursolic acid (U3), uzarigenin (U4), and withaferin A (W2).

Other less active deterrents in Figure 1 that are similar in structure to the complexity of the A and B rings of azadirachtin and warburganal (W1), muzigadial (M15), ugandensidial (U1), and polygodial (P23) include the ajugarins (A6—8), the caryoptins and clerodins (C9—18, C31, and C32), the salannins (D1 and D2), glaucolide-A (G4), and peroxyferolide (P3) — both of which are missing the bridge between the A and B rings — inflexin (I1), isomedin (I6), (−)-kaur-16-en-19-oic acid (K2), 13-keto-8(14)-podocarpen-18-oic acid (K3), santonin (S4), the schkuhrins (S6 and S7), the shiromodiols (S11 and S12), shiromool (S13), and trachyloban-19-oic acid (T8).

The prevalence of furan, lactone, ketone, and epoxide groups in these two classes of deterrents is striking. No doubt, they probably play some role in feeding deterrent activity. One also observes the common presence of terpenoid and the obvious absence of alkaloidal structures in Figure 1. All of these structural comparisons, however, are so general that it is difficult to postulate any specific structure-activity relationships for feeding deterrent activity.

It is the personal view of the author that there is a great potential for the use of insect feeding deterrents in pest management. However, the use of feeding deterrents in all probability will be in conjunction with other methods of control. For example, sex pheromones, ovipositional attractants, and feeding attractants could be used to monitor or survey the need for control of a particular pest; if control were needed, feeding deterrents, insect predators, ovipositional deterrents, molting inhibitors, confusion techniques with pheromonal substances, viruses and bacteria specific for the pest, biodegradable pesticides, etc., could be used.

At this time, several investigators and commercial companies are testing insect feeding deterrents in the field for various pests. The only limitations that exist in the development and utilization of insect feeding deterrents are time, money, and the extreme complexity of the problem. One must weigh the cost of development of such complex insect control methods vs. the cost of cleaning up the environment through the indiscriminate use of toxic chemicals that are not biodegradable or degrade into other toxic by-products. These are tough questions that must be addressed, but the author believes that the natural approach to the control of insect pests will prevail and is well worth the cost of research and development.

It is hoped that this review will stimulate the isolation and/or synthesis of more potent insect feeding deterrents, along with the development of resistant crop varieties that naturally incorporate the feeding deterrent substances. In the future, we may see feeding deterrents being used for insect pest management to control insects indirectly through starvation. From present indications, one would not anticipate harm to useful natural parasites, predators, and pollinators. Perhaps pests would even turn to weeds[1] when food crops are sprayed with yet undiscovered feeding deterrents.

PART B: INSECT FEEDING DETERRENTS (1980-1987)

E. David Morgan and J. David Warthen, Jr.

The discovery of new substances from natural sources which deter or suppress feeding in insects has continued without any signs of exhaustion, though there is now a pattern emerging of the types of compounds from plants which display activity. Many compounds from among the highly oxygenated triterpenoids, such as azadirachtin, diterpenes of the clerodin type, sesquiterpene lactones, flavones, and tannin derivatives, are listed here. This still leaves many other active compounds outside these categories. The overwhelming proportion of studies have been on higher plants, but there are reports of at least two microbiological products[209,210] and one from marine organisms,[211] in addition to the avermectins, fungal products toxic, and also feeding-deterrent, to insects. The avermectins have been widely studied in recent years. Possibly many active compounds remain to be discovered in these two rich sources of unusual natural products.

The first report of practical field use of feeding deterrents has been published,[212] though others are to be published soon. It is interesting to observe the greater participation of the manufacturers of traditional synthetic pesticides in research on feeding deterrents and the boost to research on the neurophysiology of taste reception in insects that has occurred.

The feeding deterrence of neem (*Azadirachta indica* A. Juss) and the compound azadirachtin here received increased attention, though its practical use is still very limited. There have been three international neem conferences, the first two in Germany.[213,214] The proceedings of the third, at Nairobi in 1986, are now available. A chapter in Volume III of this series[215] has been devoted largely to neem, and the first volume of a new series[216] is devoted entirely to neem.

Since preparing the first part of this survey, the correct molecular structure of azadirachtin (see structure A20) has been solved through an X-ray crystallographic study of one of its derivatives[217] and numerous NMR studies.[218-221]

No further listing of the many studies of insect deterrence by crude neem extracts are noted in the tables, chiefly because the actual contents of these extracts are uncertain. From our experience of analyzing some of these crude extracts used in feeding tests, they may contain very variable or negligible amounts of azadirachtin and compounds closely related to it.

There has been increasing activity in attempts to synthesize some of the more active naturally occurring feeding deterrents, and analogs of them, to discover something about the source of their activity. This has been reviewed elsewhere.[332] Some 63 isomeric clerodane diterpenes, some synthetic, have been surveyed with their levels of activity and the type of test used, in an effort to relate structure and activity.[314]

Many reviews have been written covering some aspect of this subject, of particular interest is one on the properties of 46 quassinoids compared with azadirachtin.[305] The subject of terpenoid antifeedants has been reviewed recently,[314] and the properties of a large number of drimane and clerodane types have been summarized.[332] Russell has written an interesting general review, though it is not easily available.[315] Phytoecdysteroids, plant steroids with the same or similar structure to insect moulting hormones, are not strictly feeding deterrents, though in many cases they affect insect growth and development and are claimed by some to be produced by plants as a protection against insects. Some hundred phytoecdysteroids are now known. They have been reviewed, together with their structure formulae and the plant families from which they have been isolated.[342]

There has been increasing activity in attempts to synthesize some of the more active naturally occurring feeding deterrents and analogs of them, to discover something about the source of their activity, e.g., in Reference 332. Ley has been active in this field and has reviewed synthetic work.[314] Synthesis of the drimane and clerodane types has also been reviewed in detail.[332] However, the subject of synthetic compounds is outside the scope of this chapter.

Table 4 lists those compounds which have been found to be active feeding deterrents, 1980 to 1987. The data is in the same format as Table 1. Some additional data on compounds described in Table 1 is added in a supplementary table. Table 5 continues the information of Table 2. Table 6 is an attempt to bring together those plant families from which active compounds have been isolated and the compounds found active in those families.

The information given on the nature of feeding tests, the insect species used, and the effective concentration are very variable in papers. We have been constrained to list compounds without much detail of this sort. Almost all of the compounds reported here have been listed in the *Dictionary of Organic Compounds*[222] and its supplements, where melting points, optical rotation, and references to spectroscopic data are given, so we have not attempted to duplicate that information here, but to give molecular structures wherever possible.

TABLE 4
Insect Feeding Deterrents, 1980—1987

Feeding deterrent	Plant source	Insect scientific name conc for deterrent activity	Ref.
Abietic Acid A19	*Larix laricina*	*Pristiphora erichsonii*	284
Abyssinin A20	*Bersama abyssinica*	*Heliothis zea*	231
12 β-Acetoxyharrisonin (see H2)	*Harrisonia abyssinica*	*Spodoptera exempta*	281
8-Acetyl-3α-hydroperoxy-2α-isohex-3-enyl-6-isopent-2-enyl-5-methoxy-2β-methylchroman A21	*Harrisonia abyssinica*	*Spodoptera eridania* *Spodoptera exempta*	280
6-O-Acetylnimbanidol A22	*Azadirachta indica*	*Heliothis virescens*	311
		Pectinophora gossypiella	338
7-Acetyltrichilin A23	*Trichilia roka*		243
Acetylvismione B (see V9, R=Ac)	*Vismia guaramirangae*	*Locusta migratoria*	329
Achillin A24	*Artemisia ludoviciana* *Achillea sp.*	*Melanoplus sanguinipes* 0.5%	298
Acrecoline A25	*Areca cathecu*	*Phormia regina*	322
Amygdalin A26	Peach	*Choristoneura rosaceana*	274
Anthranilic acid	*Alchornea triplinervia*	*Anthonomis grandis* *Heliothis virescens*	239
Arctolide A28	*Arctotis grandis*	*Trogoderma granarium* *Tribolium confusum*	242
Asimicin A29	*Rollinia sylvatica*	*Acalymma vittatum* 0.5 ppm	316
Azadirachtin A30 (Correct molecular structure)	*Azadirachta indica*	Many species	217—220
Azedarachol A31	*Melia azedarach* root bark	*Argrotis sejetum*	226
Bakkenolid A B12	*Homogyne alpina*	*Sitophilus granarius*	241
		Tribolium confusum	256
		Trogoderma granarius	257
		Leptinotarsa decemlineata 0.1%	317
Balanites saponin B13	*Balanites roxburgii*	—	300
Bilobalide B14	*Ginkgo biloba*	*Pieris rapae crucivora*	313
Bisabolangelone B15	*Angelica silvestris*	*Sitophiles granarius*	241
		Tribolium confusum	256
		Trogoderma granarium	317
		Leptinotarsa decemlineata	
		Mythimna unipuncta	
Bruceine A B16 (R=CH₂CHMe₂)	*Brucea amarissima*	*Locusta migratoria*	279
Bruceine B (see B16, R=CH₃)	*Brucea amarissima*	*Locust migratoria*	279
Cadinene C44	*Eupatorium adenophorum*		233
Camphor C45	*Alchornea triplinervia*	*Anthonomis grandis* *Heliothis virescens*	239
Capillarin C46	*Artemisia capillaris*	*Pieris brassicae*	299
Capillen C47 (Agropyrene)	*Artemisia capillaris* *Agropyron repens* *Triticum repens*	*Pieris rapae crucivora*	272
Caryophyllene oxide C48	*Melampodium divaricatum*	*Atta cephalotes*	333
Castanospermine C49	—	*Acyrthosiphon pisum*	229
Catalposide C50	*Catalpa speciosa*	"Nectar thieves"	
		Poanes hobomok 0.4%	277
Cedrolone C51	*Cedrela toona*	*Spodoptera litura* 0.1%	271
α-Chaconine C52	*Solanum chacoense* and other *S. sp*	*Choristoneura fumiferana* $10^{-3}M$	250

TABLE 4 (continued)
Insect Feeding Deterrents, 1980—1987

Feeding deterrent	Plant source	Insect scientific name conc for deterrent activity	Ref.
Chaparrinone C53 (R=H)	*Simaba multiflora*	*Locusta migratoria*	230
	Soulamea soulameoides	*Heliothis virescens*	279
		Spodoptera frugiperda	
Chicoriin C54	*Cichorium intybus*	*Schistocerca gregaria*	330
Chlorogenic acid C55	*Salix chaenomeloides,*	*Lochmaeae capreae cribrata,*	235
	S. integra, S. Koriyanagi	*Agelasa coerulea, Altica olera-*	
	(Willows)	*ceae, Atrachya menetriesi,*	
		Fleutriauxia armata, Galelu-	
		cella vittaticollis, Gallerucida	
		bifasciata, Gastrolina de-	
		pressa, Gastrophysa atrocy-	
		anea, Oulema oryzae,	
		Pyrrhalta humeralis	
		(leaf eating beetles)	
Cinnamaldehyde	*Alchornea triplinervia*	*Anthonomis grandis*	239
13-*trans*-Cinnamoyloxylupanine C56	*Lupinus polyphyllus*	*Choristoneura fumiferana*	249
(−)-Claussequinone C57	*Cyclolobium clausseni*	*Costelytra zealandica*	326
	C. vecchii, Millettia pendula		
Clerodane diterpenes C58	*Verbenaceae*	*Spodoptera sp.*	314
(60 active compounds)	*Labiatae*		
ar-Curcumene C59	*Artemisia capillaris*	*Pieris brassicae*	299
Dehydroabietic acid D19	*Larix laricina*	*Pristiphora erichsonii*	284
	Pinus, Cedrus sp.		
n-Decyl gallate D20	—	*Schizaphis graminum* 16 ppm	303
8-Deoxylactupicrin D21	*Cichorium intybus*	*Schistocerca gregaria*	330
Desacetylarctolide D22	*Arctotis grandis*	*Sitophilus granarius*	242
		Tribolium confusum	
Deacetylvismione A (see V8, R=H)	*Vismia guaramirangae*	*Locusta migratoria*	329
11,13-Dihydrovernodalin D23	*Vernonia amygdalina*	*Spodoptera exempta*	261
2,3-Dihydrowithanolide E (see W4)	*Withania Sp.*	*Spodoptera littoralis*	
		Epilachna varivestis	304
2,5-Dihydroxymethyl-3,4-dihy-droxy-pyrrolidine D24	—	*Locusta migratoria* 0.01%	236
		Prodenia litura	
		Schistocerca	
		Spodoptera exempta	
		Spodoptera littoralis	
6,7-Dimethoxyisochroman-3-one D25	—	*Achoea janata* 2%	240
Ellagic acid E9	Galls	*Schizaphis graminum* 15 ppm	303
Encecalin E10	*Encelia california*	*Peridroma saucia*	228
		Plusia gamma	290
		Oncopeltus fasciatus	342
		Heliothis zea	
Eperu-8(17)-en-15,19-dioic acid butenolide (see L12 lamber-tianic acid, furan ring con-verted to lactone)	*Gridelia humilis*	*Schizaphis graminum*	335
(4-8)Epicatechin-catechin dimer	—	*Locusta migratoria*	325
5β,6β-Epoxy-1β,14α,17β,20-tetrahydroxywith-24-enolide E11	*Physalis peruviana*	*Spodoptera littoralis*	292
Erivanin E12	*Artemisia* sp.	*Sitophilus granarius*	242

TABLE 4 (continued)
Insect Feeding Deterrents, 1980—1987

Feeding deterrent	Plant source	Insect scientific name conc for deterrent activity	Ref.
Eupatoriopicrin E13	*Eupatorium cannabinum*	*Sitophilus granarius*	
		Tribolium confusum	241
		Trogoderma granarium	257
Ferruginin A F10	*Vismia guaramirangae*	*Spodoptera exempta*	
		Heliothis virescens	329
Ferruginin B F11	Vismia guaramirangae	*Heliothis virescens*	
		Locusta migratoria	329
Geraniin G8	*Geranium thunbergii*	*Shizaphis graminum*	
		Myzus persicae	303
Ginkgolide A G9 (R=H)	*Ginkgo biloba*	*Pieris rapae crucivora*	313
Ginkgolide B G9 (R=OH)	*Ginkgo biloba*	*Pieris rapae crucivora*	313
Glaucarubinone G10	—	*Locusta migratoria*	279
2-β-D-Glucosyl-4-hydroxy-7-methoxy-1,4-benzoxazin-3-one G11	*Zea mays*	*Schizaphis graminum*	234
Grayanotoxin-III G12	*Kalmia latifolia* *Leucothoe grayana*	*Lymantria dispar*	291
Harunganin H11	*Vismia guaramirangae*	*Spodoptera exempta* *Heliothis virescens*	329
Helenalin H12	*Helenium* sp.	*Sitophilus granarius* 2%	241
		Tribolium confusum	242
		Trogoderma granarium	317
		Leptinotarsa decemlineata	
1-(1,3,4,5,6,7-Hexahydro-4-hydroxy-3,8-dimethyl-5-azulenyl)ethanone H13	*Dictyota dichotoma*	*Spodoptera littoralis*	210
Hildecarpin H14	*Tephrosia hildebrandtii*	*Maruca testulalis*	223
Hiptagin H15	*Lotus pedunculatus*	*Costelytra zealandica*	328
Hispidulin H16	*Tithonia diversifolia, Ambrosia hispida, Gaillardia fastigiata, Florensia cernua, Scutellaria przewalskii*	*Philosamia ricini*	324
Hydroxamic acid H17	*Zea mays, Triticum durum, Secale cereale, Arundo donax*	*Schizaphis graminum*	260
2'-Hydroxyformononetin (2',7-Dihydroxy-4'-methoxy-isoflavone) H18	*Bowdichia nitida*	*Costelytra zealandica*	326
2'-Hydroxygenistein H19	*Lupinus angustifolius*	*Costelytra zealandica*	326
			340
6α-Hydroxygrindelic acid	*Grindelia humilis*	*Schizaphis graminum*	335
18-Hydroxygrindelic acid H21	*Grindelia humilis*	*Schizaphis graminum* 0.002%	335
(4R,5S)-5-Hydroxy-2-hexen-4-olide H22	*Osmunda japonica*	*Eurema hecabe mandarina*	251
			268
(−)-3'-Hydroxy-4'-methoxy-7-hydroxy-8-methylflavan H23	*Lycoris radiata*	*Eurema hecabe mandarina*	266
23-(R,S)-Hydroxytoonacilide H24	*Toona ciliata*	*Epilachna varivestis*	289
21-(R,S)-Hydroxytoonacilide (see 23-Hydroxy, lactide ring reversed)	*Toona ciliata*	*Epilachna varivestis*	289
Ipolamiide I8	*Stachytarpheta mutabilis*	*Locusta migratoria*	283
		Schistocerca gregaria	
Isoalantolactone I8a	*Inula helenium*	*Sitophilus granarius*	334
		Tribolium confusum	
		Trogoderma granarium	

TABLE 4 (continued)
Insect Feeding Deterrents, 1980—1987

Feeding deterrent	Plant source	Insect scientific name conc for deterrent activity	Ref.
ent-Isoal antolactone enantioner of Isoalantolactone, I8a)	*Locopholea heterophylla*	*Tribolium confusum* *Sitophilus granarius* *Trogoderma granarium*	334
Isobrucein A (see B16, 1α-OH instead of 3-OH)	Simaba multiflora *Soulamea soulameoides*	*Heliothis virescens* *Spodoptera frugiperda*	230
Isopimaric acid I9	*Larix laricina*	*Pristiphora erichsonii*	284
(−)-3-Isothujone I10 ((−)-α-Thujone)	*Thuja plicata*	*Pissodes strobi* *Altica ambiens* *Hylemya antigua*	288
Kalmitoxin-I K5	*Kalmia latifolia*	*Lymontria dispar*	291
Kalmitoxin-IV K6	*Kalmia latifolia*	*Lymantria dispar*	291
Kievitone K7	*Phaseolus vulgaris* infected with *Sclerotinia fructicola*	*Costelytra zealandica*	326
Lactarorufin A L10	*Lactarius rufus*	*Tribolium confusum*	242
Lactucin L11	*Cichorium intybus* *Lactuca virosa*	*Schistocerca gregaria*	330
Lactupicrin (p-hydroxyphenyl-acetic ester of Lactucin)	*Cichorium intybus*	*Schistocerca gregaria*	330
Lambertianic acid L12	*Grindelia humilis*	*Schizaphis graminum* 0.006%	335
Lasiocarpine L13	*Tussilago farfara* *Heliotropium europaeum* *H. Lasiocarpum,* *H. arbainense*	*Choristoneura fumiferama* 1.2 m*M*	252
Levopimaric acid L14	*Grindelia humilis*	*Schizaphis graminum* 0.03%	335
Licoisoflavone A L15	*Lupinus angustifolius*	*Costelytra zealandica* 100 μg/ml	340
Licoisoflavone B L16	*Lupinus angustifolius* cv "Uniharvest"	*Costelytra zealandica* 2 μg/ml *Heteronychus arator* 10 μg/ml	326 340
Limonin L17	*Citrus paradisi*	*Spodoptera litura* 0.5% *Leptinotarsa decemlineata* 100 μg/cm²	271 273, 308
Limonim diosphenol L18 (Evodol)	*Evodia glauca*	—	301
Linifoline A L19	*Helenium linifolium*	*Sitophilus granarius* *Tibolium confusum* *Trogoderma granarium*	241
Luteone L20	*Lupinus angustifolius*	*Costelytra zealandica* 2 μg/ml	326, 340
(−)-Maackiain M19	*Maackia amurensis* *Sophora tomentosa*	*Costelytra zealandica*	326
(−)-Medicarpin M20	—	*Costelytra zealandica*	326
Melampodin A M21	*Melampodium americanum* *M. leucanthrum*	*Spodoptera frugiperda*	264
Melampodinin A M22	*Melampodium americanum* *M. leucanthrum*	*Spodoptera frugiperda*	264
Melicopicine M23	*Teclea trichocarpa*	*Spodoptera exempta*	227, 232
12-Methoxyabietic acid M24	*Larix laricina*	*Pristiphora erichsonii*	284
5-Methoxy-N,N-dimethyl tryptamine M25	Gramineae *Virola calophylla*	*Schizaphis graminum* 0.5 m*M* *Anthonomis grandis*	258 306
6-Methoxytecleanthine M26	*Teclea trichocarpa*	*Spodoptera exempta*	227, 232
5-Methoxy tryptamine	Graminae	*Schizaphis graminum* 0.5 m*M*	258
4'-O-Methyl(−)-epigallicatechin M28	—	*Locusta migratoria* 0.1%	325
Methyleugenol M29	*Artemisia capillaris*	*Pieris brassicae*	299
7-Methylhydroxamic acid M30	*Zea mays, Triticum durum, Elymus gayanus, Chusquea cumingii*	*Schizaphis graminum*	260

TABLE 4 (continued)
Insect Feeding Deterrents, 1980—1987

Feeding deterrent	Plant source	Insect scientific name conc for deterrent activity	Ref.
N-Methyl-*trans*-4-hydroxy-L-proline	*Copaifera* sp.	*Spodoptera littoralis*	296
O-Methyllycorenine M32	*Lycoris radiata*	*Eurema hecabe mandarina*	266
2-Methyl-6-methoxytetrahydro-β-carboline M33	*Virola calophylla*	*Anthonomis grandis*	306
(−)-7-O-Methylphaseolin (see P34, R=Me)	*Phaseolus vulgaris*	*Costelytra zealandica*	326
(+)-2′-O-Methylphaseoliniso-flavan (see P35, R=Me)	*Phasolus vulgaris*	*Costelytra zealandica*	326
Methyl sainfuran M34	*Onobrychis viciifolia*	—	244
N-Methyltryptamine M35	Gramineae	*Schizaphis graminum* 0.5 m*M*	258
Miconidin M36	*Miconia* sp.	*Pieris brassicae* 0.05%	248
		Spodoptera exempta 0.05%	
		Spodoptera littoralis 0.01%	
		Heliothis armigera 0.2%	
		Schistocerca gregaria 0.05%	
		Locusta migratoria 0.05%	
Melletia chalcone M37	*Milletia pachycarpa*	*Philosamia ricini* 1%	263
Momordicine II M38	*Momordica charantia*	*Aulacophora foveicollis*	302
Nicalbin A N9	*Nicandra physaloides*	*Epilachna varivestis*	282
		Tribolium castaneum	304
Nicalbin B N10	*Nicandra physaloides*	*Epilachna varivestis*	282
		Tribolium castaneum	304
Nicandrenone N-11 (Nic-1)	*Nicandra physaloides*	*Epilachna varivestis*	282
			292
Nimbolidin A (see N12, R=benzoyl)	*Melia azedarach*	*Epilachna varivestis* 0.01%	247
Nimbolidin B N12 (R=tigloyl)	*Melia azedarach*	*Epilachna varivestis* 0.005%	247
Nimbolinin B N13	*Melia azedarach*	*Epilachna varivestis*	312
Nomilin N14	Citrus sp.	*Spodoptera frugiperda*	262
		Heliothis zea	273
Obacunone 03	Citrus sp.	*Spodoptera frugiperda*	273
	Phellodendron amurense	*Heliothis zea*	
Oblongolide 04	*Phomopsis oblonga*	*Scolytus scolytus*	235, 321
Ochinolide A 05 (R=benzoyl)	*Melia azedarach*	*Epilachna varivestis* 0.01%	247
Ochinolide B (see 05, R=Tigloyl)	*Melia azedarach*	*Epilachna varivestis* 0.01%	247
n-Octyl gallate 06	—	*Schizaphis pisum* 182 ppm	
		Myzus persicae 56 ppm	303
(4R,5S)-Osmundalactone 07	*Osmunda japonica*	*Eurema hecabe mandarina*	251, 268
Palustric acid P28	*Pinaceae palustris*	*Neodiprion dubiosus* 0.1%	
		N. leconti, N. rugifrons	253
Papaverine P29	*Papaver somnifera*	*Phormia regina*	322
Pedonin P30	*Harrisonia abyssinica*	—	309
Pedunculagin P31	Gall nuts	*Schizaphis graminum*	
		Myzus persicae	303
Peramine P32	*Lolium perenne* infected with *Acremonium loliae*	*Listronotus bonariensis*	238
			318
(−)-Phaseollidin P33	*Phaseolis vulgaris* infected with *Sclerotinia fructicola*	*Costelytra zealandica*	326
(−)-Phaseolin P34 (R=H)		*Costelytra zealandica*	293
	Phaseolus vulgaris	*Heteronychus arator*	326
(−)-Phaseolinisoflavan P35 (R=H)	*Phaseolus vulgaris*	*Costelytra zealandica*	326
1-Phenylheptatriyne P36	*Asteraceae* sp.	*Euxoa messoria* 10 ppm	275

TABLE 4 (continued)
Insect Feeding Deterrents, 1980—1987

Feeding deterrent	Plant source	Insect scientific name conc for deterrent activity	Ref.
1-Phenyl-2,4-pentadiyne P37	*Artemisia capillaris*	*Pieris rapae crucivora*	272
Phloretin 4'-O-β-D-glucopyranoside (see P12)	*Malus pumila*	*Schizaphis graminum* 0.02%	225
Piceatannol P38	*Scirpus maritimus*	—	307
(+)-Pisatin P39	*Pisum sativum*	*Costelytra zealandica*	326
	Lathyrus spp.		
Plagiochilin A P40	*Plagiochila ovalifolia*	*Spodoptera exempta*	287
Plectrin P41	*Plectranthus barbatus*	*Schizaphis graminum*	246
Precocene I P42	*Ageratum houstonianum*	*Costelytra zealandica*	326
Precocene II P43	*Ageratum houstonianum*	*Rhodnius prolixus* 0.5%	276
	A. conyzoides, verbesina sp.	*Costelytra zealandica*	326
Prieurianin P44	*Nymania capensis*	*Heliothis virescens*	
	Trichilia prieuriana	*Epilachnis varivestis*	224
Primin P45 (see Miconidin)	*Miconia* sp.	*Pieris brassicae* 0.005%	248
		Heliothis armigera 0.1%	
		Spodoptera littoralis 0.01%	
		Spodoptera exempta 0.01%	
		Schistocerca gregaria 0.01%	
		Locusta migratoria 0.1%	
Proanthocyanidin A P46	*Cola acuminata*	*Locusta migratoria*	325
Proanthocyanidin B P46	—	*Locusta migratoria*	325
Psilotin P47	*Psilotum nudum*	*Ostrinia nubilalis* 0.1%	323
	Tmesipteris tannensis		
Pungenin P48	*Picea pungens*	*Choristoneura fumiferana*	327
Quassin Q3	*Quassia amara*	*Epilachna varivestis*	
	Picraena excelsa	*Spodoptera eridania*	245
Rohituka 7 R2	*Aphanamixis polystacha*	*Heliothis virescens*	224
(−)-Rotenone R3	*Derris elliptica*	*Costelytra zealandica*	326
Sainfuran S22	*Onobrychis viciifolia*	—	244
Salannolactam — (21) S23 (-(23) with lactam ring reversed)	*Azadirachta indica*	*Epilachna varivestis*	310
Salicortin S24	*Populus tremuloides*	*Papilio glaucus glaucus*	344
Sandaracopimaric acid S25	*Larix laricina*	*Pristiphora erichsonii*	284
	Pinus, Juniperus, Cupressus sp.	*Schizaphis graminum* 0.1%	335
Schkuhriolide S26	*Schkuhria schkuhrioides*	*Sitophilus granarius*	242
Scirpusin B S27	*Scirpus maritimus*	—	307
Senecioic acid S28	*Alchornea triplinervia*	*Anthonomis grandis*	239
6α-Senecioyloxy chaparrinone	*Simaba multiflora*	*Heliothis virescens*	230
	Soulamea soulameoides	*Spodoptera frugiperda*	
Senkirkine S29	*Tussilago farfara*	*Choristoneura fumiferana* 1 mM	252
	Nardosmia laevigata		
	Farfugium japonicum		
	Crotalaria laburnifolia, senecio sp.		
Simalikalactone D S30	*Quassia africana*	*Locusta migratoria*	279
Soulameanone S31	*Soulamea muelleri*	*Locusta migratoria*	279
Spathulenol S32	*Melampodium divaricatum*	*Atta cephalotes*	333
Specionin S33	*Catalpa speciosa*	*Choristoneura fumiferana* 100 ppm	237 265, 320
Steviol S34	*Stevia rebaudiana*	*Schizaphis graminum*	343
18-Succinoyloxygrindelic acid (see 18-Hydroxygrindelic acid)	*Grindelia humilis*	*Schizaphis graminum* 0.003%	335
Tafricanin A T13	*Teucrium africanum*	—	278
Tafricanin B T14	*Teucrium africanum*	—	278

TABLE 4 (continued)
Insect Feeding Deterrents, 1980—1987

Feeding deterrent	Plant source	Insect scientific name conc for deterrent activity	Ref.
Tagitinin A T15	*Tithonia diversifolia*	*Philosamia ricini*	324
Tagitinin C T16	*Tithonia diversifolia*	*Philosamic ricini*	324
Tecleanthine T17	*Teclea trichocarpa*	*Spodoptera exempta*	227, 232
Tenulin T18	*Helenium amarum*	*Astrinia nubilalis*	297
	H. tenuifolium, H. elegans	*Peridroma saucia*	
α-Terthienyl T19 (Terthiophene)	*Asteraceae*	*Euxoa messoria*	255
(+)-3-Thujone T20	*Thuja plicata*	*Pissodes strobi*	288
		Altica ambiens	
		Hylemya antiqua	
3-Tigloylazadirachtol T21	*Azadirachta indica*	*Heliothis virescens*	336
		Pectinophora gossypiella	337, 338
13-Tigloyloxylupanine T22	*Lupinus polyphyllus*	*Choristoneura fumiferana*	249
Tr-A T23 (R=Et)	*Trichilia roka*	*Agrotis sejetum*	294
Tr-B T24	*Trichilia roka*	*Agrotis sejetum*	294
Tr-C (see T23, R=Me)	*Trichilia roka*	*Agrotis sejetum*	294
Tremulacin T25	*Populus tremuloides*	*Papilio glaucus glaucus*	344
Tremuloidin T26	*Populus tremuloides*	*Papilio glaucus glaucus*	344
Trewiasine T27	*Trewia nudiflora*	*Acalymma vittatum* 0.1%	267
		Cydia pomonella	
Trichilin A T28	*Trichilia roka*	*Spodoptera eridania*	
		Epilachna varivestis	270
		Manduca sexta	285
Trichilin B (12-epimer of trichilin A)	*Trichilia roka*	*Spodoptera eridania*	270, 285
Trichilin C (As trichilin A but 11β–OH, 12C=O)	*Trichilia roka*	*Spodoptera eridania*	270, 285
Trichilin D (as trichilin A, but 12-C=O)	*Trichilia roka*	*Spodoptera eridania*	270, 285
2,3,4-Trihydroxybenzaldehyde T29	—	*Locusta migratoria*	325
2,4,6-Trihydroxybenzaldehyde T30	—	*Locusta migratoria*	325
1-(2,4,6-Trimethoxyphenyl)but-2-en-1-one T31	*Arachnoides standishii*	*Eurema hecabe mandarina*	259
Vasicine V4	*Adhatoda vasica*	*Aulacophora foveicollis* 0.05%	
		Epilachna vijintioctopunctata 0.1%	319
Vasicinol V5	*Adhatoda vasica*	*Aulacophora foveicollis* 0.05%	
		Epilachna vijintioctopunctata 0.1	319
Vasicinone V6	*Adhatoda vasica*	*Aulacophora foveicollis* 0.05%	
		Epilachna vijintioctopunctata	319
Vismin V7	*Vismia guaramirangae*	*Spodoptera exempta*	
		Heliothis virescens	
		Locusta migratoria	329
Vismione A V8	*Vismia guaramirangae*	*Spodoptera littoralis*	
		Spodoptera exempta	
		Locusta migratoria	329
Vismione B V9	*Vismia guaramirangae*	*Spodoptera exempta*	329
		Heliothis virescens	
		Locusta migratoria	
Wilforine W3	*Maytenus rigida*	*Pieris rapae*	254
	Tripterygium wilfordii	*Locusta migratoria*	
	Euonymus alatus		
Withanolide E W4	*Physalis peruviana*	*Epilachna varivestis*	282
	Withania	*Spodoptera littoralis*	292

TABLE 4 (continued)
Insect Feeding Deterrents, 1980—1987

Feeding deterrent	Plant source	Insect scientific name conc for deterrent activity	Ref.
Yatein Y1	*Libocedrus yateenis*	*Sitophilus granarius*	241
		Tribolium confusum	257
		Trogoderma granarium	317

Compounds already reported in Table 1 which have been tested and found active on further insect species

Ajugarin I A6	*Ajuga remota*	*Pieris brassicae*	269
		Prodenia litura	
		Spodoptera exigua	
Alantolactone A9	*Inula helenium*	*Trogoderma granarium*	334
		Sitophilus granarius	
Atropine A16	—	*Phormia regina*	322
Carvone	—	*Sitophilus granarius*	241
Cinnamic acid C26	*Alchornea triplinervia*	*Anthonomus grandis*	239
Gentisic acid G1	*Alchornea triplinervia*	*Anthonomus grandis*	239
Gramine G7	*Gramineae*	*Shizaphis graminum* 0.5 mM	238, 258
Quinine Q2	—	*Phormia regina*	322
Senecionine S10	*Senecio* anonymous	*Choristoneura fumiferana*	252
Solanidine S16	*Solanum* sp.	*Choristoneura fumiferana* $10^{-3}M$	250
Solanine S17	*Solanum* sp.	*Choristoneura fumiferana* $10^{-3}M$	250
Sparteine S18	—	*Phormia regina*	322
Strychnine S20	—	*Phormia regina*	322
Tomatidine T6	*Solanum* sp.	*Choristoneura fumiferana* $10^{-3}M$	250
Tomatine T7	*Solanum* sp.	*Phormia regina*	322, 250
		Choristoneura fumiferana $10^{-3}M$	
Vanillic acid V1	—	*Locusta migratoria*	325
Veratrine	—	*Sitophilus granarius*	241
		Tibolium confusum	
		Trogoderma granarium	

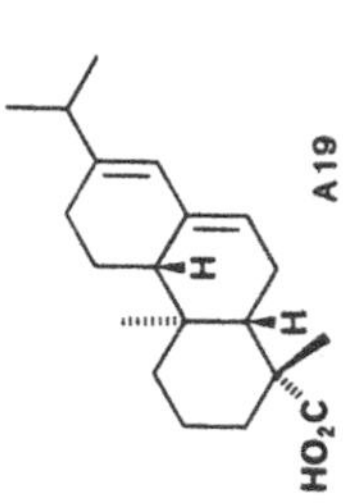

A26

A28

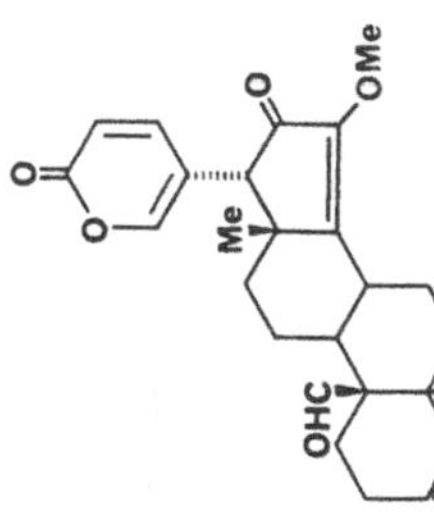

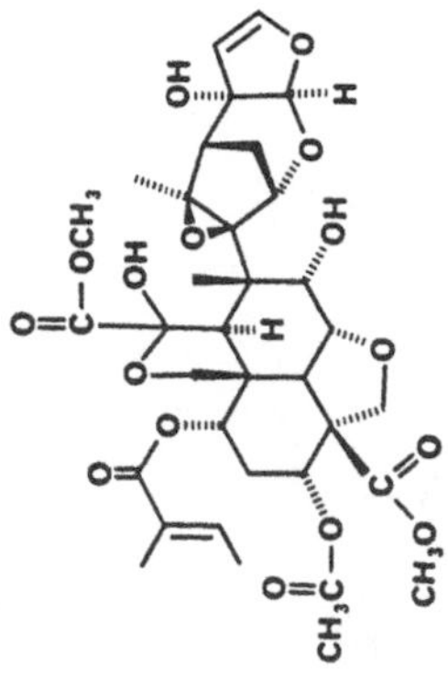

A30
Azadirachtin

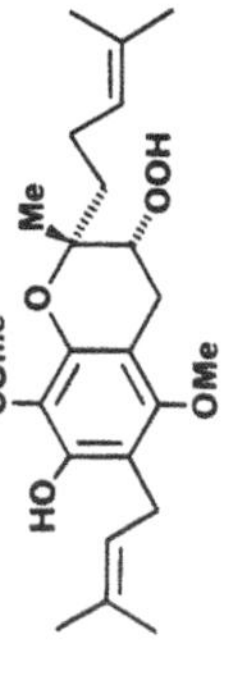

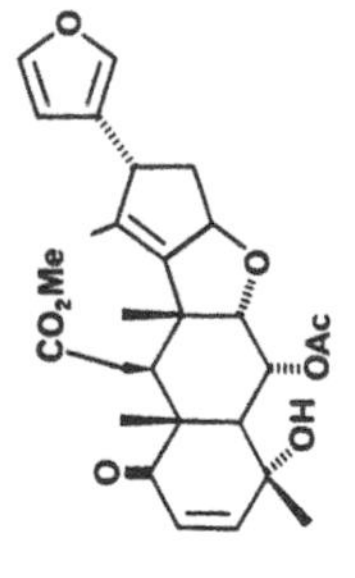

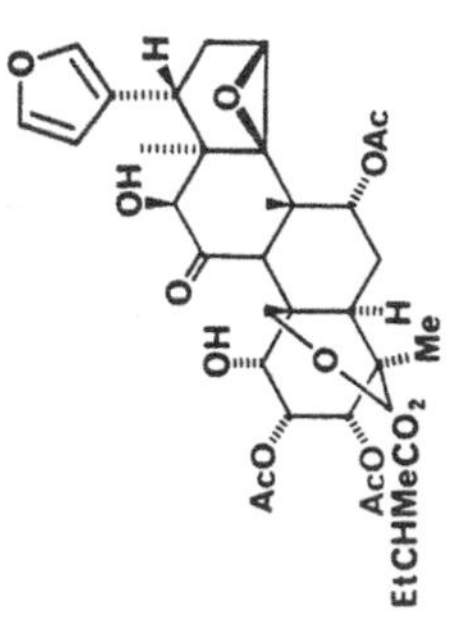

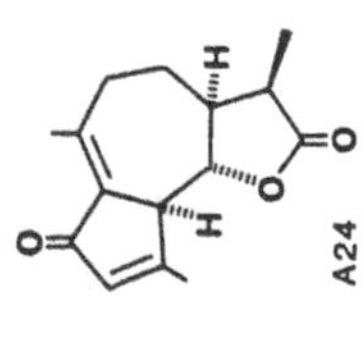

A25

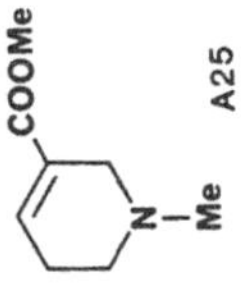

COOMe

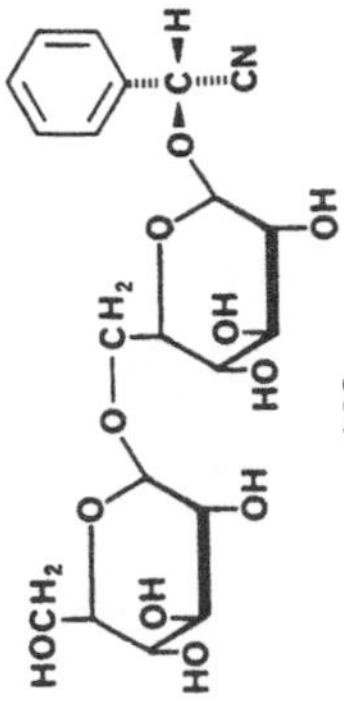

N—Me

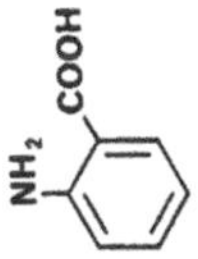

A27

NH₂ COOH

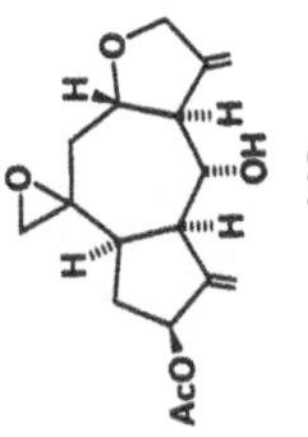

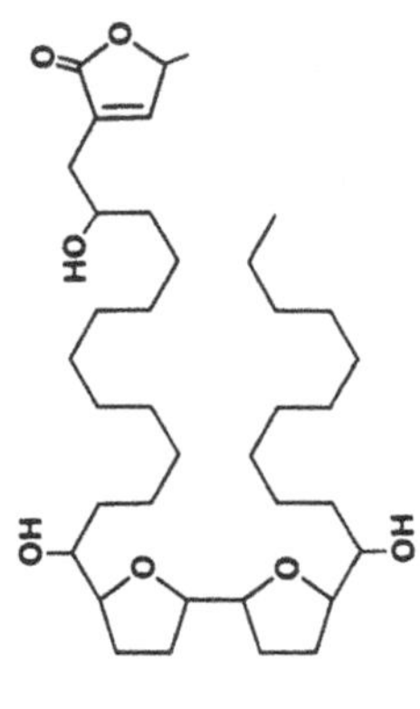

A29

A20

A22

A24

A19

A21

A23

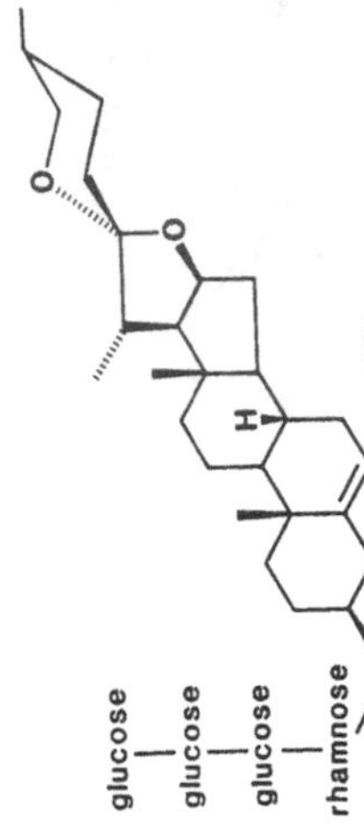

glucose—glucose—glucose—rhamnose

B13

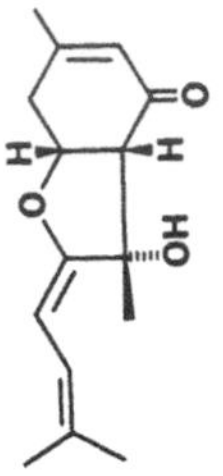

B15

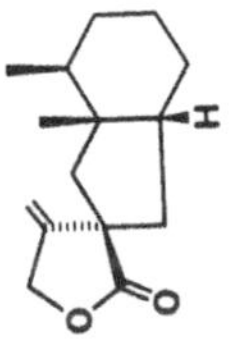

B12

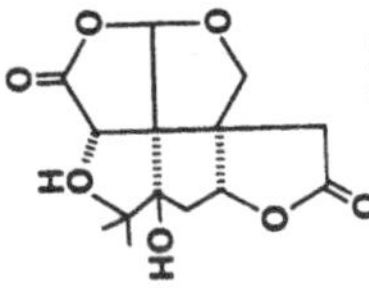

B14

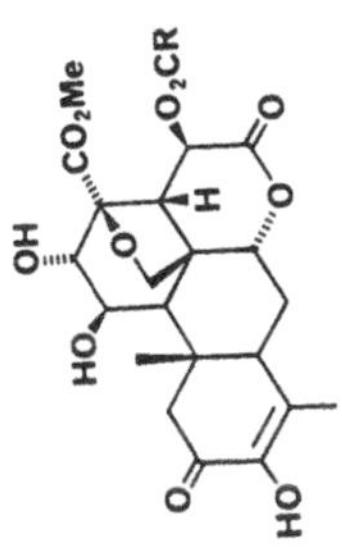

B16

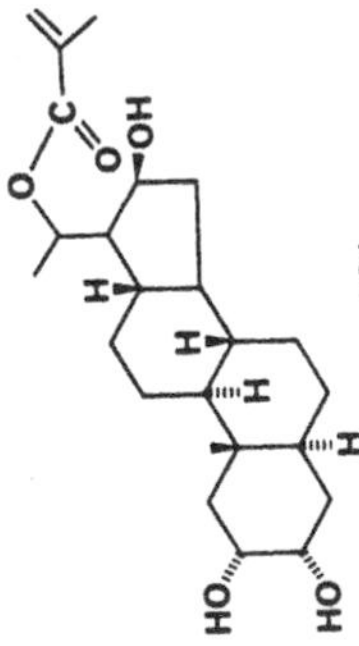

A31

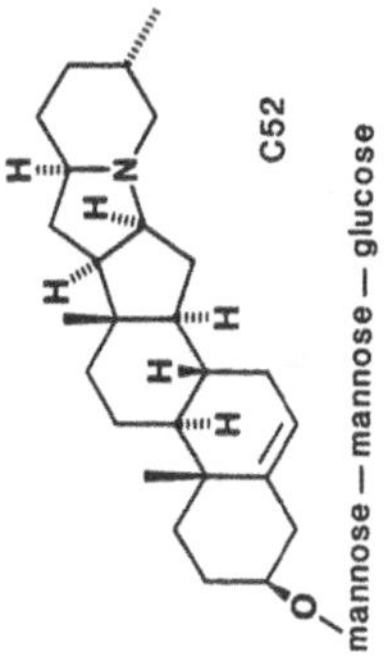

C52

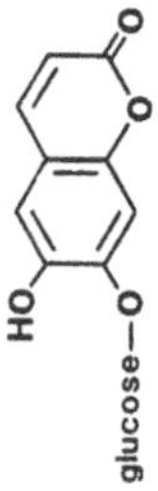

C54

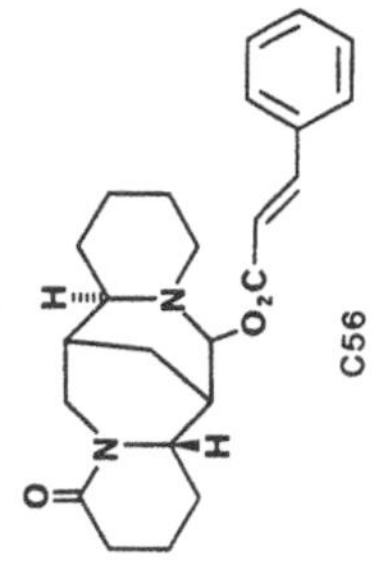

C56

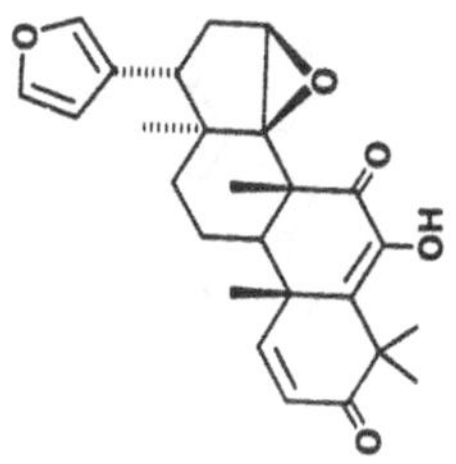

C51

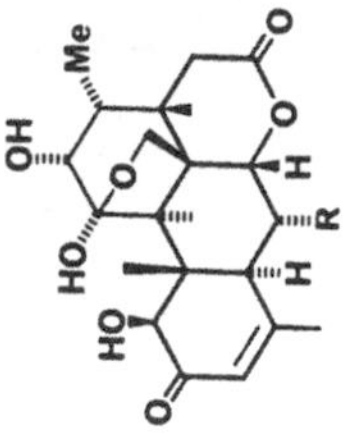

C53

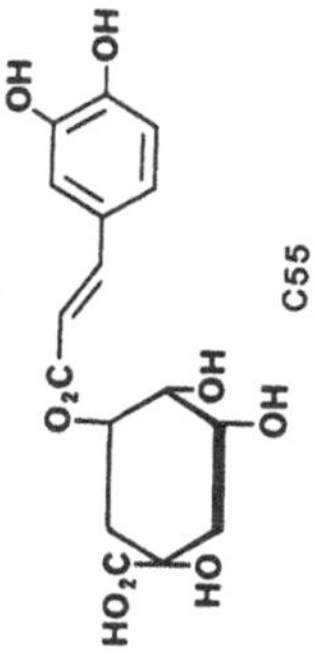

C55

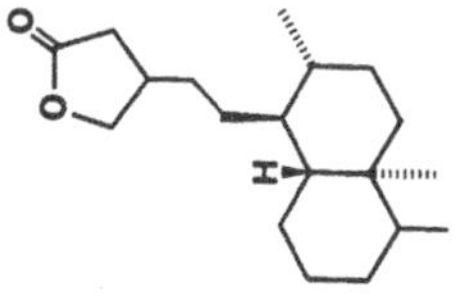

C58

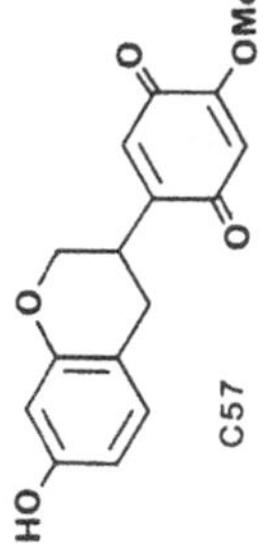

C57

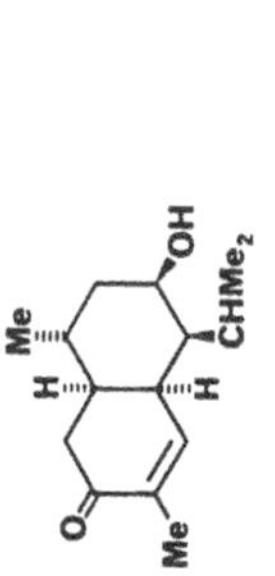

C44

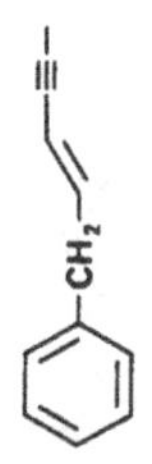

C47

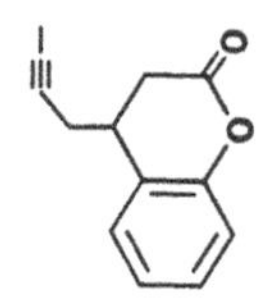

C46

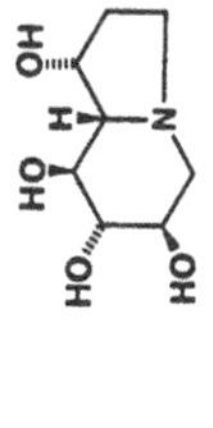

C49

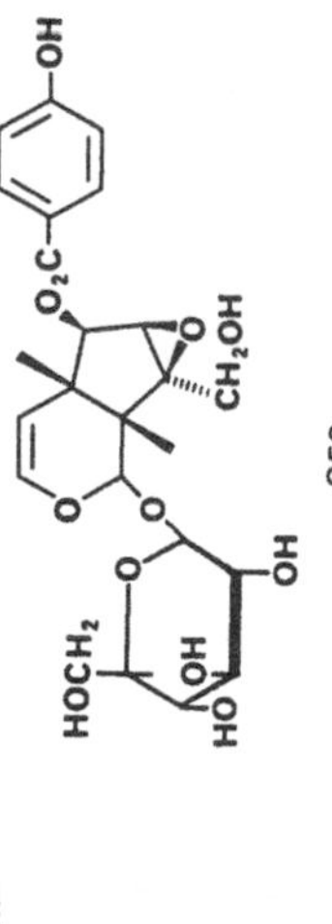

C50

α-Chaconine

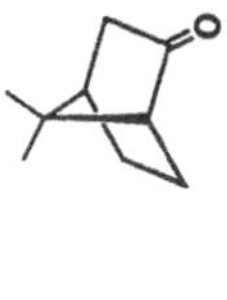

C45

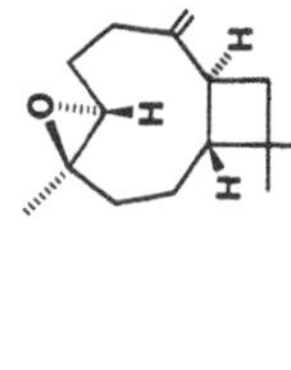

C48

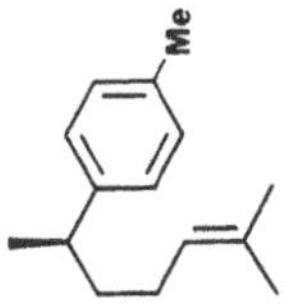

D23

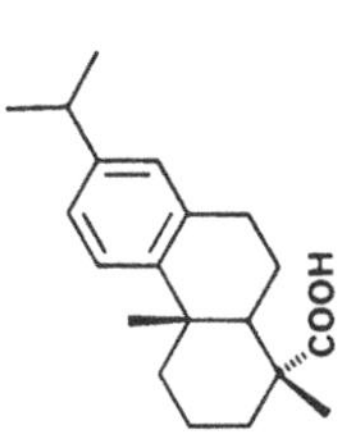

D25

E12

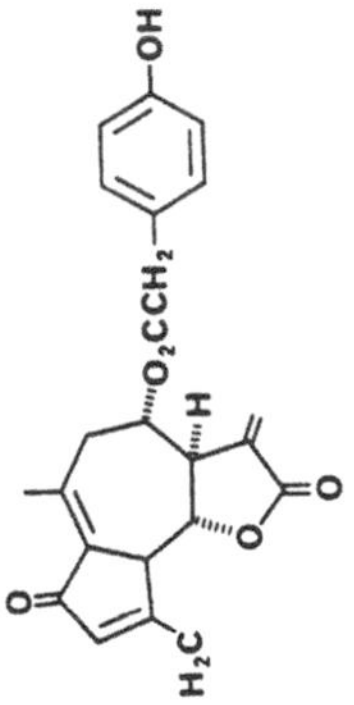

D22

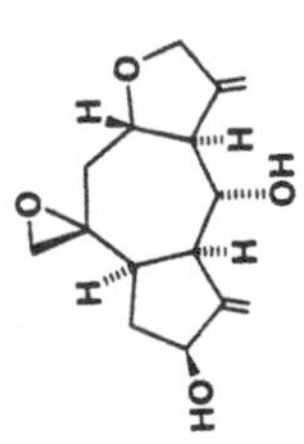

D24

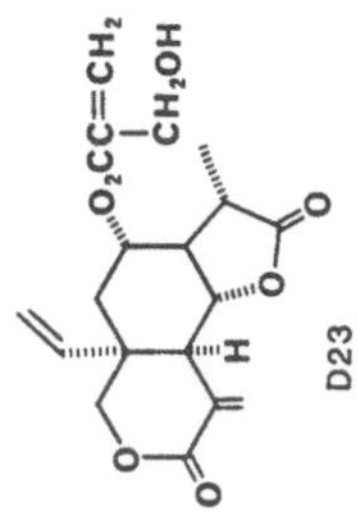

E11

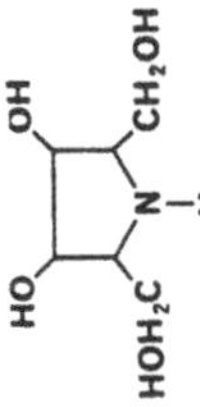

E13

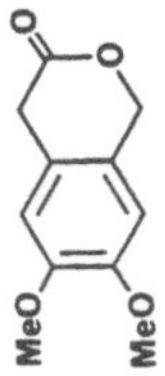

D20

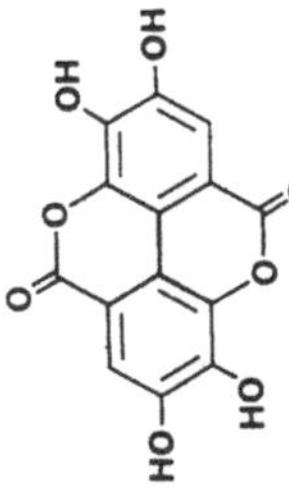

D21

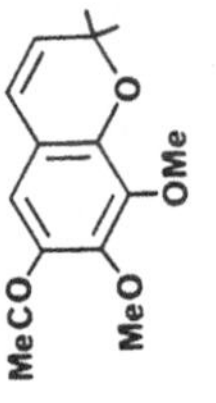

E10

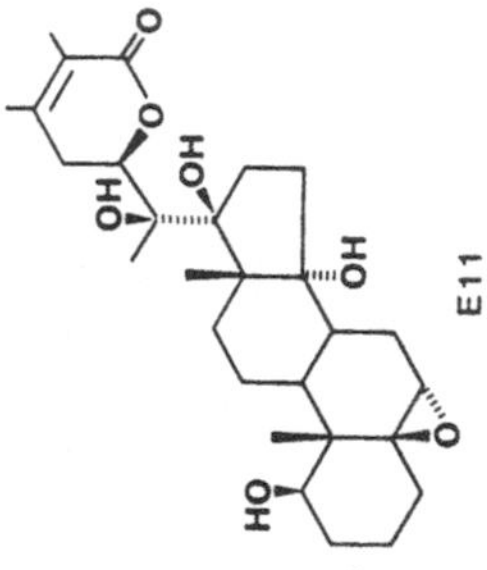

C59

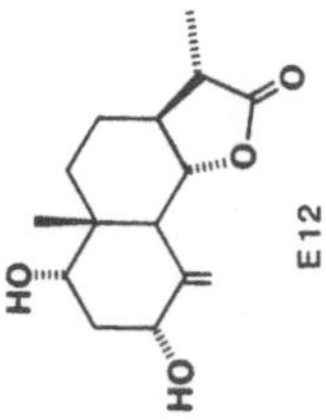

C60

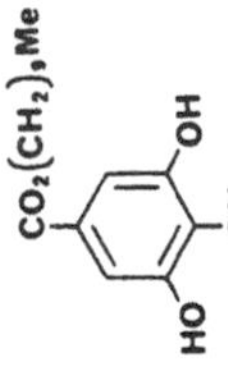

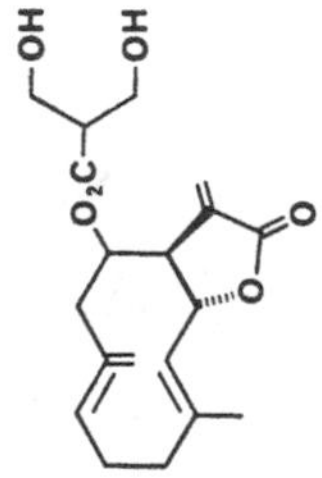

E9

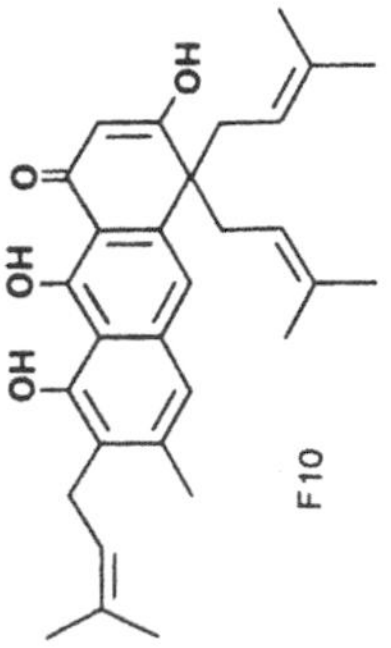

F10

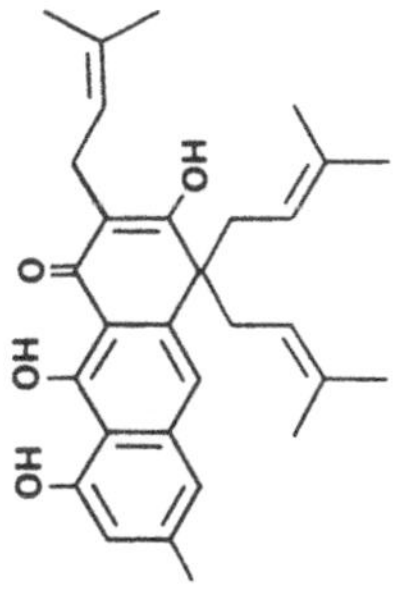

F11

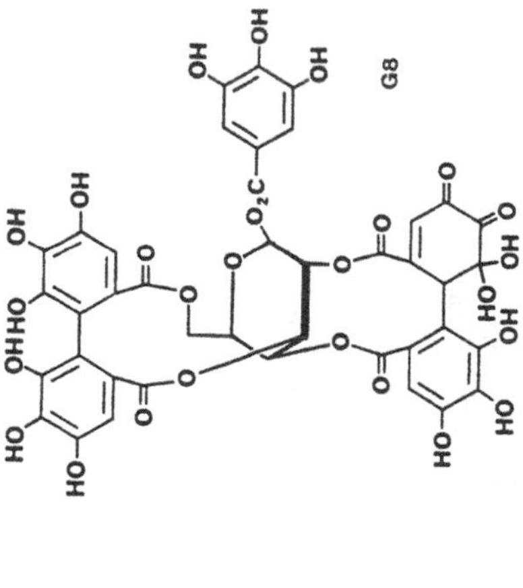

G8

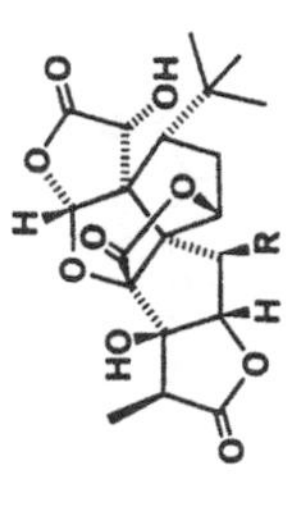

G9

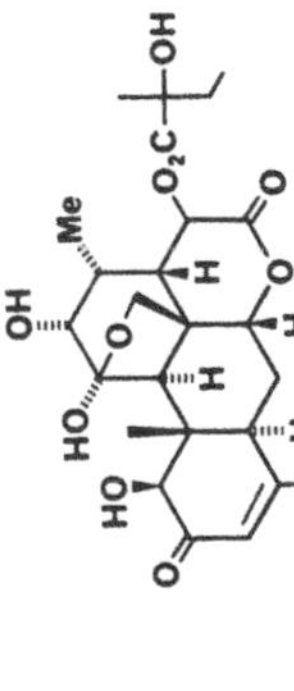

G10

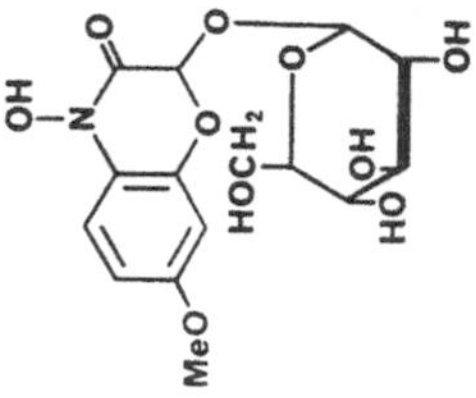

G11

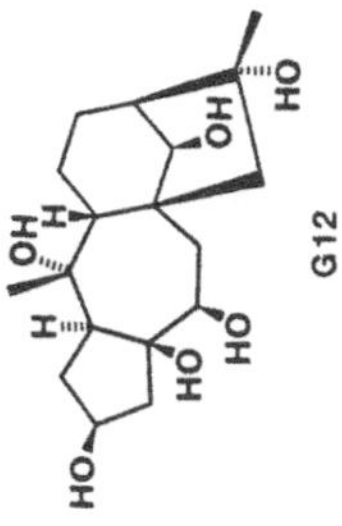

G12

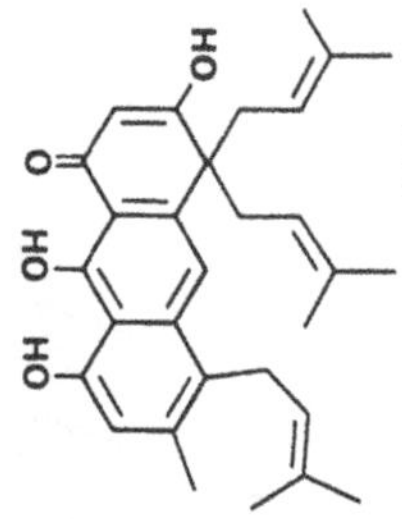

H11

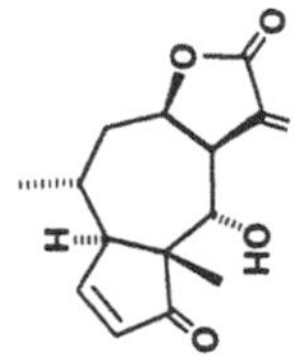

H12

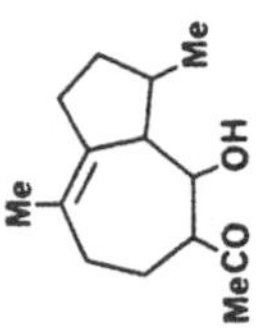

H13

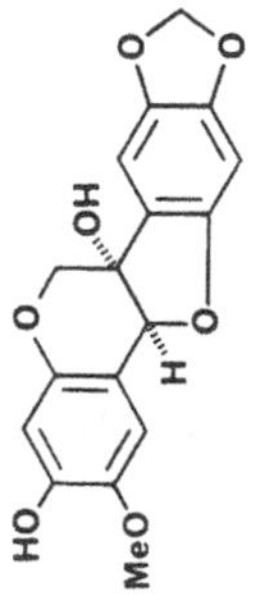

H14

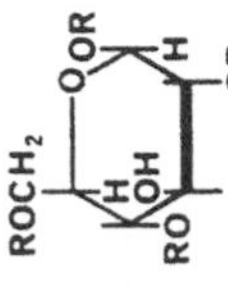

$R=NO_2CH_2CH_2CO-$

H15

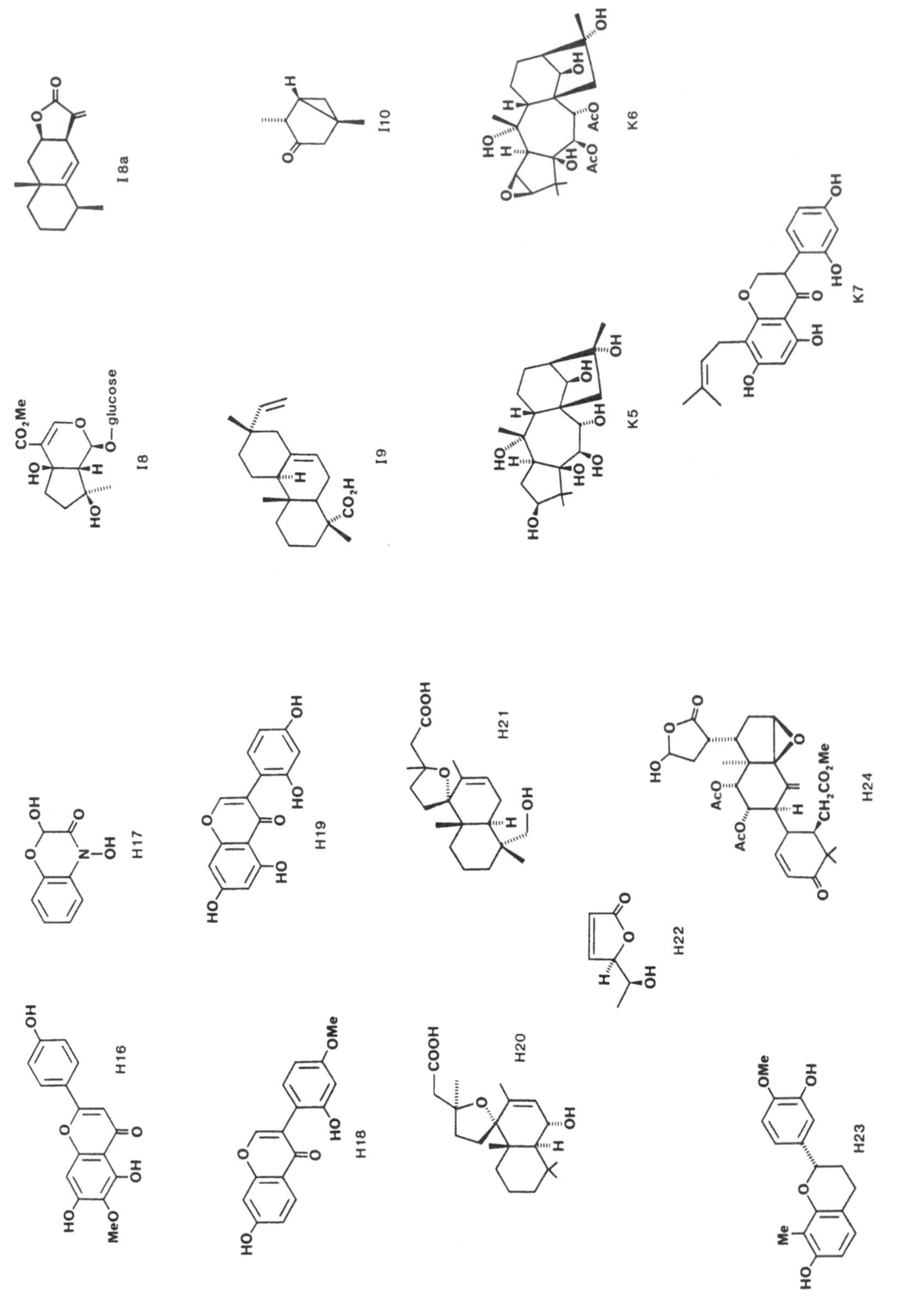

I8a

I10

K6

I8

I9

K5

K7

H17

H19

H21

H24

H22

H16

H18

H20

H23

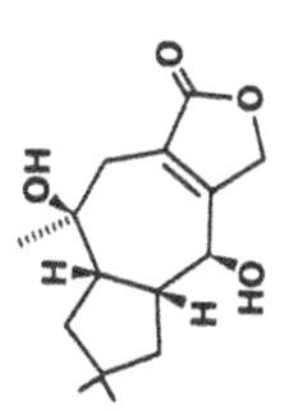

L18

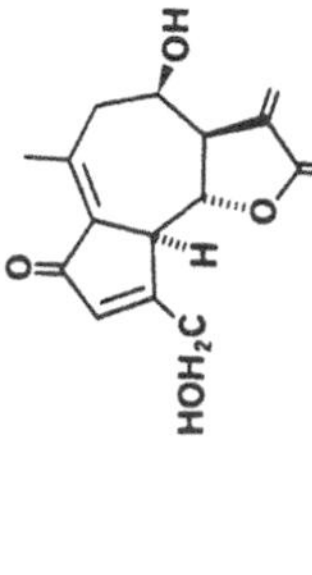

L20

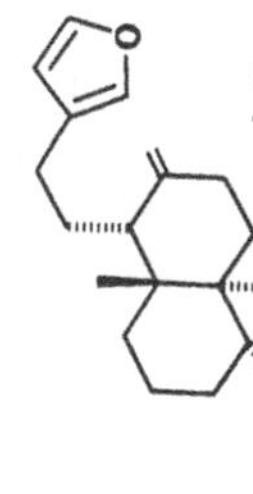

M20

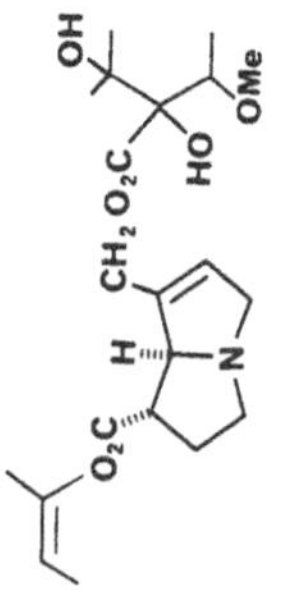

M22

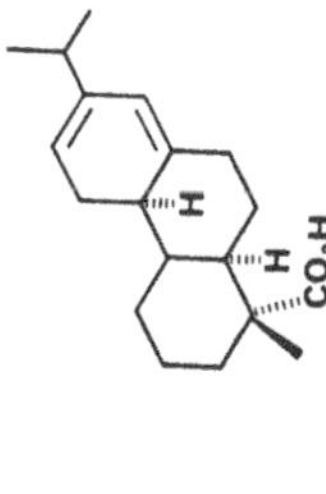

L17

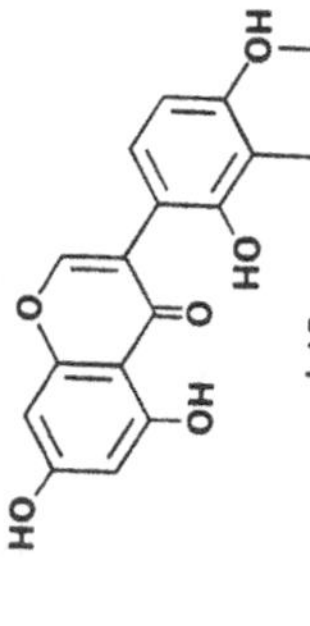

L19

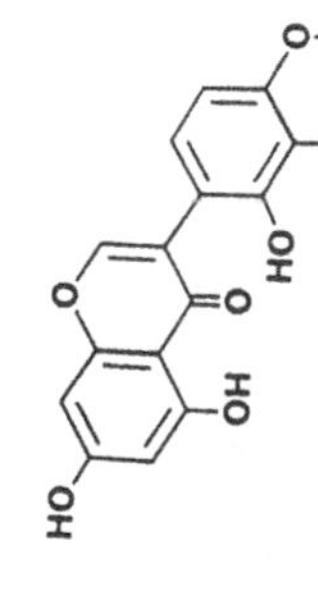

M19

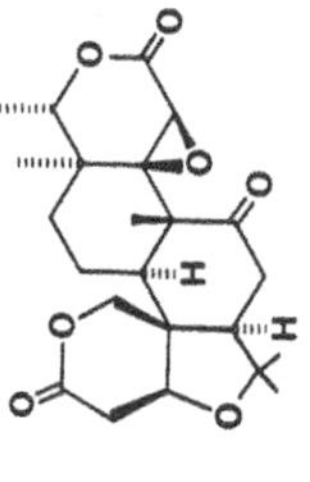

M21

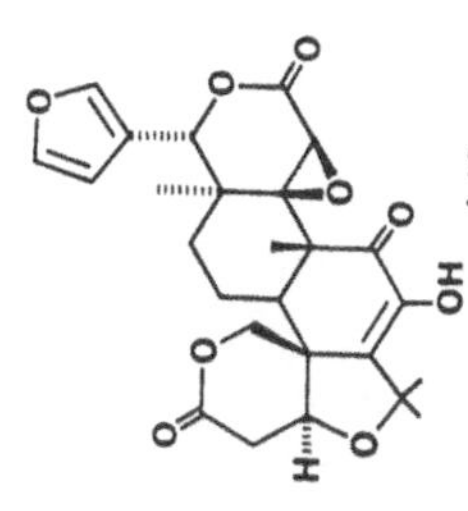

L12

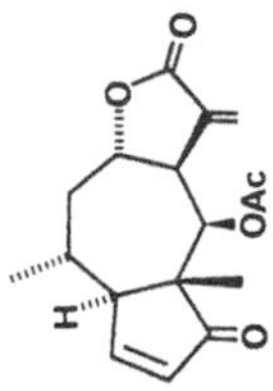

L14

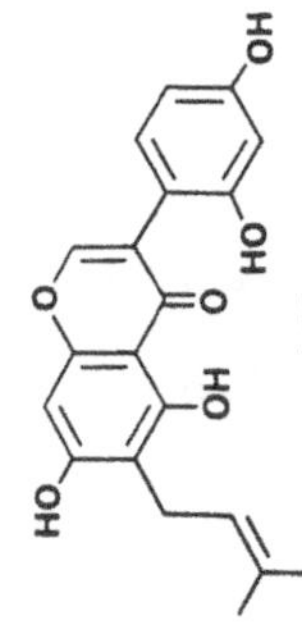

L16

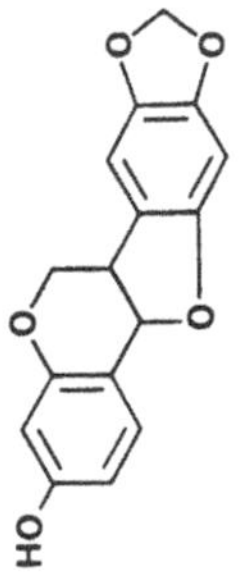

L10

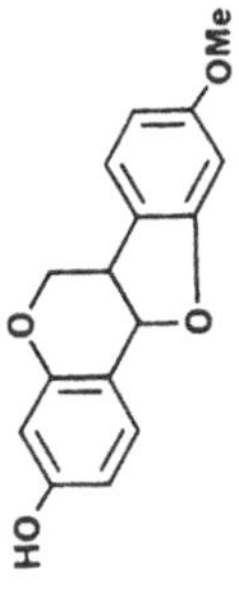

L13

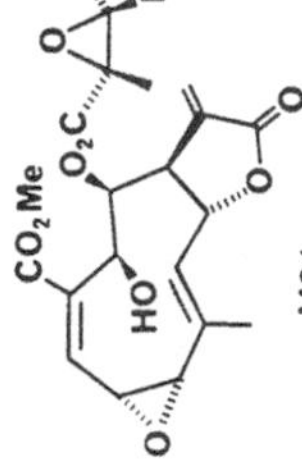

L11

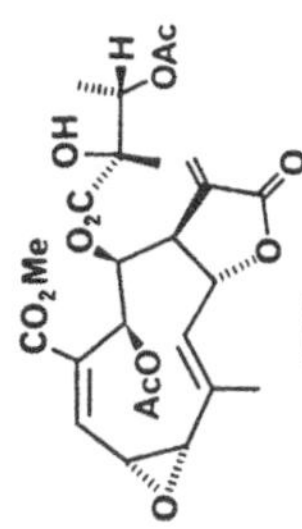

L15

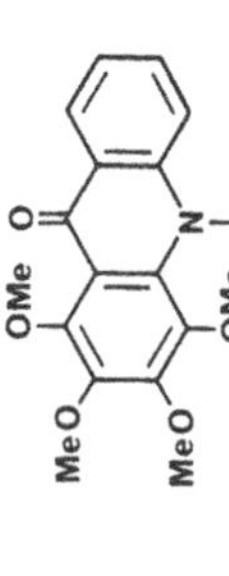
M23

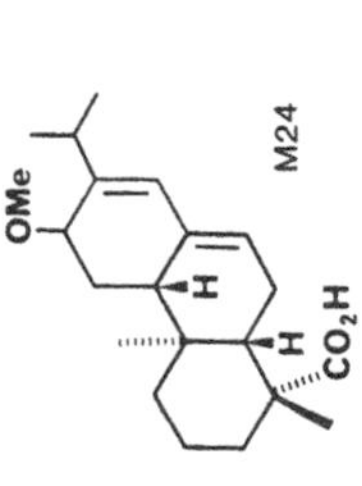
M24

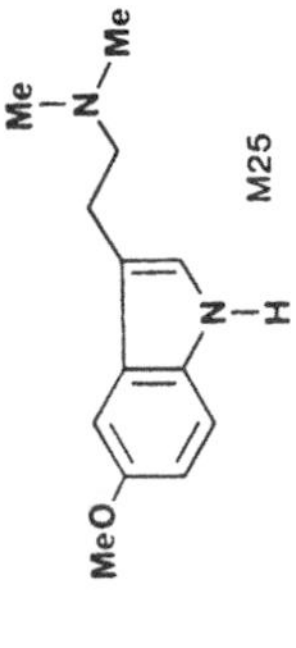
M25

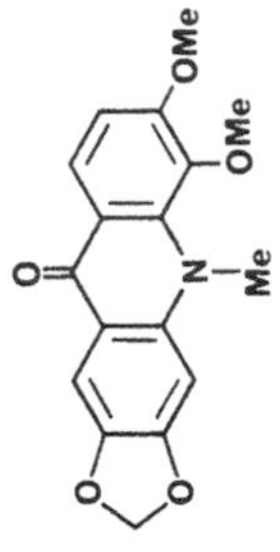
M26

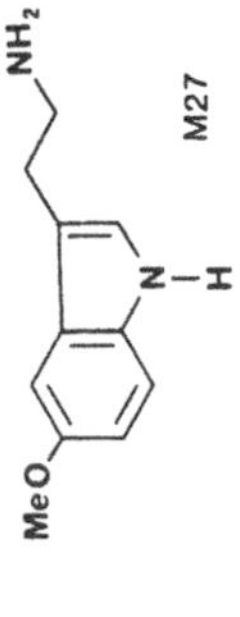
M27

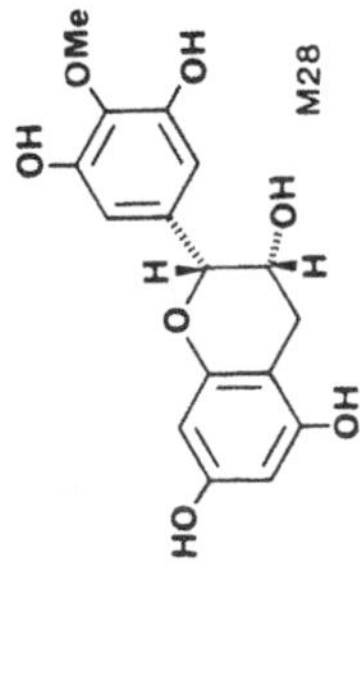
M28

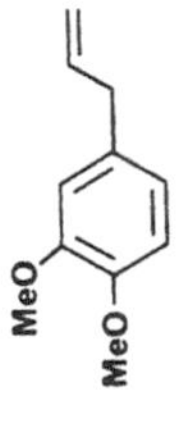
M29

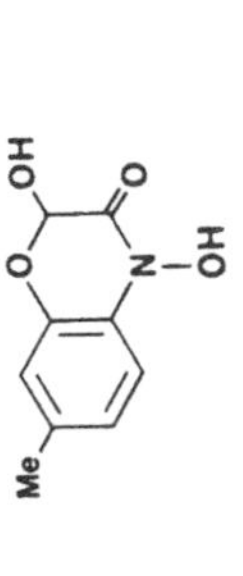
M30

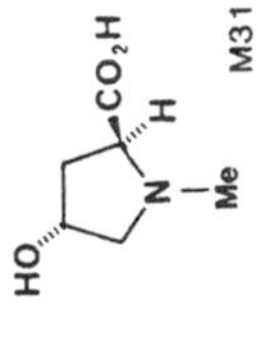
M31

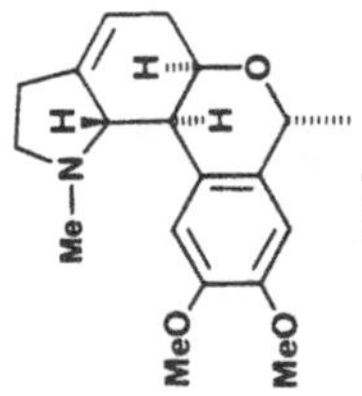
M32

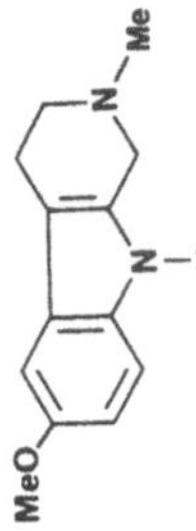
M33

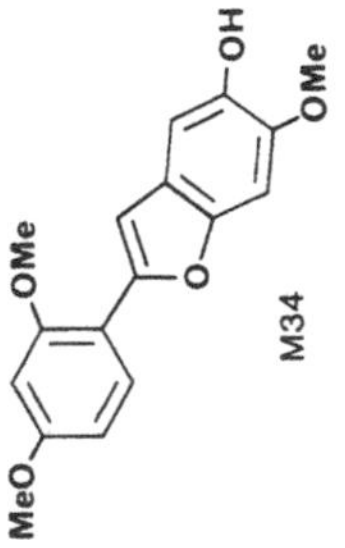
M34

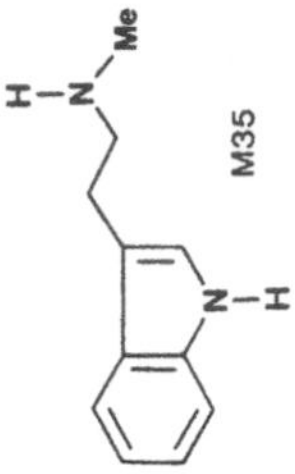
M35

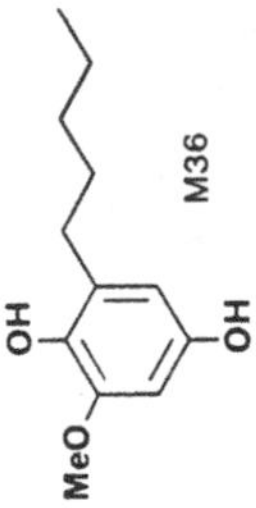
M36

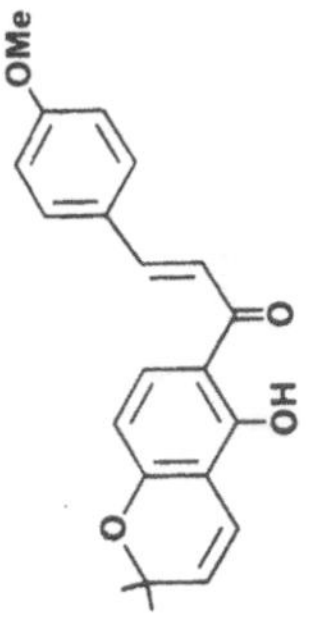
M37

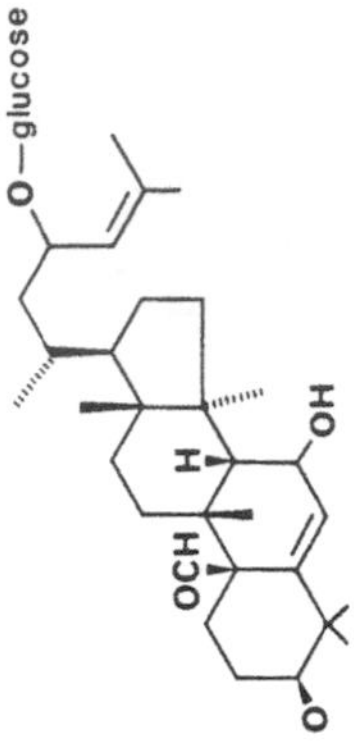
M38

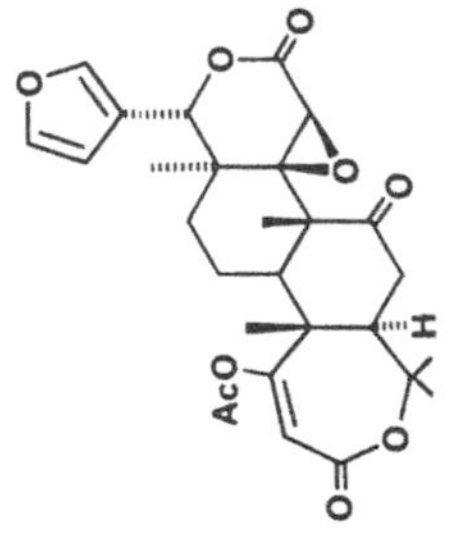

O3

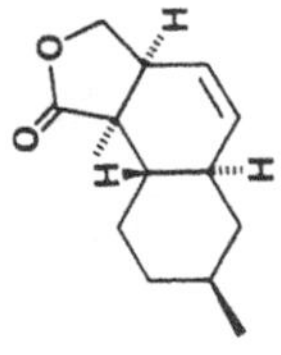

O4

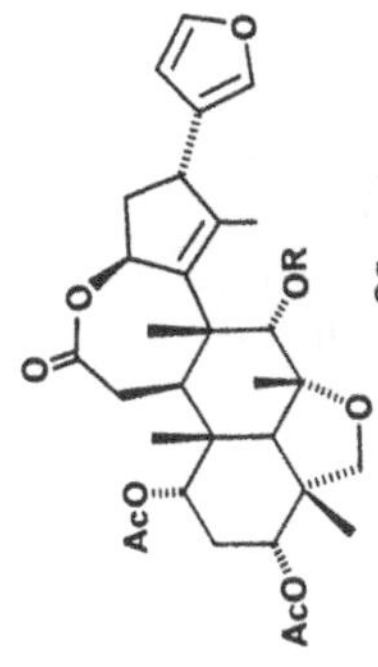

O5

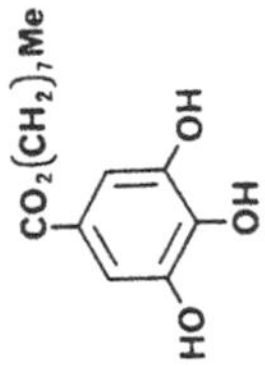

O6

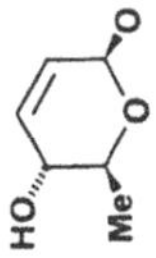

O7

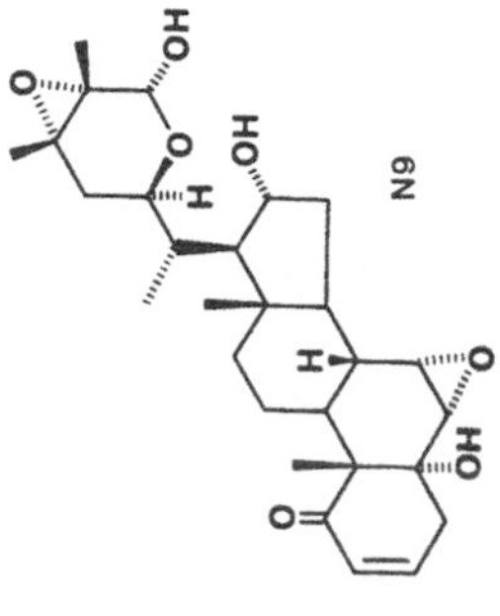

N9

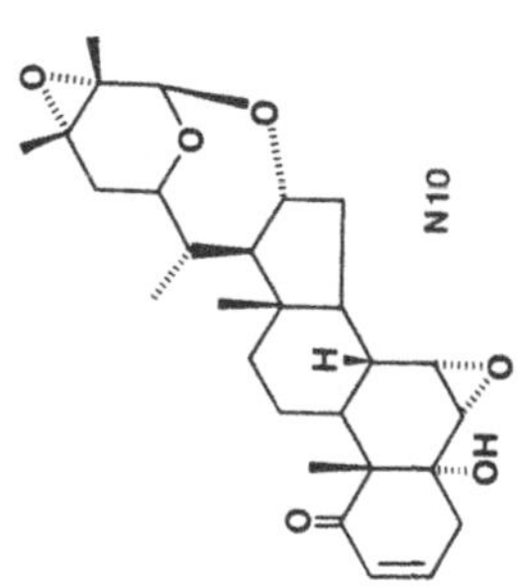

N10

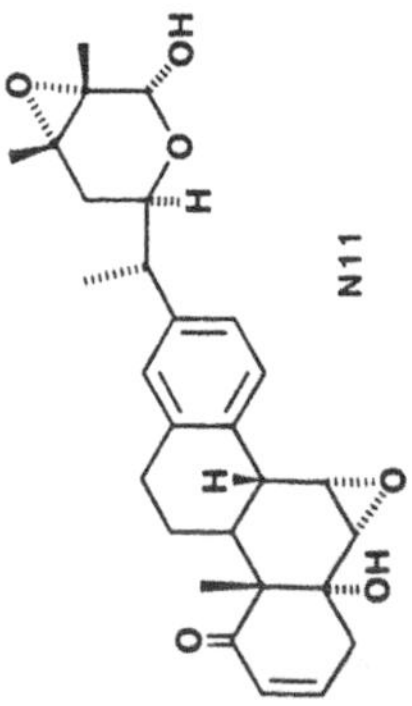

N11

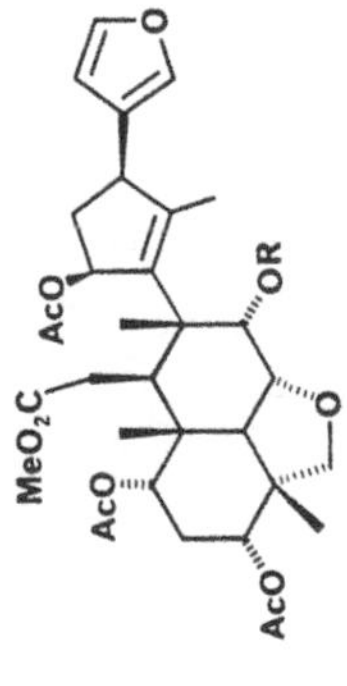

N12

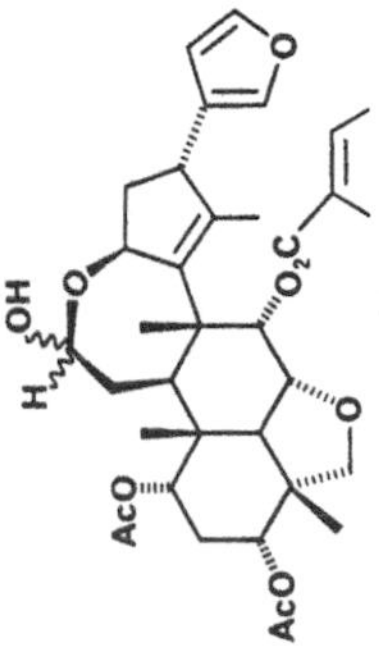

N13

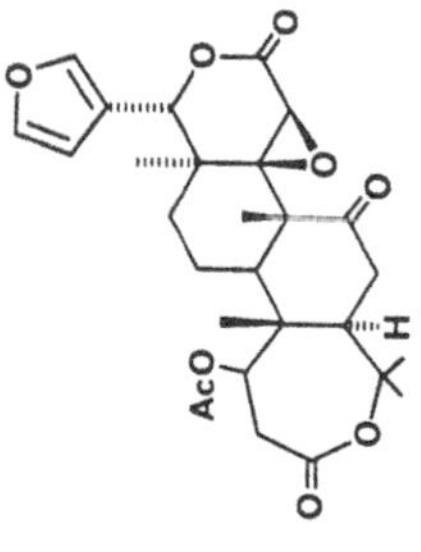

N14

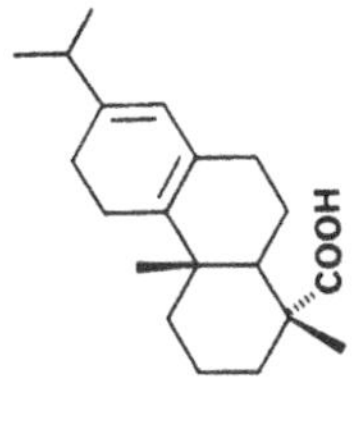

P35

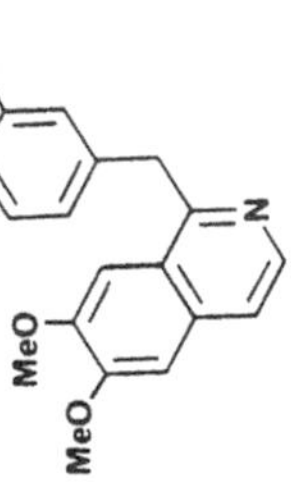

P37

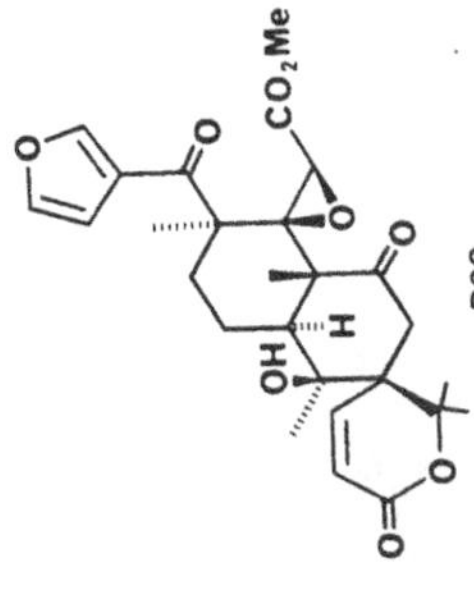

P39

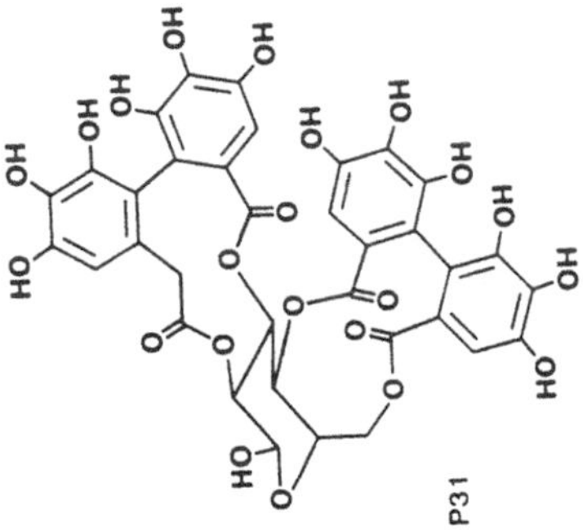

P34

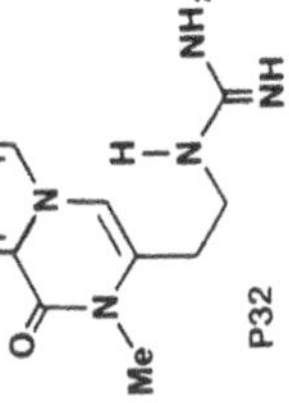

P36

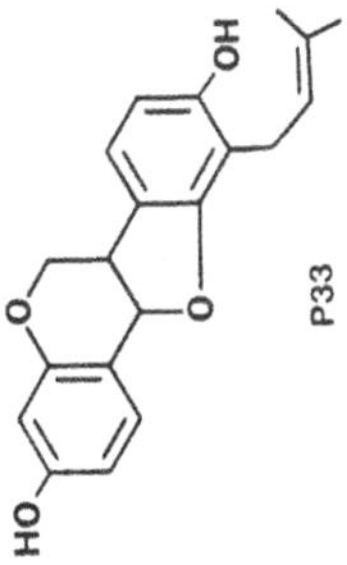

P36

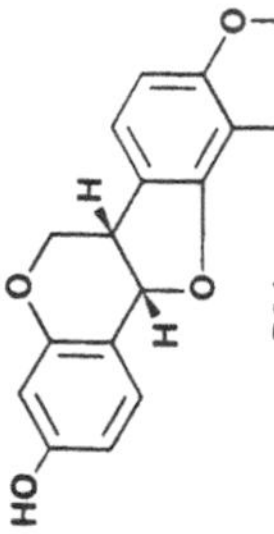

P29

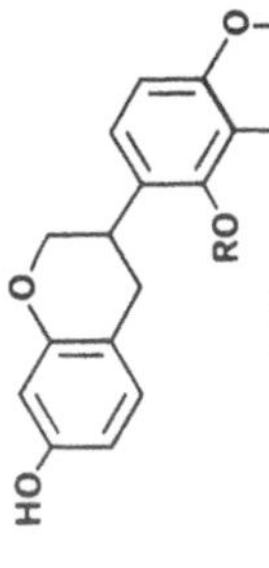

P31

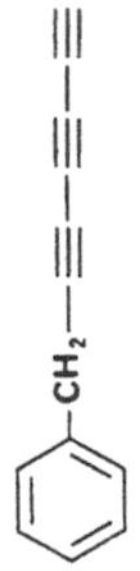

P33

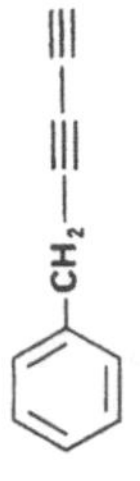

P28

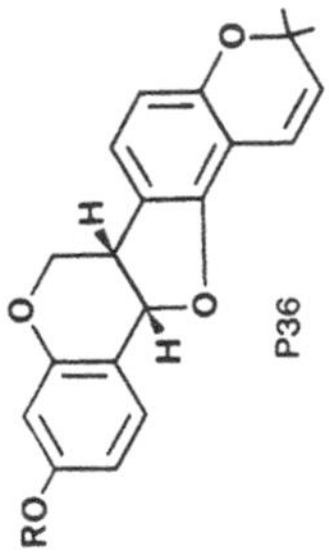

P30

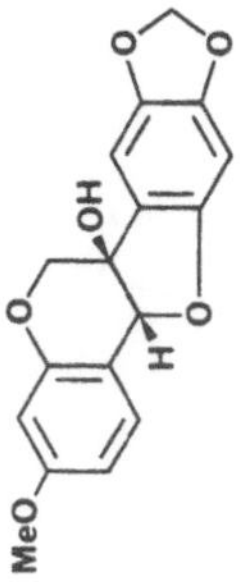

P32

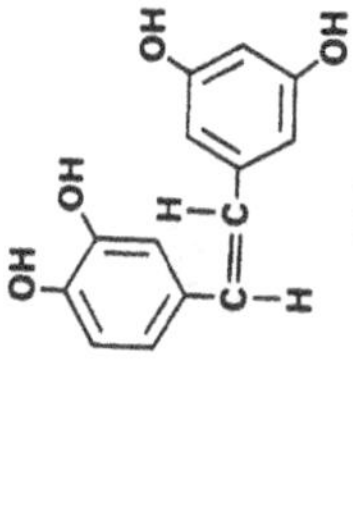

P38

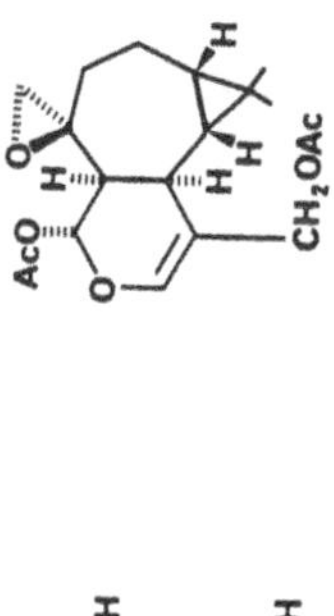

P40

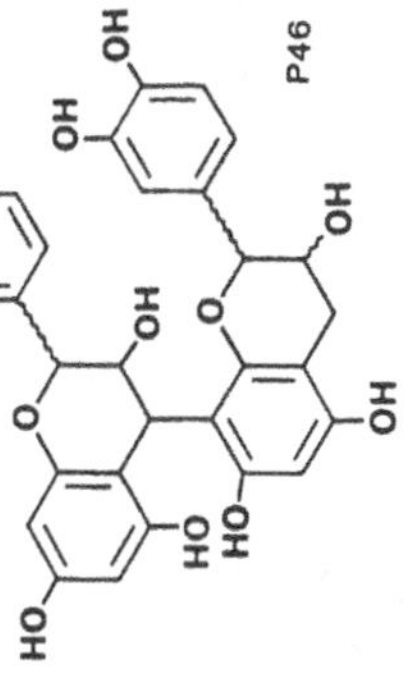

P46

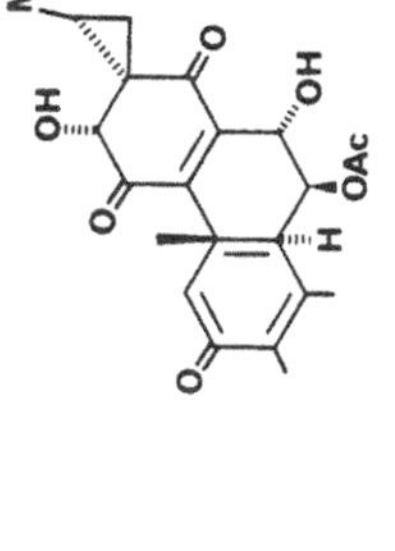

P41

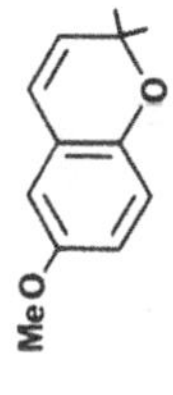

P42

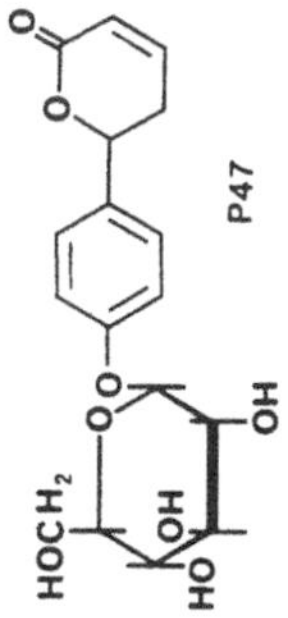

P47

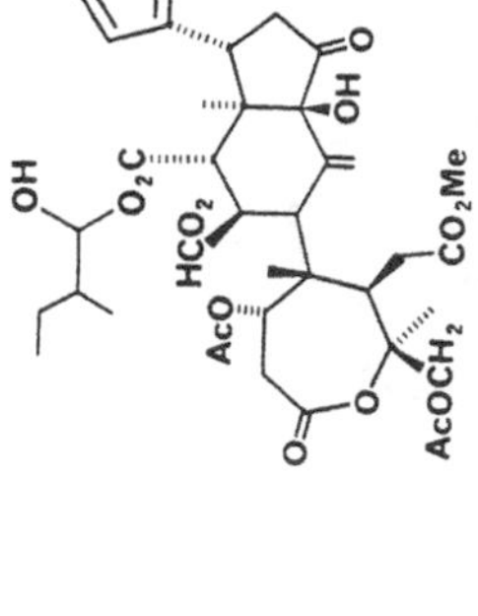

P44

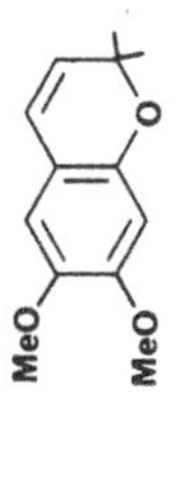

P43

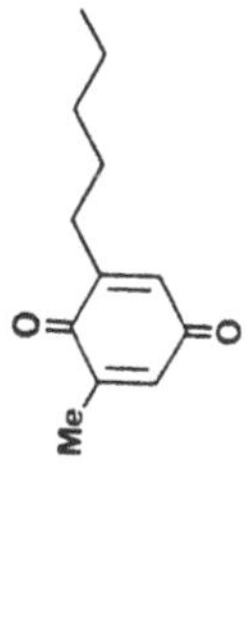

P45

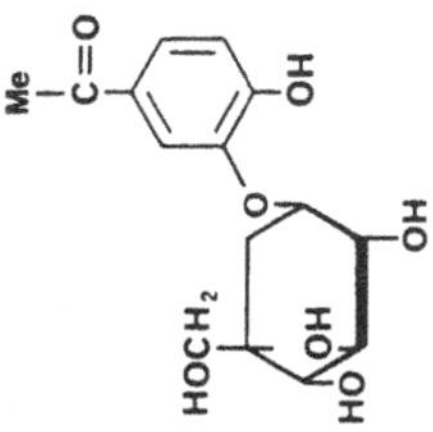

P48

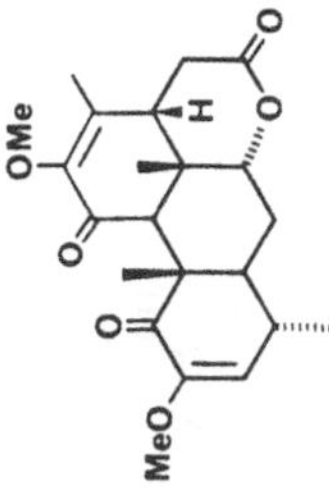

Q3

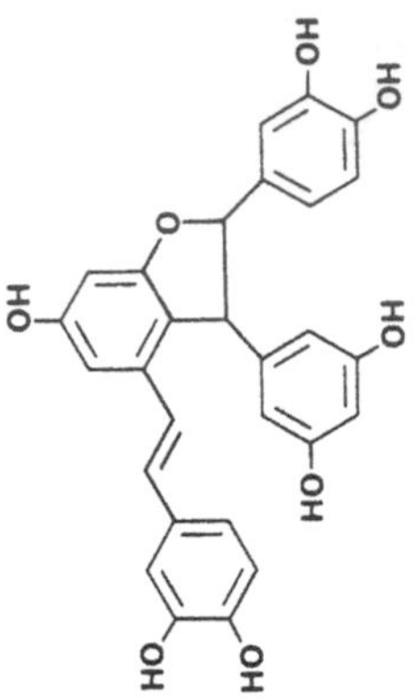

S27

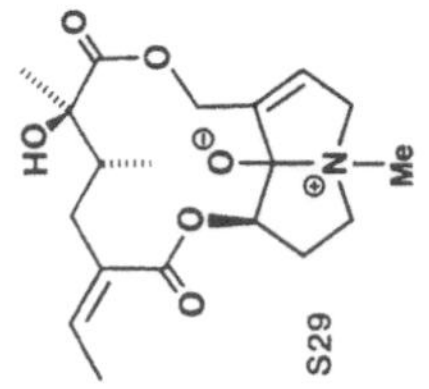

S29

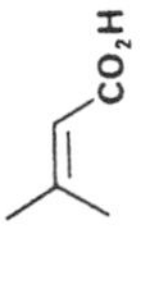

S28

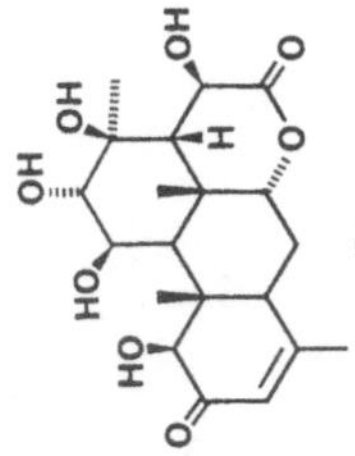

S31

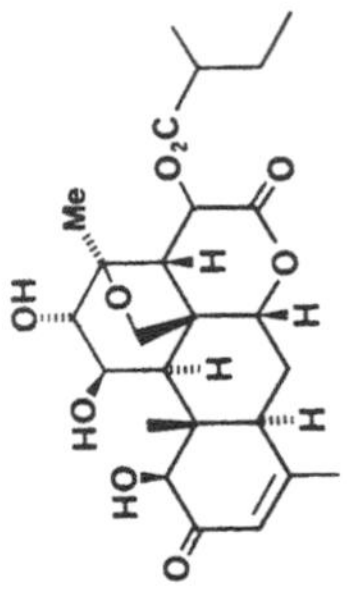

S30

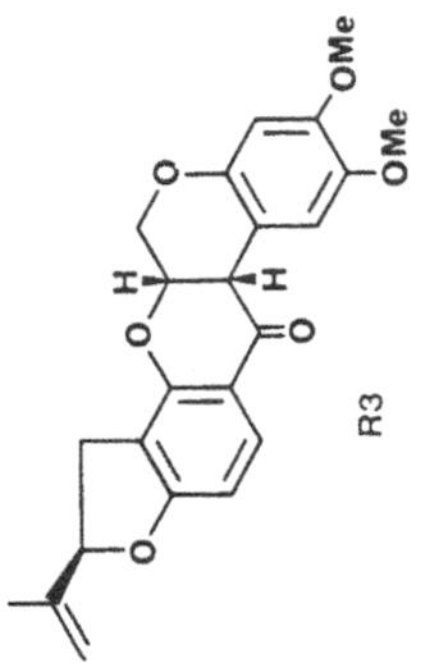

R3

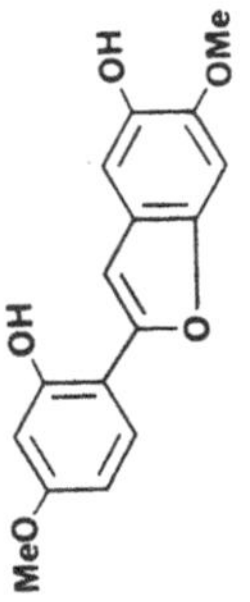

S22

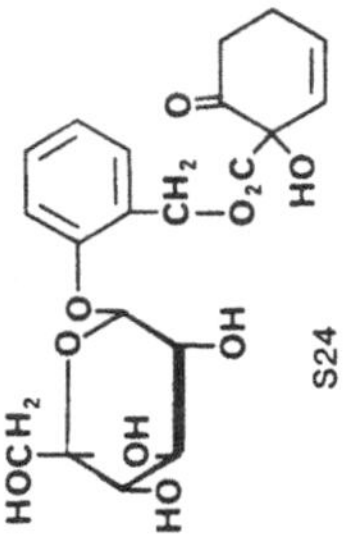

S24

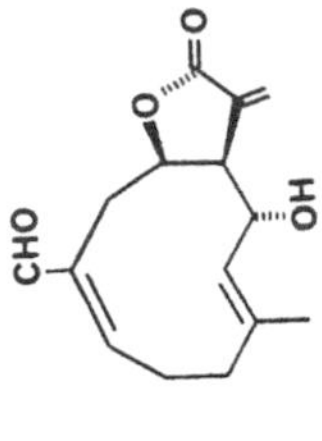

S26

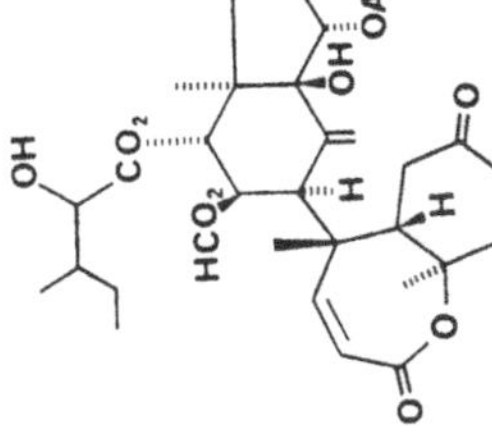

R2

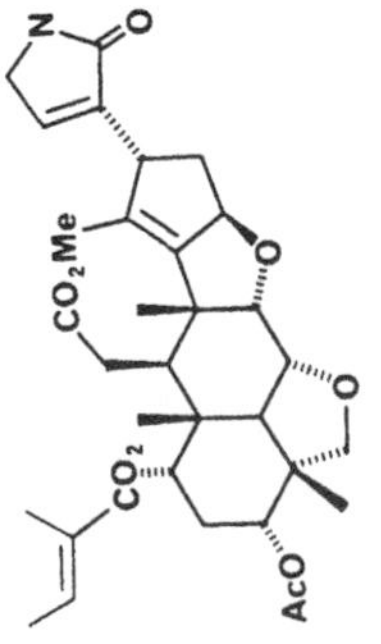

S23

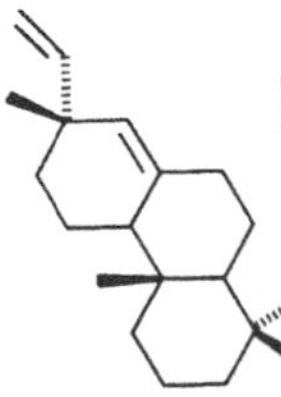

S25

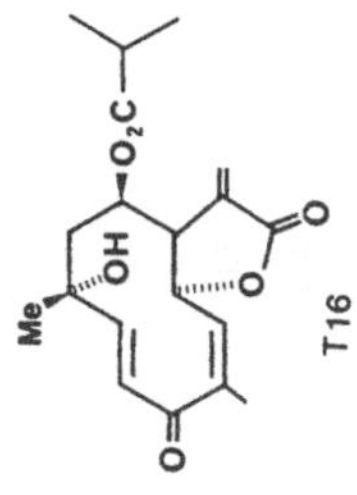

T16

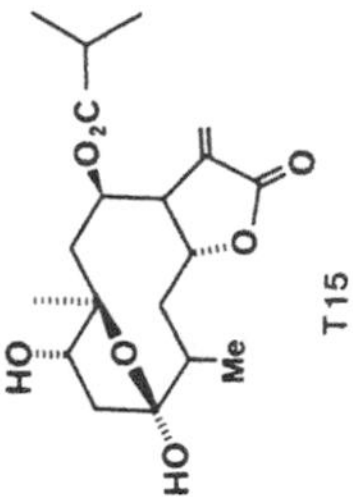

T15

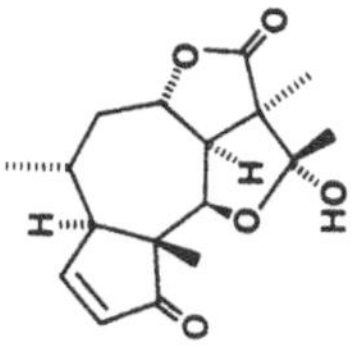

T18

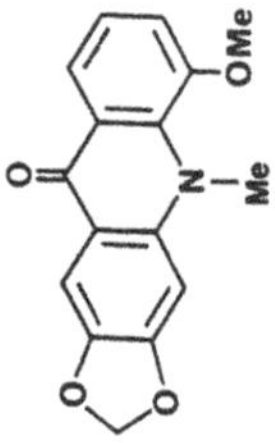

T17

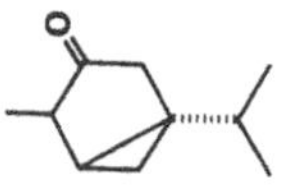

T20

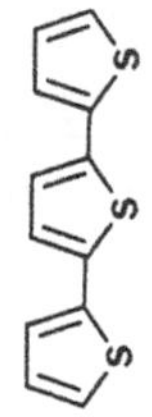

T19

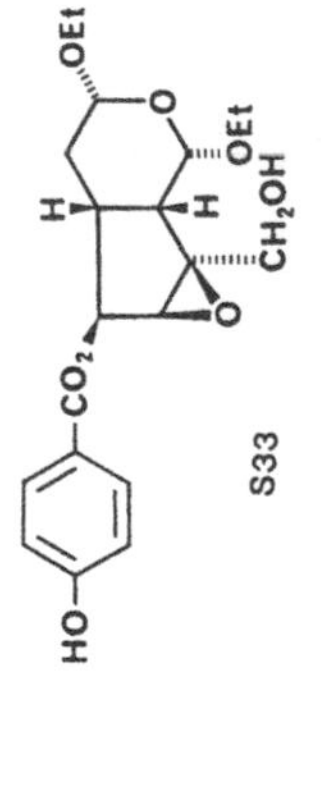

S33

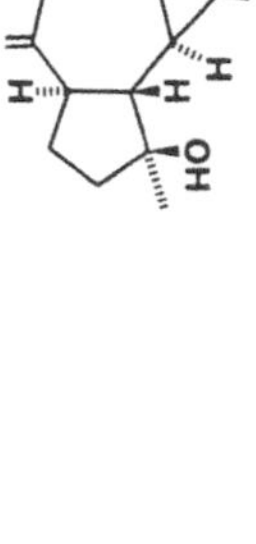

S32

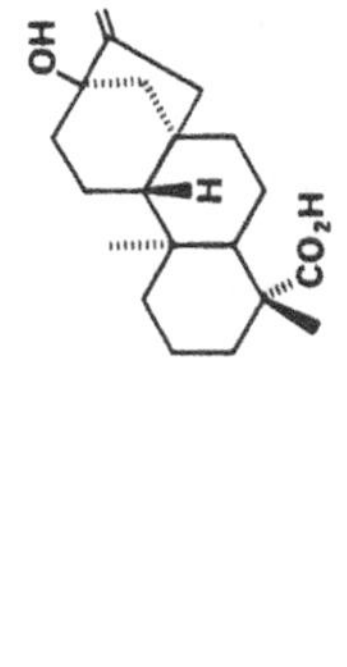

S34

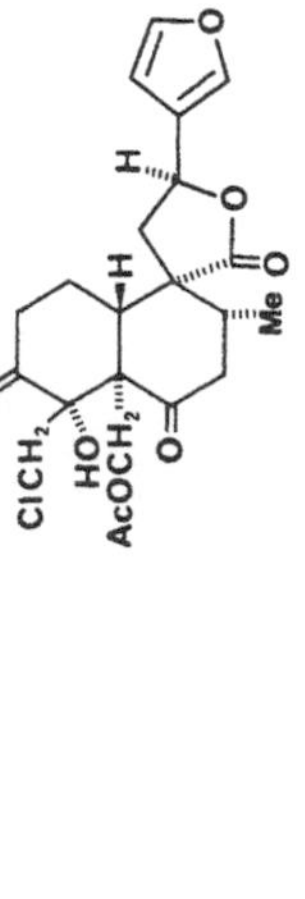

T13

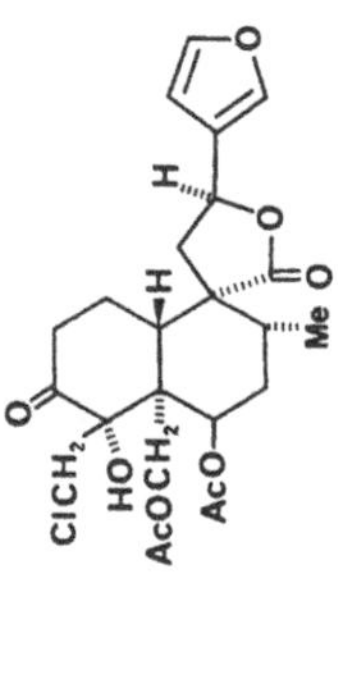

T14

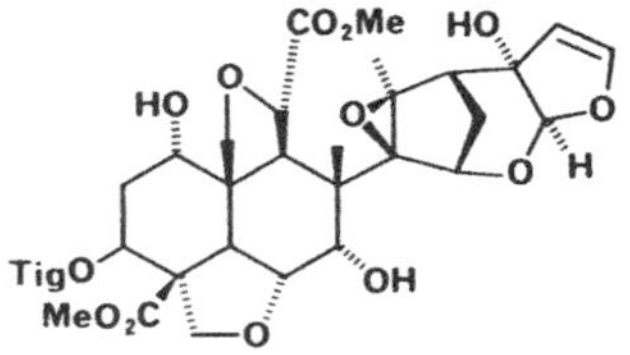

T29

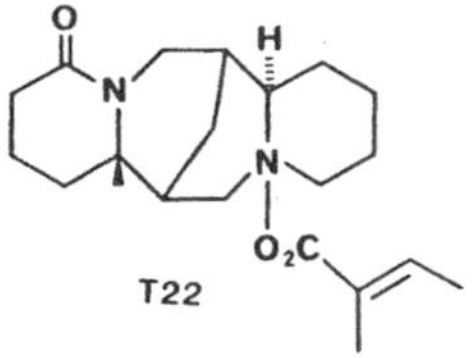

T31

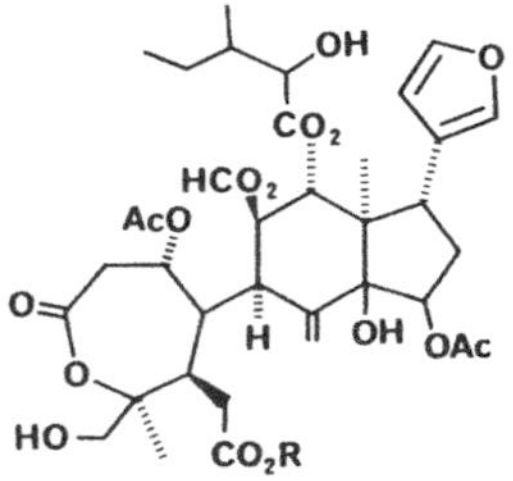

T28

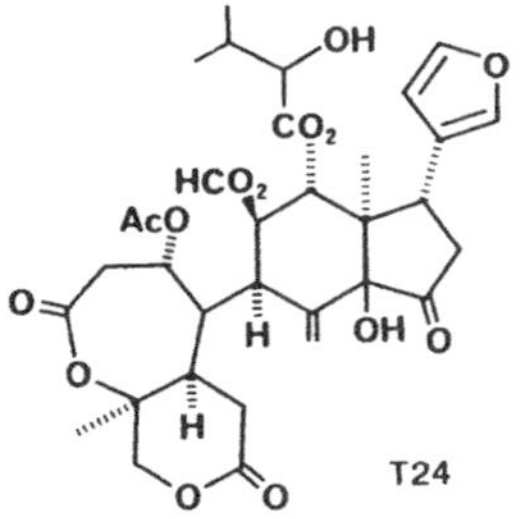

T30

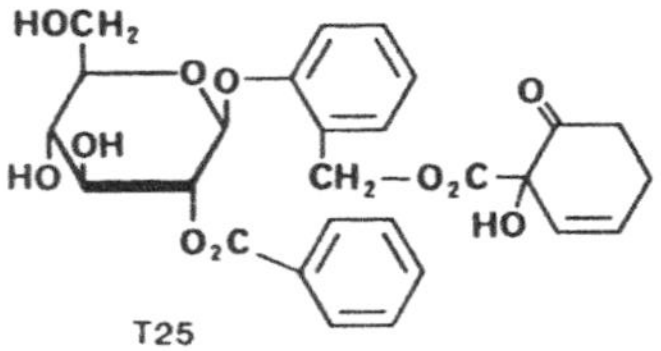

T27

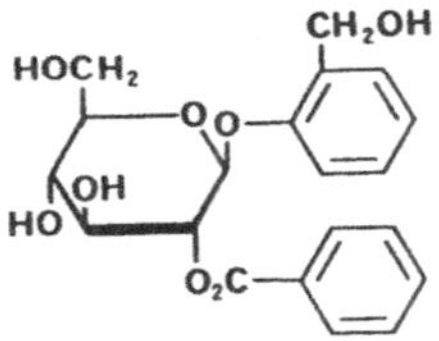

T22

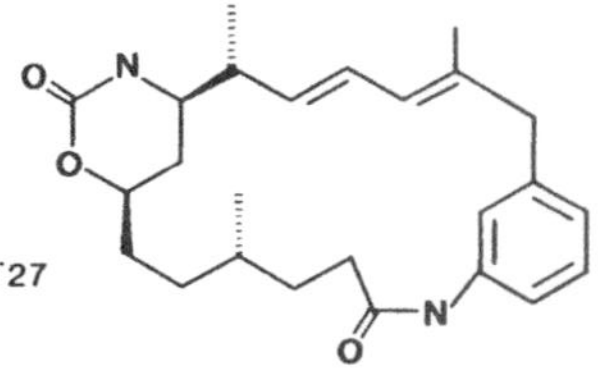

T24

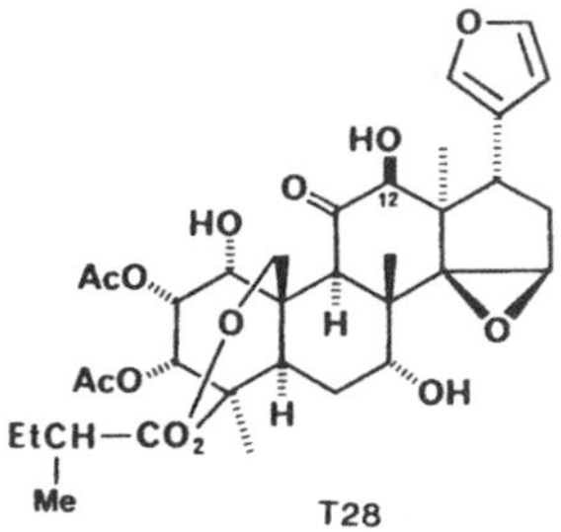

T26

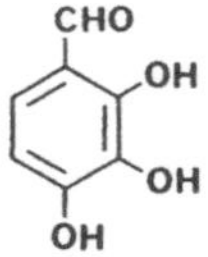

T21

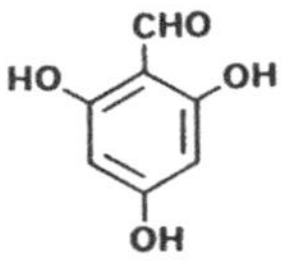

T23

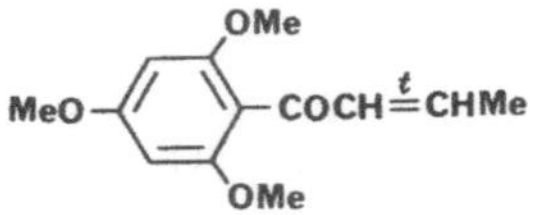

T25

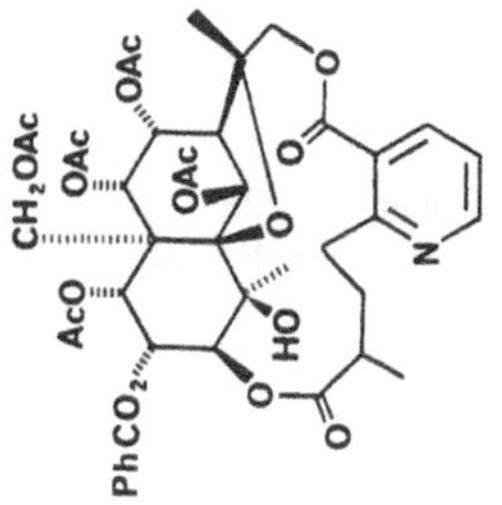

W4

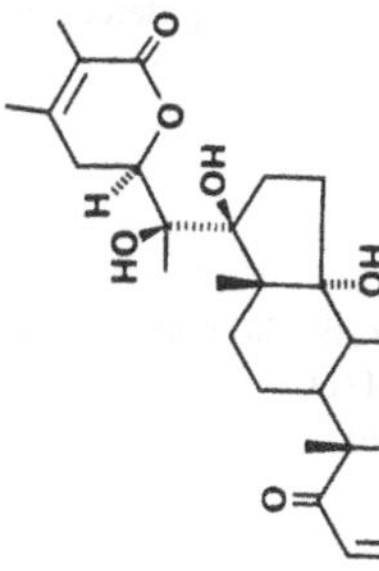

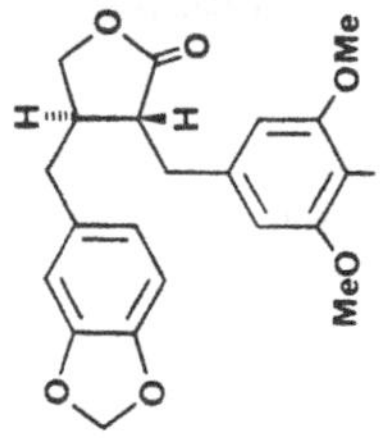

Y1

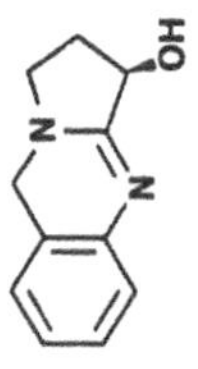

W3

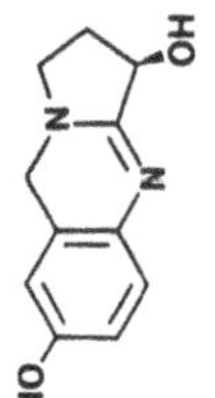

V5

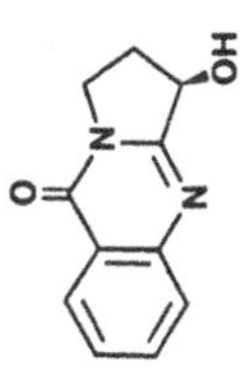

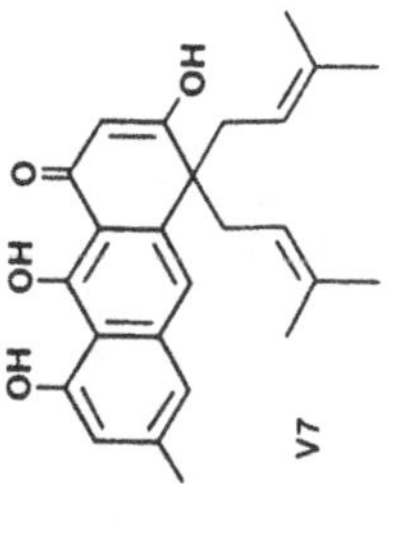

V7

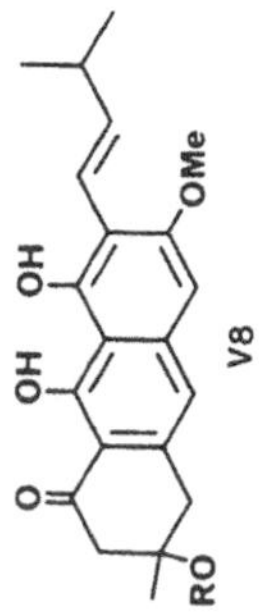

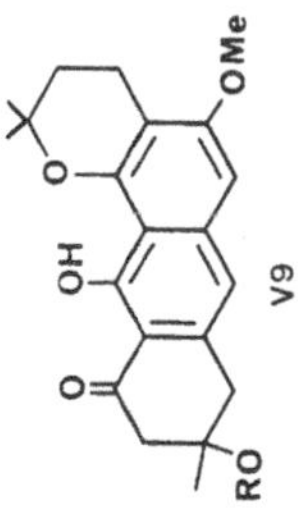

V9

V4

V6

V8

TABLE 5
Order, Scientific Name, and Feeding Deterrent Cross Reference for Tables 1 and 4

Insect		Feeding deterrent
Scientific name	Common name	

COLEOPTERA

Acalymma vittata (Fabricius)	Striped cucumber beetle	Asimicin A29
		Trewiasine T27
Achea janata (Linnaeus)	Croton caterpillar	6,7-Dimethoxyisochromanone D28
Altica ambiens Lecoute	Alder flea beetle	3-Isothujone I1O
		3-Thujone T20
Altica oleraceae	—	Chlorogenic acid C55
Agelasa coerulea	—	Chlorogenic acid C55
Authonomis grandis	Boll weevil	Anthranilic acid A27
		Camphor C45
		Cinnamaldehyde
		Trans-Cinnamic acid C26
		Gentisic acid G1
		5-Methoxy-N,N-dimethyltryptamine M25
		2-Methyl-6-methoxytetrahydro-β-carboline M33
		Senecioic acid S28
Altrachya menetriesi	—	Chlorogenic acid C55
Aulacophora foveicollis	Red pumpkin beetle	Momordicine II M38
		Vasicine V4
		Vasicinol V5
		Vasicinone V6
Costelytra zealandica	—	Claussequinone C57
		Hiptagin H15
		Hydroxyformononetin H18
		2′-Hydroxygenistein H19
		Kievitone K7
		Licoisoflavone A L15
		Licoisoflavone B L16
		Luteone L20
		Maackiain M19
		Medicarpin M20
		7-O-Methylphaseolin P34, R=Me
		7-O-Methylphaseolinisoflavan P35, R=Me
		Phaseolidin P33
		Phaseolin P34
		Phaseolinisoflavan P35
		Pisatin P39
		Precocene I P42
		Precocene II P43
		Rotenone R3
Epilachna varivestis Mulsant	Mexican bean beetle	3-Desacetylsalannin D1
		1,3-Diacetylvilasinin D4
		2,3-Dihydrowithanolide E
		21-Hydroxytoonacilide H24
		23-Hydroxytoonacilide
		Nicalbin A N9
		Nicalbin B N10
		Nicandrenone N11
		Nimbolidin A N12 R=benzyl
		Nimbolidin B N12 R=tigloyl
		Nimbolinin B N13
		Ochinolide A O5 R=benzoyl
		Ochinolide B 05 R=tigloyl
		Prieurianin P44

TABLE 5 (continued)
Order, Scientific Name, and Feeding Deterrent Cross Reference for Tables 1 and 4

Insect		Feeding deterrent
Scientific name	**Common name**	
		Quassin Q3
		Salannol S1 S23
		Salannolactam-(21) S23
		Salannolactam-(23)
		Withanolide E W4
Epilachna vijintioctopunctata	—	Vasicine V4
		Vasicinol V5
		Vasicinone V6
Fleutiauxia armata	—	Chlorogenic acid C55
Galelucella vittaticollis	—	Chlorogenic acid C55
Gallerucida bifasciata	—	Chlorogenic acid C55
Gastrolina depressa	—	Chlorogenic acid C55
Gastrophysa atrocyanea	—	Chlorogenic acid C55
Heteronychus arator	—	Licoisoflavone B 46
		Phaseolin P34
Leptinotarsa decemlineata (Say)	Colorado potato beetle	Bakkenolide A B12
		Bisabolangelone B15
		Helenalin H12
		Limonin L17
Lochmaeae capreae cribrata	—	Chlorogenic acid C55
Oulema oryzae	—	Chlorogenic acid C55
Pissodes strobi (Peck)	Engelmann spruce weevil	3-Isothujone I10
		3-Thujone T20
Pyrrhalta humeralis	—	Chlorogenic acid C55
Scolytus scolytus	European bark beetle	Oblongolide O4
Sitophilus granarius (Linnaeus)	Granary weevil	Bakkenolide A B12
		Bisabolangelone B15
		Carvone C7
		Desacetylarctolide D22
		Erivanin E12
		Eupatoriopicrin E13
		Helenalin H12
		Isoalantolactone I8a
		ent-Isoalantolactone
		Linifoline A L19
		Schkuhriolide S26
		Veratrine
		Yatein Y1
Tribolium castenaeium (Herbst)	Red flour beetle	Nicalbin A N9
		Nicalbin B N10
Tribolium confusum Jacquelin du Val	Confused flour beetle	Arctolide A28
		Bakkenolide A B12
		Bisabolangelone B15
		Desacetylarctolide D22
		Eupatoriopicrin E13
		Helenalin H12
		Isoalantolactone I8a
		ent-Isoalantolactone
		Lactarorufin A L10
		Linifoline A L19
		Veratrine
		Yatein Y1
Trogoderma granarium Everts	Khapra beetle	Arctolide A28
		Bakkenolide A B12

TABLE 5 (continued)
Order, Scientific Name, and Feeding Deterrent Cross Reference for Tables 1 and 4

Insect		
Scientific name	**Common name**	**Feeding deterrent**
		Bisabolangelone B15
		Eupatoriopicrin E13
		Helanalin H2
		Isoalantolactone
		ent-Isoalantolactone
		Lactarorufin A, L10
		Linifoline A, L19
		Veratrine
		Yatein Y1

DIPTERA

Hylemya antiqua (Meigen)	Onion maggot	3-Isothujone I10
		3-Thujone T20
Phormia regina (Meigen)	Black blowfly	Acrecoline A25
		Atropine A16
		Papaverine P29
		Quinine Q2
		Sparteine S18
		Strychnine S20
		Tomatine T7

HEMIPTERA

Oncopeltus fasciatus	Large milkweed bug	Encecalin E10
Rhodnius prolixus	—	Precocene II P43

HOMOPTERA

Acyrthosiphon pisum (Harris)	Pea aphid	Castanospermine C49
Listronotus bonariensis	—	Peramine P32
Myzus persicae (Sulzer)	Green peach aphid	Geraniin G8
		n-Octyl gallate O6
		Pedunculagin P31
Pristiphora erichsonii (Hartig)	Larch sawfly	Abietic acid A19
		Dehydro abietic acid D19
		Isopimaric acid I9
		12-Methoxyabietic acid M24
		Sandaracopimaric acid S25
Schizaphis graminum (Rondani)	Green bug	n-Decyl gallate D20
		Eperu-8(17)-en-15,19-dioic acid butenolide
		Ellagic acid E9
		Gramine G7
		Geraniin G8
		2-β-D-Glucosyl-4-hydroxy-7-methoxy-1,4-benzoxazin-3-one G11
		Hydroxamic acid H17
		6α-Hydroxygrindelic acid H20
		18-Hydroxygrindelic acid H21
		Lambertianic acid L12
		Levopimaric acid L14
		5-Methoxy-N,N-dimethyltryptamine M25
		5-Methoxytryptamine M27
		7-Methylhydroxamic acid M30
		N-Methyltryptamine M35

TABLE 5 (continued)
Order, Scientific Name, and Feeding Deterrent Cross Reference for Tables 1 and 4

Insect		
Scientific name	**Common name**	**Feeding deterrent**
		Pedunculagin P31
		Plectrin P41
		Phloretin 4′-O-β-D-glucopyranoside (see P12)
		Sandaracopimaric acid S25
		Steviol S34
		18-Succinoyloxygrindelic acid (see H21)
Schizaphis pisum	—	n-Octyl gallate O6

HYMENOPTERA

Atta cephalotes	Leafcutter ant	Caryophyllene oxide C48
		Spathulenol S32
Neodiprion dubiosus (Schell)	Brownheaded jackpine sawfly	Palustric acid P28
Neodiprion leconti (Fitch)	Redheaded pine sawfly	Palustric acid P28
Neodiprion rugifrons (Middleton)	Redheaded jackpine sawfly	Palustric acid P28

LEPIDOPTERA

Agrotis sejetum	—	Azedarachol A31
		Tr-A, Tr-B, Tr-C T23, T24
Choristoneura fumiferana (Clemens)	Spruce budworm	α-Chaconine C52
		13-*trans*-Cinnamoyllupanine C56
		Lasiocarpine L13
		Pungenin P48
		Senecionine S10
		Senkirkine S29
		Solanidine S16
		Solanine S17
		Specionin S33
		13-Tigloyloxylupanine T22
		Tomatidine T6
		Tomatine T7
Choristoneura rosaceana (Harris)	Obliquebanded leafroller	Amygdalin A26
Cydia pomonella	Codling moth	Trewiasine T27
Eurema hecabe mandarina	Yellow butterfly	5-Hydroxy-2-hexen-4-olide H22
		3′Hydroxy-4′-methoxy-7-hydroxy-8-methylflavan H23
		O-Methyllycorenine M32
		Osmundalactone O7
		1-(2,4,6-Trimethoxyphenyl)-but-2-en-1-one T31
Euxoa messoria (Harris)	Darksided cutworm	1-Phenylheptatriyne P36
		α-Terthienyl T19
Heliothis armigera	Cotton boll worm	Miconidin M36
		Primin P45
Heliothis virescens (Fabricius)	Tobacco budworm	6-O-Acetylnimbandiol A22
		Anthranilic acid A27
		Camphor C45
		Chaparrinone C35
		Ferruginin A F10
		Harunganin H11
		Isobrucein A (see B16)
		Prieurianin P44

TABLE 5 (continued)
Order, Scientific Name, and Feeding Deterrent Cross Reference for Tables 1 and 4

Scientific name	Common name	Feeding deterrent
		Rohituka-7 R2
		6-Senecioyloxychapparinone (see C53)
		3-Tigloylazadirachtol T21
		Vismin V7
		Vismione B V9
Heliothis zea (Boddie)	Corn earworm	Abyssinin A20
		Encecalin E10
		Obacunone O3
Lymantria dispar	Gypsy moth	Grayanotoxin-I G12
		Kalmitoxin-I K5
		Kalmitoxin-IV K6
Manduca sexta	Tobacco hornworm	Trichilin A T28
Maruca testulalis (Geyer)	Bean pod borer	Hildecarpin H14
Mythimna unipuncta	—	Bisabolangelone B15
Ostrinia nubilalis (Hübner)	European corn borer	Psilotin P47
		Tenulin T18
Papilio glaucus glaucus	Eastern tiger swallowtail	Salicin S2
Pectinophora gossypiella (Saunders)	Pink bollworm	6-O-Acetylnimbandiol A22
		3-Tigloylazadirachtol T21
Peridroma saucia	Variegated cutworm	Encecalin E10
		Tenulin T18
Philosamia ricini	—	Hispidulin H16
		Milletia chalcone M37
		Tagitinin A T15
		Tagitinin C T16
Pieris brassicae	Cabbage moth	Ajugarin I A6
		Capillarin C46
		ar-Curcumene C59
		Methyleugenol M29
		Miconidin M36
		Primin P45
Pieris rapae	Imported cabbage worm	Wilforine W3
Pieris rapae crucivora	—	Bilobalide B14
		Capillen C47
		Ginkgolide A G9
		Ginkgolide B (see G9)
		1-Phenyl-2,4-pentadiyne P37
Poanes hobomok	—	Catalposide C50
Plusia gamma	—	Encecalin E10
Prodenia litura	—	Ajugarin I A6
		2,5-Dihydroxymethyl-3,4-dihydroxypyrrolidine D24
Spodoptera spp.	—	Clerodane diterpenes C58 (60 active compounds)
Spodoptera eridania (Cramer)	Southern armyworm	8-Acetyl-3α-hydroperoxy-2α-isohex-3-enyl-6-isopent-2-enyl-5-methoxy-2β-methylchroman A22
		Quassin Q3
		Trichilin A T28
		Trichilin B
		Trichilin C
		Trichilin D
Spodoptera exempta (Walker)	African armyworm Nutgrass armyworm	12β-Acetoxyharrisonin (see H2)
		8-Acetyl-3α-hydroxyperoxy-2α-isohex-3-enyl-6-isopent-2-enyl-5-methoxy-2β-methylchroman A22
		11,13-Dihydrovernodalin D23

TABLE 5 (continued)
Order, Scientific Name, and Feeding Deterrent Cross Reference for Tables 1 and 4

Insect		Feeding deterrent
Scientific name	**Common name**	
		2,5-Dihydroxymethyl-3,4-dihydroxypyrrolidine D24
		Ferruginin A F10
		Harunganin H11
		6-Methoxytecleanthine M26
		Melicopicine M23
		Miconidin M36
		Plagiochilin A P40
		Primin P45
		Tecleanthine T17
		Vismin V7
		Vismione A V8
		Vismione B
Spodoptera exigua (Hübner)	Beet armyworm	Ajugarin I A6
Spodoptera frugiperda (J. E. Smith)	Fall armyworm	Chaparrinone C53
		Isobrucein A (see B16)
		Melampodin A M21
		Melampodinin A M22
		Nomilin N14
		Obacunone O3
		6α-Senecioyloxychaparrinone (see C53)
Spodoptera littoralis	Egyptian cotton leafwork African armyworm	2,3-Dihydrowithanolide E
		2,5-Dihydroxymethyl-3,4-dihydroxypyrrolidine D24
		5β,6β-Epoxy-1β,14α,17β,20-tetrahydroxywith-24-enolide E11
		1-(1,3,4,5,6,7-Hexahydro-4-hydroxy-3,8-dimethyl-5-azulenyl-ethanone H13
		N-Methyl-*trans*-4-hydroxy-L-proline M31
		Miconidin M36
		Primin P45
		Vismione A V8
		Withanolide E W4
Spodoptera litura	Tobacco cutwork	Cedrolone C51
		Limonin L17

ORTHOPTERA

Locusta migratoria	African migratory locust	Acetylvismione B (see V9)
		Bruceine A B16
		Bruceine B
		Chaparrinone C53
		Deacetylvismione A (see V8)
		2,5-Dihydroxymethyl-3,4-dihydroxypyrrolidine D24
		(4-8)Epicatechin-catechin dimer
		Ferruginin A F10
		Ferruginin B F11
		Glaucarubinone G10
		Ipolamiide I8
		4-O'Methyl(−)epigallocatechin M28
		Miconidin M36
		Primin P45
		Proanthocyanidin A P46
		Proanthocyanidin B P46

TABLE 5 (continued)
Order, Scientific Name, and Feeding Deterrent Cross Reference for Tables 1 and 4

Insect		Feeding deterrent
Scientific name	Common name	
		Simalikalactone D S30
		Soulameanone S31
		2,3,4-Trihydroxybenzaldehyde T29
		2,4,6-Trihydroxybenzaldehyde T30
		Vanillic acid V1
		Vismin V7
		Vismione A V8
		Vismione B V9
		Wilforine W3
Melanoplus sanguinipes (Fabricius)	Migratory grasshopper	Achillin A24
Schistocera gregaria (Förskel)	Desert locust	Chicoriin C54
		8-Deoxylactupicrin D21
		2,5-Dihydroxymethyl-3,4-dihydroxypyrrolidine D24
		Ipolamiide I8
		Lactucin L11
		Lactupicrin
		Miconidin M36
		Primin P45

TABLE 6
Insect Feeding Deterrents Assembled Under the Plant Families[a] in Which They Have Been Found. Compounds from Tables 1 and 4.

Division	Family	Approx no. species known	No. of species examined	Compounds
Thallophyta	Cyanophyceae (Alga)	—	1	1-(1,3,4,5,6,7-Hexahydro-4-hydroxy-3,8-dimethyl-5-aznlenyl)ethanone H13
	Agaricaceae (Fungus)	—	1	Lactarorufin A L10
	Usneaceae (Lichen)	—	1	Vulpinic acid V3
Bryophyta	Plagiochilaceae (Liverwort)	—	5	Plagiochilin A P40 *ent*-Isoalantolactone 18a
Pteridiophyta (ferns)	Aspidiaceae	—	1	1-(2,4,6-Trimethoxyphenyl)-but-2-en-1-one T31
	Osmundaceae	—	1	(4R,5S)-5-Hydroxy-2-hexen-4-olide H22 (4R,5S)-Osmundalactone O7
	Psilotaceae	—	1	Psilotin P47
Spermatophyta (Gymnospermae)	Ginkgoaceae	1	1	Bilobalide B14 Ginkgolide A G9 Ginkgolide B
	Pinaceae	240	7	Abietic acid A19 Borneol B8 Borneol acetate B9 Δ3-Carene C6 Chrysin C23 Dehydroabietic acid D19 Isopimaric acid I9 Keto-8(14)-podocarpen-18-oic acid K

TABLE 6 (continued)
Insect Feeding Deterrents Assembled Under the Plant Families[a] in Which They Have Been Found. Compounds from Tables 1 and 4.

Division	Family	Approx no. species known	No. of species examined	Compounds
				Limonene L2
				12-Methoxyabietic acid M24
				Palustric acid P28
				α-Pinene P17
				Pungenin P48
				Sandaracopimaric acid S25
				α-Terpineol T2
				Terpineol acetate T3
	Cupressaceae	120	4+	3-Isothujone T10
				Sandarocopimaric acid
				3-Thujone T20
				Yatein Y1
Angiospermae (Monocotoledonae)	Cyperaceae	4000	1	Piceatannol P38
				Scirpusin B S27
	Graminae	8000	10+	(E)-Aconitic acid A4
				2-Benzoxazolinone B3
				Caffeic acid C1
				Capillen C47
				Cholesterol acetate C20
				Cholesterol myristate C21
				Cholesterol oleate C22
				o-Coumaric acid C35
				p-Coumaric acid C36
				Dhurrin D3
				2,4-Dihydroxy-7-methoxy-2H-1,4-benzoxazin-3(44)-one (DIMBOA)
				Ferulic acid F6
				Gentisic acid G1
				2-β-D-Glucosyl-4-hydroxy--7-methoxy-1-4-benzoxazin--3-one G11
				Gramine G7
				Hordenine H4
				Hydroxamic acid H17
				o-Hydroxybenzaldehyde H7
				m-Hydroxybenzaldehyde H8
				p-Hydroxybenzaldehyde H9
				Isoferulic acid I5
				6-Methoxy-2-benzoxazolinone M4
				5-Methoxy-N,N-dimethyltrypt-amine M25
				5-Methoxytryptamine M27
				7-Methylhydroxamic acid M30
				N-Methyltryptamine M35
				Peramine P32
				Protocatechnic acid P26
				Salicylic acid S3
				Syringic acid S21
				Vanillic acid V1
	Palmae	3500	1	Acrecoline A25
	Araceae	1800	1	Asarone A15

TABLE 6 (continued)
Insect Feeding Deterrents Assembled Under the Plant Families[a] in Which They Have Been Found. Compounds from Tables 1 and 4.

Division	Family	Approx no. species known	No. of species examined	Compounds
	Liliaceae	4200	2	Colchicine C34 Scillaren S8 Veratrine
	Amaryllidaceae	—	2	3'-Hydroxy-4'-methoxy-7-hydroxy-8-methylflavan H23 Lycorine L9 O-Methyl lycorenine M32
	(46 other families)	—	—	—
(Dicotoledonae) Subclass Magnoliidae	Magnoliaceae	200	2	Epitulipinolide diepoxide Lipiferolide Magnoline M1 Peroxyferolide P3 Tulirinol T11
	Annonaceae	2100	1	Asimicin A29
	Myristaceae	250	1	5-Methoxy-N,N-dimethyltryptamine M25 2-Methyl-6-methoxytetrahydro-β-carboline M33
	Canellaceae	—	2	Muzigadial M15 Polygodial P23 Ugandensidial U1 Warburganal W1
	Lauraceae	2200	2	Epieudesmin E6 Eudesmin E7 Shiromodiol diacetate S11 Shirmodiol monoacetate S12
	Piperaceae	1400	1	Isoasarone I2 Piperenone P20
	Aristolochiaceae	600	1	Aristolochic acid A14
	Menispermaceae	425	1	Cocculolidine C33 Isoboldine I4 (E,S)-Nerolidol N4
	Papaveraceae	300	1	Papaverine P29
	(27 other families)	—	—	
Subclass Hamamelidae	Moraceae	1500	1	Morin M13 Osajin O2 Pomiferin
	Juglandaceae	50	1	Juglone J1
	Fagaceae	600	1	Quercitin Q1
	(20 other families)	—	—	
Subclass Caryophyllidae	Cactaceae	2000	1	Lophocereine Pilocereine P16
	Plumbaginaceae	—	1	Plumbagin P22
	(12 other families)			
Subclass Dilleniidae	Guttiferae	900	1	Acetylvismione B (see V9) Deacetylvismione A (see V8) Ferruginin A F10 Ferruginin B F11 Harunganin H11 Vismin V7 Vismione A V8 Vismione B V9
	Sterculiaceae	1000	1	Proanthocyanidin A P46

TABLE 6 (continued)
Insect Feeding Deterrents Assembled Under the Plant Families[a] in Which They Have Been Found. Compounds from Tables 1 and 4.

Division	Family	Approx no. species known	No. of species examined	Compounds
	Malvaceae	1500	1	Gossypol G6 Vasicinone V6
	Cucurbitaceae	850	3	Cucurbitacins Momordicine II M38
	Salicaceae	350	4	Chlorogenic acid C55 2',6'-Dihydroxy-4'-methoxy-dihy- drochalcone D14
	Cruciferae	3000	2	Cucurbitacins 2-Phenylethylisothiocyanate P11 Sinigrin S14
	Ericaceae	2500	3	Grayanotoxin-III G12 Kalmitoxin-I K5 Kalmitoxin-IV K6 Unedoside U2
	(62 other families)	—	—	—
Subclass Rosidae	Rosaceae	3000	2	Amygdalin A26 Phloretin 4'-O-β-D- glucopyranoside (see P12)
	Leguminosae	5000	26+	Albizziin A10 Ammonium nitrate L-canavanine C3 13-*trans*-Cinnamoyloxylupanine C56 Claussequinone C57 (Z)-*o*-Coumaric acid glucoside (see C35) Coumarin C37 *p*-Cymene C43 Djenkolic acid L-Dopa D18 Hildecarpin H14 Hiptagin H15 2'-Hydroxyformononetin H18 2'-Hydroxygenistein H19 Kaempferol K1 Kievitone K7 Licoisoflavone A L15 Licoisoflavone B L16 Luteone L20 Maackiain M19 N-Methyl-*trans*-4-Hydroxy-L- proline M31 7-O-Methylphaseolin (see P34) 2'-O-Methylphaseolinisoflavan (see P35) Methyl sainfuran M34 *Milletia* chalcone M37 Phaseolidin P33 Phaseolin P34 Phaseolinisoflavan P35 Pipecolic acid P19 Pisatin P39 Rotenone R3

TABLE 6 (continued)
Insect Feeding Deterrents Assembled Under the Plant Families[a] in Which They Have
Been Found. Compounds from Tables 1 and 4.

Division	Family	Approx no. species known	No. of species examined	Compounds
				Sainfuran S22
				Sativan S5
				Senkirkine S29
				Terpinen-4-ol T1
				α-Terpineol T2
				13-Tigloyloxylupanine T22
				(3R)-Vestitol V2
	Myrtaceae	3000	1	1,4-Cineole C24
				1,8-Cineole C25
	Melastomataceae	4000	1	Miconidin M36
				Primin P45
	Celastraceae	850	3	Wilforine W3
	Euphorbiaceae	7500	4	Anthranilic acid A27
				Camphor C45
				Cinnamaldehyde
				trans-Cinnamic acid C26
				Crotepoxide C39
				9,10-Dihydroxy-1-octadecanol acetate
				α-Eleostearic acid E4
				Gentisic acid G1
				Senecioic acid S28
				Trewiasine T27
	Rhamnaceae	900	1	Emodin E5
	Melianthaceae	35	1	Abyssinin A20
	Hippocastanaceae	—	1	Aesculitin A5
	Aitonaceae	1	1	Prieurianin P44
				Rohituka 7
	Aceraceae	150	2	Gramine G7
	Anacardiaceae	600	1	Cardol
	Simaroubaceae	170	7	12β-Acetoxy harrisonin (see H2)
				8-Acetyl-3α-hydroxyperoxy-2α-isohex-3-enyl-6-isopent-2-enyl-5-methoxy-2β-methyl-chroman A22
				Bruceine A B16
				Bruceine B (see B16)
				Chaparrinone C53
				Harrisonin H2
				Isobruceine A (see B16)
				Pedonin P30
				Quassin Q3
				6α-Senecioyloxychaparrinone C53
				Simalikalactone D S30
				Soulameanone S31
	Balanitaceae	—	1	*Balanites* saponin B13
	Rutaceae	1600	7	Bergapten B6
				Evoxine E8
				Isopimpenillin I7
				Japonin
				Kokusagine K4
				Limonin L17
				Limonin diosphenol L18
				Melicopicine M23

TABLE 6 (continued)
Insect Feeding Deterrents Assembled Under the Plant Families[a] in Which They Have Been Found. Compounds from Tables 1 and 4.

Division	Family	Approx no. species known	No. of species examined	Compounds
				6-Methoxytecleanthine M26
				N-Methylflindersine M8
				Nomilin N14
				Obacunone O3
				Pilocarpine P15
				Tecleanthine T17
				Xanthotoxin X1
				Zanthophylline Z1
	Meliaceae	600	9	6-Acetoxytoonacilin A2
				6-O-Acetylnimbandiol A22
				7-Acetyltrichilin A23
				Azedarachol A31
				Azadirachtin A30
				Cedrolone C51
				Cycloeucalenol
				3-Desacetylsalannin D1
				1,3-Diacetylvilasinin D4
				2-Formylanthraquinone F7
				23-(R,S)-Hydroxytoonacilide H24
				21-(R,S)-Hydroxytoonacilide
				Meliantriol M2
				Meliatin
				N-Methylflindersine M8
				Nimbandiol N6
				Nimbolidin A (see N12)
				Nimbolidin B N12
				Nimbolinin B N13
				Ochinolide A 05
				Ochinolide B 05
				Prieurianin P44
				Salannin D2
				Salannol S1
				Salannolactam-(21) S23
				3-Tigloylazadirachtol T21
				Toonacilin A3
				Tr-A, Tr-B, Tr-C T23, T24
				Trichilin A, B, C and D T28
				Xylomollin X2
	Zygophyllaceae	250	2	Vasicinone V6
	Geraniaceae	750	1	Geraniin G8
	Umbelliferae	3000	4	Angelicin A12
				Bisabolangelone B15
				Myristicin M18
				Peucedanin P4
	(92 other families)	—	—	—
Subclass Asteridae	Strychnaceae	500	1	Brucine B10
	Apocynaceae	2000	1	Friedelin F9
	Asclepiadaceae	2000	2	Gymnemic acids
				Uzarigenin U4
				Uzarigenin glycoside (see U4)
	Solanaceae	2300	13+	α-Chaconine C52
				Demissidine
				Demissine

TABLE 6 (continued)
Insect Feeding Deterrents Assembled Under the Plant Families[a] in Which They Have
Been Found. Compounds from Tables 1 and 4.

Division	Family	Approx no. species known	No. of species examined	Compounds
				Dihydro-α-solanin
				2,3-Dihydrowithanolide E (see W4)
				5β,6β-Epoxy-1β,14α,17β,20-te-trahydrowith-24-enolide E11
				Leptine I
				Leptine III
				Leptinine I
				Leptinine II
				Nicalbin A N9
				Nicalbin B N10
				Nicandrenone N11
				Nicotine N5
				Nornicotine N8
				Scopolamine S9
				Solacaulin
				Solanidine S16
				Solanine S17
				Tomatidine
				Tomatine T7
				Withaferin W2
				Withanolide E W4
	Boraginaceae	2000	2	Lasiocarpin L13
	Verbenaceae	2600	7	Caryopterin C9
				Caryopterin hemiacetal C13
				Caryoptionol C10
				Clerodendrin-A C31
				Clerodendrine B C32
				Clerodin C11
				Clerodin hemiacetal C14
				Dihydrocaryoptin C15
				Dihydrocaryoptionol C16
				Dihydroclerodin-I C17
				Epicaryoptin C12
				Ipolamiide I8
				7-Methoxyjuglone M5
				2-Methylanthraguinone M7
				Phytol P14
	Labiatae	3200	7	Ajugarin I, II, III A6, A7, A8
				Clerodane diterpenes C58
				Effusin E2
				Hecogenin H3
				Hispidulin H16
				Indican dioegin
				Inflexin I1
				Isomedin I6
				Lucenine
				Nepetalactone N2
				Plectrin P41
				Santonin S4
				Tafricanin A T13
				Tafricanin B T14
				Tomatin T7

TABLE 6 (continued)
Insect Feeding Deterrents Assembled Under the Plant Families[a] in Which They Have
Been Found. Compounds from Tables 1 and 4.

Division	Family	Approx no. species known	No. of species examined	Compounds
	Oleaceae	600	1	Fraxetin F8
	Scrophulariaceae	2700	1	Digitonin D9
	Bignoniaceae	800	6	Catalposide C50
				Lapachol L1
				Lapachonone
				Specionin S33
	Acanthaceae	2600	1	Vasicine V4
				Vasicinol V5
				Vasicinone V6
	Campanulaceae	2000	1	Lobeline L5
	Rubiaceae	6500	2	Quinine Q2
				Unedoside U2
	Compositae	19000	40+	Achillin A24
				Alantolactone A9
				Arctolide A28
				Bakkenolide A B12
				Cadinene C44
				Capillarin C46
				Capillen C47
				Caryophyllene oxide C48
				Chicoriin C54
				ar-Curcumene C59
				8-Deoxylactupicrin D21
				Desacetylarctolide D22
				11,13-Dihydrovernodalin D23
				Encecalin E10
				Eperu-8(17)-en-15,19-dioic acid-butenolide (see lambertianic acid)
				Erivanin E12
				Eupatoriopicrin E13
				Glancolide-A G4
				Helenalin H12
				Hispidulin H16
				6α-Hydroxygrindelic acid H20
				18-Hydroxygrindelic acid H21
				Isoalantolactone I8a
				Kaur-16-en-oic acid K2
				Lactucin L11
				Lactupicrin
				Lambertianic acid L12
				Lasiocarpine L13
				Levopimaric acid L14
				Linifoline A L19
				Melampodin A M21
				Melampodinin A M22
				Methyl (Z)-10-acetoxy-8,9-epoxy-2-decen-4,6-diynoate
				Methyl (Z,Z)-10-acetoxymatric-ariate M6
				Methyleugenol M29
				Parthenin P1
				1-Phenylheptatriyne P36
				1-Phenyl-2,4-pentadiyne P37

TABLE 6 (continued)
Insect Feeding Deterrents Assembled Under the Plant Families[a] in Which They Have Been Found. Compounds from Tables 1 and 4.

Division	Family	Approx no. species known	No. of species examined	Compounds
				Precocene I P42
				Precocene II P43
				Schkuhrin-I S6
				Schkuhrin-II S7
				Schkuhriolide S26
				Senecionine S10
				Seneciphylline
				Senkirkine S29
				Spathulenol S32
				Steviol S34
				18-Succinoyloxygrindelic acid
				Tagitinin A T15
				Tagitinin C T16
				Tenulin T18
				Terthienyl T19
				Toxyl angelate
				Trachyloban-19-oic acid T8
	(29 other families)	—	—	—

[a]   Classification of families based upon that of Cronquist,[345] with genera alloted to families according to Willis.[346]

REFERENCES TO PART A

1. **Munakata, K.,** Insect feeding deterrents in plants, in *Chemical Control of Insect Behavior*, Shorey, H. H. and McKelvey, J. J., Jr., Eds., John Wiley & Sons, New York, 1977, chap 6.
2. **Slansky, F., Jr.,** Effect of the lichen chemicals atranorin and vulpinic acid upon feeding and growth of larvae of the yellow-striped armyworm, *Spodoptera ornithogalli, Environ. Entomol.*, 8, 865, 1979.
3. **Warthen, J. D., Jr.,** *Azadirachta indica*: A source of insect feeding inhibitors and growth regulators, *Agric. Rev. & Manuals*, Northeast. Ser., No. 4, U.S. Department of Agriculture Science and Education Administration, Beltsville, MD, 1979.
4. **Munakata, K.,** Insect antifeedants of *Spodoptera litura* in plants, in *Host Plant Resistance To Pests*, Hedin, P. A., Ed., ACS Symp. Ser., 62, 185, 1977.
5. **Kubo, I. and Nakanishi, K.,** Insect antifeedants and repellents from African plants, in *Host Plant Resistance To Pests*, Hedin, P. A., Ed., ACS Symp. Ser., 62, 165, 1977.
6. **Dethier, V. G., Browne, L. B., and Smith, C. N.,** The designation of chemicals in terms of the responses they elicit from insects, *J. Econ. Entomol.*, 53, 134, 1960.
7. **Adams, C. M. and Bernays, E. A.,** The effect of combinations of deterrents on the feeding behaviour of *Locusta migratoria, Entomol. Exp. Appl.*, 23, 101, 1978.
8. **Chapman, R. F. and Bernays, E. A.,** The chemical resistance of plants to insect attack, in *Pontif. Accad. Sci. Scr. Varia*, 41, 603, 1976.
9. **Munakata, K.,** Insect antifeeding substances in plant leaves, *Pure Appl. Chem.*, 42, 57, 1975.
10. **Chapman, R. F.,** The chemical inhibition of feeding by phytophagous insects: A review, *Bull. Entomol. Res.*, 64, 339, 1974.
11. **Hedin, P. A., Maxwell, F. G., and Jenkins, J. N.,** Insect plant attractants, feeding stimulants, repellents, deterrents, and other related factors affecting insect behavior, in *Proc. Summer Inst. Biol. Control of Plant Insects and Diseases*, University Press, Mississippi, 1974, 494.
12. **Wright, D. P., Jr.,** Antifeedants, in *Pest Control*, Kilgore, W. W. and Doutt, R. L., Eds., Academic Press, Orlando, 1967, chap. 8.

13. **Schoonhoven, L. M. and Jermy, T.,** A behavioural and electrophysiological analysis of insect feeding deterrents, in Crop Prot. Agents — Their Biol. Evaluation, Proc. Int. Conf. Evaluation Biol. Act., 1977, 133.

14. **Ascher, K. R. S.,** Fifteen years (1963-1978) of organotin antifeedants — a chronological bibliography, *Phytoparasitica,* 7, 117, 1979.

15. **Kubo, I., Miura, I., Pettei, M. J., Lee, Y. -W., Pilkiewicz, F., and Nakanishi, K.,** Muzigadial and warburganal, potent antifungal antiyeast, and African army worm antifeedant agents, *Tetrahedron Lett.,* 4553, 1977.

16. **Peterse, A. J. G. M., Roskam, J. H., and de Grott, Ae.,** A model synthesis of ring B of warburganal and muzigadial, *J. Netherlands Chem. Soc.,* 97, 277, 1978.

17. **Nakanishi, K.** Insect growth regulators from plants, in *Pontif. Accad. Sci. Scr. Varia,* 41, 185, 1976.

18. **Kraus, V. W., Grimminger, W., and Sawitzki, G.,** Toonacilin und 6-Acetoxy-toonacilin, zwei neue B-*seco*-Tetranortriterpenoide mit frasshemmender Wirkung, *Angew. Chem.,* 90, 476, 1978.

19. **Koh, H., Kim, M., Obata, T., Fukami, H., and Ishii, S.,** Antifeedant in barnyard grass against the brown planthopper — *trans*-aconitic acid, *Pap. Semin.,* 117, 1977.

20. **Norris, D. M.,** Role of repellents and deterrents in feeding of *Scolytus multistriatus,* in *Host Plant Resistance To Pests,* Hedin, P. A., Ed., ACS Symp. Ser. 62, 215, 1977.

21. **Goldsmith, D. J., Srouji, G., and Kwong, C.,** Insect antifeedants, 1. Diels-Alder approach to the synthesis of ajugarin I, *J. Org. Chem.,* 43, 3152, 1978.

22. **Kubo, I., Lee, Y., Balogh-Nair, V., Nakanishi, K., and Chapya, A.,** Structure of ajugarins, *J. Chem. Soc. Chem. Commun.,* 22, 949, 1976.

23. **Meinwald, J., Prestwich, G. D., Nakanishi, K., and Kubo, I.,** Chemical ecology: Studies from East Africa, *Science,* 199, 1167, 1978.

24. **Ascher, K. R. S., Meisner, J., and Flowers, H. M.,** Effects of amino acids on the feeding behavior of the larva of the Egyptian cotton leafworm, *Spodoptera littoralis* Boisd., *Phytoparasitica,* 4, 85, 1976.

25. **Kogan, M.,** The role of chemical factors in insect/plant relationships, in Proc. 15th Int. Congr. Entomol., 1977, 211.

26. **Picman, A. K., Elliott, R. H., and Towers, G. H. N.,** Insect feeding deterrent property of alantolactone, *Biochem. Syst. Ecol.,* 6, 333, 1978.

27. **Jacobson, M.,** Insecticides from plants, a review of the literature, 1954-1971, in *Agriculture Handbook,* U.S. Department of Agriculture, Aricultural Research Service, 461, 1975.

28. **Evans, C. S. and Bell, E. A.,** Non-protein amino acids of *Acacia* species and their effect on the feeding of the acridids *Anacridium melanorhodon* and *Locusta migratoria, Phytochemistry,* 18, 1807, 1979.

29. **Navon, A. and Bernays, E. A.,** Inhibition of feeding in acridids by nonprotein amino acids, *Comp. Biochem. Physiol.,* 59A, 161, 1978.

30. **Clift, A. D.,** Activity of chlordimeform hydrochloride and amitraz mixtures with *Bacillus thuringiensis* against *Heliothis armigera* (Hübner) (Lepidoptera: Noctuidae), *Gen. Appl. Entomol.,* 11, 21, 1979.

31. **Schoonhoven, L. M.,** Secondary plant substances and insects in *Structural and Functional Aspects of Phytochemistry, Recent Adv. Phytochem.,* 5, 197, 1972.

32. **Hedin, P. A., Jenkins, J. N., and Maxwell, F. G.,** Behavioral and developmental factors affecting host plant resistance to insects, in *Host Plant Resistance To Pests,* Hedin, P. A., Ed., ACS Symp. Ser., 62, 231, 1977.

33. **Beeman, R. W. and Matsumura, F.,** Anorectic effect of chlordimeform in the American cockroach, *J. Econ. Entomol.,* 71, 859, 1978.

34. **Yajima, T. and Munakata, K.,** Phloroglucinol-type furocoumarins, a group of potent naturally-occurring insect antifeedants, *Agric. Biol. Chem.,* 43, 1701, 1979.

35. **Bernays, E. A. and Chapman, R. F.,** Deterrent chemicals as a basis of oligophagy in *Locusta migratoria* (L.), *Ecol. Entomol.,* 2, 1, 1977.

36. **Jacobson, M.,** Allelopathy and entomology research, in Rep. Res. Plann. Conf. Role Secondary Compounds Plant Interactions (Allelopathy), Mississippi State, State College, March 15 to 16, 1977, 33.

37. **Reed, D. K., Jacobson, M., Warthen, J. D., Jr., Uebel, E. C., and Tromley, N.,** Laboratory screening of natural products for feeding suppression of cucumber beetles, *Technical Bulletin* No. 1641 U.S. Department of Agriculture Science and Education Administration, Beltsville, MD, 1981.

38. **Reed, D. K., Warthen, J. D., Jr., and Uebel, E. C.,** Effects of two triterpenoids from neem on feeding by cucumber beetles, *J. Econ. Entomol.* 75, 1109, 1982.

39. **Meisner, J., Ascher, K. R. S., Aly, R., and Warthen, J. D., Jr.,** Response of *Spodoptera littoralis* (Boisduval) and *Earias insulana* (Boisduval) larvae to azadirachtin and salannin, *Phytoparasitica,* 9, 27, 1981.

40. **Ruscoe, C. N. E.,** Growth disruption effects of an insect antifeedant, *Nat. New Biol,* 236, 159, 1972.

41. **Zanno, P. R., Miura, I., Nakanishi, K., and Elder, D. L.,** Structure of the insect phagorepellent azadirachtin. Application of PRFT/CWD carbon-13 nuclear magnetic resonance, *J. Am. Chem. Soc.,* 97, 1975, 1975.

42. **Butterworth, J. H. and Morgan, E. D.,** Isolation of a substance that suppresses feeding in locusts, *Chem. Commun.,* 23, 1968.

43. **Warthen, J. D., Jr., Redfern, R. E., Uebel, E. C., and Mills, G. D., Jr.,** An antifeedant for fall armyworm larvae from neem seeds, *Agric. Res. Results,* Northeast. Ser., No. 1, U.S. Department of Agriculture Science and Education Administration, Beltsville, MD 1978.

44. **Russell, G. E.,** Some effects of benzimidazole compounds on the transmission of beet yellows virus by *Myzus persicae,* in *Proc. Br. Crop Prot. Conf. — Pests and Dis.,* 3, 831, 1977.

45. **Norris, D. M.,** Physico-chemical aspects of the effects of certain phytochemicals on insect gustation, *Symp. Biol. Hung.,* 16, 197, 1976.

46. **Juneja, P. S., Gholson, R. K., Burton, R. I., and Starks, K. J.,** The chemical basis for greenbug resistance in small grains. I. Benzyl alcohol as a possible resistance factor, *Ann. Entomol. Soc. Am.,* 65, 961, 1972.

47. **Yajima, T., Kato, N., and Munakata, K.,** Isolation of insect anti-feeding principles in *Orixa japonica* Thunb., *Agric. Biol. Chem.,* 41, 1263, 1977.

48. **Dethier, V. G.,** Evolution of receptor sensitivity to secondary plant substances with special reference to deterrents, *Am. Nat.,* 115, 45, 1980.

49. **Tamaki, G.,** TH 6041: Knockdown and feeding inhibition of the zebra caterpillar and the Colorado potato beetle, *J. Econ. Entomol.,* 69, 644, 1976.

50. **Smelyanets, V. P.,** Mechanisms of plant resistance in Scotch pine *(Pinus silvestris)* 3. Phase of secondary insect choice of pine trees (contact-gustatory preferendum), *Z. Angew Entomol.,* 84, 113, 1977.

51. **Nielsen, J. K.,** Host plant discrimination within Cruciferae: Feeding responses of four leaf beetles (Coleoptera: Chrysomelidae) to glucosinolates, cucurbitacins and cardenolides, *Entomol. Exp. Appl.,* 24, 41, 1978.

52. **Rosenthal, G. A.,** The biological effects and mode of action of L-canavanine, a structural analogue of L-arginine, *Q. Rev. Biol.,* 52, 155, 1977.

53. **Carrel, J. E. and Eisner, T.,** Cantharidin: potent feeding deterrent to insects, *Science,* 183, 755, 1974.

54. **Harada, N. and Uda, H.,** Absolute stereochemistries of 3-epicaryoptin, caryoptin, and clerodin as determined by chiroptical methods, *J. Am. Chem. Soc.,* 100, 8022, 1978.

55. **Hosozawa, S., Kato, N., and Munakata, K.,** Absolute configuration of caryoptin and 3-epicaryoptin — an exception in the exciton chirality method, *Tetrahedron Lett.,* 3753, 1974.

56. **Hirata, M. and Sogawa, K.,** Antifeeding activity of chlordimeform for plant-sucking insects, *Appl. Entomol. Zool.,* 11, 94, 1976.

57. **Higgans, R. A.,** Fundal and SN 513 as antifeeding agents for the green cloverworm when applied to soybean leaflets, in *Proc. North Cent. Branch Entomol. Soc. Am.,* 32, 20, 1977.

58. **Nakamuta, K. and Saito, T.,** The antifeeding effects of chlordimeform on the silkworm larvae, *Bombyx mori* L., *Jpn. J. Appl. Entomol. Zool.,* 21, 158, 1977.

59. **Chippendale, G. M. and Reddy, G. P. V.,** Dietary sterols: role in larval feeding behavior of the southwestern corn borer, *Diatraea grandiosella, Experientia,* 28, 485, 1974.

60. **Wolcott, G. N.,** Organic termite repellents tested against *Cryptotermes brevis* Walker, *P. R. Univ. J. Agric.,* 39, 115, 1955.

61. **Dässler, H. -G.,** Chlorierungsprodukte von Terpenoxyden und deren insecticide Wirkung, *Pharmazie,* 13, 404, 1958.

62. **Floyd, M. A., Evans, D. A., and Howse, P. E.,** Electrophysiological and behavioural studies on naturally occurring repellents to *Reticulitermes lucifugus, J. Insect Physiol.,* 22, 697, 1976.

63. **Descoins, C.,** New developments in agrochemistry, *Can. Nutr. Diet,* 14, (Suppl. 4, Pesticides), 93, 1979.

64. **Kojima, Y. and Kato, N.,** Synthesis of clerodin homologue. A synthetic approach of structure-activity relationships, *Chem. Abstr.,* 93, 955, 1980.

65. **Munakata, K.,** Insect antifeeding substances in plant leaves, *Pure Appl. Chem.,* 42, 57, 1975.

66. **Wada, K. and Munakata, K.,** An insecticidal alkaloid, cocculolidine from *Cocculus trilobus* DC. Part I. The isolation and the insecticidal activity of cocculolidine, *Agric. Biol. Chem.,* 31, 336, 1967.

67. **Nielsen, J. K., Larsen, L. M., and Sorensen, H.,** Cucurbitacin E and I in *Iberis amara*: Feeding inhibitors for *Phyllotreta nemorum, Phytochemistry,* 16, 1519, 1977.

68. **Carroll, C. R. and Hoffman, C. A.,** Chemical feeding deterrent mobilized in response to insect herbivory and counteradaption by *Epilachna tredecimnotata, Science,* 209, 414, 1980.

69. **Gombos, M. A. and Gasko, K.,** Extraction of natural antifeedants from the fruits of *Amorpha fruticosa* L., *Acta Phytopathol. Acad. Sci. Hung.,* 12, 349, 1977.

70. **Gombos, M., Szendrei, K., Feuer, L., Toth, G., and Kecskes, M.,** Environmental aspects in the evaluation of the antifeedants extracted from *Amorpha fructicosa* L., *Proc. 18th Hung. Annu. Meet. Biochem.,* 18, 23, 1978.

71. **Jackson, W. P. and Ley, S. V.,** Synthesis of a substituted *cis*-decalin as a potential insect antifeedant, *J. Chem. Soc. Chem. Commun.,* 732, 1979.

72. **Kraus, W., Bokel, M., Cramer, R., and Sawitzki, G.,** New biologically active compounds from *Azadirachta indica* and *Melia azedarach,* presented at Int. Res. Congr. Nat. Prod. Med. Agents, Abstr. II, Strasbourg, France, July 6 to 11, 1980, 30.

73. **Woodhead, S. and Bernays, E.,** Changes in release rates of cyanide in relation to palatability of *Sorghum* to insects, *Nature,* 270, 235, 1977.

74. **All, J. N. and Benjamin, D. M.,** Potential of antifeedants to control larval feeding of selected *Neodiprion* sawflies (Hymenoptera: Diprionidae), *Can. Entomol.,* 108, 1137, 1976.

75. **Yazima, T. and Munakata, K.,** Synthesis of furoquinolines, *Agric. Biol. Chem.,* 44, 235, 1980.

76. **Klun, J. A., Tipton, C. L., and Brindley, T. A.,** 2,4-Dihydroxy-7-methoxy-1,4-benzoxazin-3-one (DIMBOA), an active agent in the resistance of maize to the European corn borer, *J. Econ. Entomol.,* 60, 1529, 1967.

77. **Jacobson, M., Crystal, M. M., and Warthen, J. D., Jr.,** Boll weevil feeding deterrents from tung oil, *J. Agric. Food Chem.,* 29, 591, 1981.

78. **Rajan Asari, P. A. and Dale, D.,** Studies on the use of antifeedants for protecting stored paddy from Angoumois grain moth, *Sitotroga cerealella, Bull. Grain Technol.,* 15, 123, 1977.

79. **Harley, K. L. S. and Thorsteinson, A. J.,** The influence of plant chemicals on the feeding behavior, development, and survival of the two-striped grasshopper, *Melanoplus bivattatus* (Say), Acrididae: Orthoptera, *Can. J. Chem.,* 45, 305, 1967.

80. **Rehr, S. S., Janzen, D. H., and Feeny, P. P.,** L-Dopa in legume seeds: a chemical barrier to insect attack, *Science,* 181, 81, 1973.

81. **Wei-Chun, M.,** Some properties of gustation in the larva of *Pieris brassicae, Entomol. Exp. Appl.,* 12, 584, 1969.

82. **Kubo, I., Kamikawa, T., Isobe, T., and Kubota, T.,** Structure of effusin, *J. Chem. Soc. Chem. Commun.,* 1206, 1980.

83. **Trial, H., Jr. and Dimond, J. B.,** Emodin in buckthorn: A feeding deterrent to phytophagous insects, *Can. Entomol.,* lll, 207, 1979.

84. **Doskotch, R. W., Fairchild, E. H., Huang, C., Wilton, J. H., Beno, M. A., and Christoph, G. G.,** Tulirinol, an antifeedant sesquiterpene lactone for the gypsy moth larvae from *Liriodendron tulipifera, J. Org. Chem.,* 45, 1441, 1980.

85. **Kubo, I.,** Insect feeding inhibitors in plants of East Africa, *Chem. Abstr.* 92, 141, 1980, Hidaka, T., Takahashi, S., and Isoe, Y., Eds., 153, 1979.

86. **Doskotch, R. W., Cheng, H. -Y., Odell, T. M., and Girard, L.,** Nerolidol: An antifeeding sesquiterpene alcohol for gypsy moth larvae from *Melaleuca leucadendron, J. Chem. Ecol.,* 6, 845, 1980.

87. **Abdul Kareem, A. and Subramaniam, T. R.,** Antifeeding effects of two organotin compounds on *Stomopteryx subsecivella* Zell. (Lepidoptera), *Madras Agric. J.,* 63, 354, 1976.

88. **Joshi, B. G., Ramaprasad, G., and Satyanarayana, S. V. V.,** Relative efficacy of neem kernel, fentin acetate and fentin hydroxide as antifeedants against tobacco caterpillar, *Spodoptera litura* Fabricius, in the nursery, *Indian J. Agric. Sci.,* 48, 19, 1978.

89. **Rudman, P. and Gay, F. J.,** The causes of natural durability in timber. Pt. VI. Measurement of antitermitic properties of anthraquinones from *Tectona grandis* L. f. by a rapid semi-micromethod, *Holzforschung,* 15, 117, 1961.

90. **Abbassy, M. A., El-Shazli, A., and El-Gayar, F.,** A new antifeedant to *Spodoptera littoralis* Boisd. (Lepid., Noctuidae) from *Acokanthera spectabilis)* Hook. *(Apocynaceae), Z. Angew. Entomol.,* 83, 317, 1977.

91. **Mabry, T. J., Gill, J. E., Burnett, W. C., Jr., and Jones, S. B., Jr.,** Antifeedant sesquiterpene lactones in the Compositae, in *Host Plant Resistance to Pests,* Hedin, P. A., Ed., ACS Symp. Ser., 62, 179, 1977.

92. **Meisner, J., Wysoki, M., and Telzak, L.,** Gossypol as phagodeterrent for *Boarmia (Ascotis) selenaria* larvae, *J. Econ. Entomol.,* 69, 683, 1976.

93. **Stipanovic, R. D., Bell, A. A., and Lukefahr, M. J.,** Natural insecticides from cotton *(Gossypium),* in *Host Plant Resistance To Pests,* Hedin, P. A., Ed., ACS Symp. Ser., 62, 197, 1977.

94. **Meisner, J., Ascher, K. R. S., and Zur, M.,** Phagodeterrency induced by pure gossypol and leaf extracts of a cotton strain with high gossypol content in the larva of *Spodoptera littoralis, J. Econ. Entomol.,* 70, 149, 1977.

95. **Chalfant, R. B., Todd, J. W., Taylor, W. K., and Mullinin, B.,** Laboratory studies on the antifeeding effect of a fungicide, guazatine, on eleven species of phytophagous insects, *J. Econ. Entomol.,* 70, 513, 1977.

96. **Higgins, R. A. and Pedigo, L. P.,** Evaluation of guazatine triacetate as an antifeedant/feeding deterrent for the green cloverworm on soybeans, *J. Econ. Entomol.,* 72, 680, 1979.

97. **Higgins, R. A. and Pedigo, L. P.,** A laboratory antifeedant simulation bioassay for phytophagous insects, *J. Econ. Entomol.,* 72, 238, 1979.

98. **Backman, P. A., Harper, J. D., Hammond, J. M., and Clark, E. M.,** Antifeeding effects of the fungicide guazatine triacetate on insect defoliators of soybeans and peanuts, *J. Econ. Entomol.,* 70, 374, 1977.

99. **Granich, M. S., Halpern, B. P., and Eisner, T.,** Gymnemic acids: secondary plant substances of dual defensive action, *J. Insect Physiol.*, 20, 435, 1974.

100. **Kubo, I., Tanis, S. P., Lee, Y., Miura, I., Nakanishi, K., and Chapya, A.,** The structure of harrisonin, *Heterocycles*, 5, 485, 1976.

101. **Kurata, S. and Sogawa, K.,** Sucking inhibitory action of aromatic amines for the rice plant- and leafhoppers (Homoptera: Delphacidae, Deltocephalidae), *Appl. Entomol. Zool.*, 11, 89, 1976.

102. **Kubo, I., Nakanishi, K., Kamikawa, T., Isobe, T., Kubota, T.,** The structure of inflexin, *Chem. Lett.*, 99, 1977.

103. **Gilbert, B. L., Baker, J. E., and Norris, D. M.,** Juglone (5-hydroxy-1,4-naphthoquinone) from *Carya ovata*, a deterrent to feeding by *Scolytus multistriatus*, *J. Insect Physiol.*, 13, 1453, 1967.

104. **Waiss, A. C., Jr., Chan, B. G., Elliger, C. A., and Garrett, V. H.,** Larvicidal factors contributing to host-plant resistance against sunflower moth, *Naturwissenschaften*, 64, 341, 1977.

105. **Ikeda, T., Matsumura, F., and Benjamin, D. M.,** Chemical basis for feeding adaptation of pine sawflies *Neodiprion rugifrons* and *Neodiprion swainei*, *Science*, 197, 497, 1977.

106. **Ikeda, T., Matsumura, F., and Benjamin, D. M.,** Mechanism of feeding discrimination between matured and juvenile foliage by two species of pine sawflies, *J. Chem. Ecol.*, 3, 677, 1977.

107. **Singh, R. P. and Pant, N. C.,** Lycorine — a resistance factor in the plants of subfamily Amaryllidoideae (Amaryllidaceae) against desert locust, *Schistocerca gregaria* F., *Experientia*, 36, 552, 1980.

108. **Lavie, D., Jain, M. K., and Shpan-Gabrielith, S. R.,** A locust phagorepellent from two Melia species, *Chem. Commun.*, 910, 1967.

109. **Rose, A. F., Butt, B. A., and Jermy, T.,** Polyacetylenes from the rabbitbrush, *Chrysothamnus nauseosus*, *Phytochemistry*, 19, 563, 1980.

110. **Capinera, J. L. and Stermitz, F. R.,** Laboratory evaluation of zanthophylline as a feeding deterrent for range caterpillar, migratory grasshopper, alfalfa weevil, and greenbug, *J. Chem. Ecol.*, 5, 767, 1979.

111. **Chou, F. Y., Hostettmann, K., Kubo, I., Nakanishi, K., and Taniguchi, M.,** Isolation of an insect antifeedant *N*-methylflindersine and several benz[C]phenanthridine alkaloids from East African plants, a comment on chelerythrine, *Heterocycles*, 7, 969, 1977.

112. **Hedin, P. A., Thompson, A. C., and Minyard, J. P.,** Constituents of the cotton bud. III. Factors that stimulate feeding by the boll weevil, *J. Econ. Entomol.*, 59, 181, 1966.

113. **Nakanishi, K. and Kubo, I.,** Studies on warburganal, muzigadial and related compounds, *Isr. J. Chem.*, 16, 28, 1977.

114. **Tanis, S. P. and Nakanishi, K.,** Stereospecific total synthesis of (±)-warburganal and related compounds, *J. Am. Chem. Soc.*, 101, 4398, 1979.

115. **Cooper, R., Solomon, P. H., Kubo, I., Nakanishi, K., Shoolery, J. N., and Occolowitz, J. L.,** Myricoside, an African armyworm antifeedant: Separation by droplet countercurrent chromatography, *J. Am. Chem. Soc.*, 102, 7953, 1980.

116. **Hsiao, T. H. and Fraenkel, G.,** The role of secondary plant substances in the food specificity of the Colorado potato beetle, *Ann. Entomol. Soc. Am.*, 61, 485, 1968.

117. **Ishikawa, S.,** Electrical response and function of a bitter substance, receptor associated with the maxillary sensilla of the larva of the silkworm, *Bombyx mori* L., *J. Cell Physiol.*, 67, 1, 1966.

118. **Sharma, R. N. and Joshi, V. N.,** Allomonic principles in *Parthenium hysterophorus*: Potential as insect control agents and role in the weed's resistance to serious insect depredation. Part II: The biological activity of parthenin on insects, *Biovigyanam*, 3, 225, 1977.

119. **Zalkow, L. H., Gordon, M. M., and Lanir, N.,** Antifeedants from rayless goldenrod and oil of pennyroyal: Toxic effects for the fall armyworm, *J. Econ. Entomol.*, 72, 812, 1979.

120. **Doskotch, R. W., El-Feraly, F. S., Fairchild, E. H., and Huang, C.,** Isolation and characterization of peroxyferolide, a hydroperoxy sesquiterpene lactone from *Liriodendron tulipifera.*, *J. Org. Chem.*, 42, 3614, 1977.

121. **Beck, S. D.,** The European corn borer, *Pyrausta nubilalis* (Hubn.), and its principal host plant. VII. Larval feeding behavior and host plant resistance, *Ann. Entomol. Soc. Am.*, 53, 206, 1960.

122. **Montgomery, M. E. and Arn, H.,** Feeding responses of *Aphis pomi, Myzus persicae* and *Amphorophora agathonica* to phlorizin, *J. Insect Physiol.*, 20, 413, 1974.

123. **Kubo, I., Taniguchi, M., Chapya., A., and Tsujimoto, K.,** An insect antifeedant and antimicrobial agent from *Plumbago capensis, Planta Medica*, 1980 (Suppl.), 185.

124. **Kubo, I., Lee, Y., Pettei, M., Pilkiewicz, F., and Nakanishi, K.,** Potent army worm antifeedants from the East African *Warburgia* plants, *J. Chem. Soc. Chem. Commun.*, 1013, 1976.

125. **Warthen, J. D., Jr., Uebel, E. C., Dutky, S. R., Lusby, W. R., and Finegold, H.,** Adult house fly feeding deterrent from neem seeds, *Agric. Res. Results*, Northeast. Ser., No. 2, U.S. Department of Agriculture Science and Education Administration, Beltsville, MD, 1978.

126. **Pappas, L. G.,** Gustatory salt rejection by *Culiseta inornata, J. Insect Physiol.*, 24, 429, 1978.

127. **Russell, G. B., Sutherland, O. R. W., Hutchins, R. F. N., and Christmas, P. E.,** Vestitol: A phytoalexin with insect feeding-deterrent activity, *J. Chem. Ecol.*, 4, 571, 1978.

128. **Pettei, M. J., Miura, I., Kubo, I., and Nakanishi, K.,** Insect antifeedant sesquiterpene lactones from *Schkuhria pinnata*: The direct obtention of pure compounds using reverse-phase preparative liquid chromatography, *Heterocycles,* 11, 471, 1978.

129. **Wada, K., Enomoto, Y., Matsui, K., and Munakata, K.,** Insect antifeedants from *Parabenzoin trilobum* (I) two new sesquiterpenes, shiromodiol-diacetate and -monoacetate, *Tetrahedron Lett.,* 4673, 1968.

130. **Wearing, C. H.,** Responses of aphids to pressure applied to liquid diet behind parafilm membrane. Longevity and larviposition of *Myzus persicae* (Sulz.) and *Brevicoryne brassicae* (L.) (Homoptera: Aphididae) feeding on sucrose and sinigrin solutions, *N. Z. J. Sci.,* 11, 105, 1968.

131. **Erickson, J. M. and Feeny, P.,** Sinigrin: a chemical barrier to the black swallowtail butterfly, *Papilio polyxenes, Ecology,* 55, 103, 1974.

132. **Ishaaya, I., Yablonski, S., Ascher, K. R. S., and Casida, J. E.,** Triphenyl and tetraphenyl derivatives of group V elements as inhibitors of growth and digestive enzymes of *Tribolium confusum* and *Tribolium castaneum* larvae, *Pestic. Biochem. Physiol.,* 13, 164, 1980.

133. **Dale, D., Saradamma, K., and Chandrika, S.,** Studies on the insect antifeedant action of tricyclohexyltin hydroxide, *Pesticides,* 12, 36, 1978.

134. **Ishaaya, I., Holmstead, R. L., and Casida, J. E.,** Triphenyl derivatives of group IV elements as inhibitors of growth and digestive enzymes of *Tribolium castaneum* larvae, *Pestic. Biochem. Physiol.,* 7, 573, 1977.

135. **Chhibber, R. C.,** Note on field evaluation of antifeedant activity of triphenyltin compounds against *Diacrisia obliqua* Walker on sugarbeet, *Indian J. Agric. Sci.,* 50, 176, 1980.

136. **Nakata, T., Akita, H., Naito, T., and Oishi, T.,** A total synthesis of (±)-warburganal, *J. Am. Chem. Soc.,* 101, 4400, 1977.

137. **Kubo, I., Miura, I., and Nakanishi, K.,** The structure of xylomollin, a secoiridoid hemiacetal acetal, *J. Am. Chem. Soc.,* 98, 6704, 1976.

138. *CRC Atlas of Spectral Data and Physical Constants for Organic Compounds,* Grasselli, J. G., Ed., CRC Press, Boca Raton, FL, 1974.

139. *The Merck Index,* Windholz, M., Ed., Merck, Rahway, NJ, 1976.

140. **Novotny, L., Herout, V., and Sorm, F.,** A contribution to the structure of absinthin and anabsinthin, *Chem. Ind.,* 19, 465, 1958.

141. **Novotny, L., Herout, V., and Sorm, F.,** On terpenes. CIX. A contribution to the structure of absinthin and anabsinthin, *Collect. Czech. Chem. Commun.,* 25, 1492, 1960.

142. **Vokac, K., Samek, Z., Herout, V., and Sorm, F.,** The structure of artabsin and absinthin, *Tetrahedron Lett.,* 3855, 1968.

143. **Marshall, J. A. and Cohen, N.,** The structure of alantolactone, *J. Org. Chem.,* 29, 3727, 1964.

144. **Kjaer, A., Larsen, P. O., and Gmelin, R.,** Structure of albizziine [L-(−)-2-amino-3-ureidopropionic acid], an amino acid from higher plants *(Mimosaceae), Experientia,* 15, 253, 1959.

145. **Haegele, W., Kaplan, F., und Schmid, H.,** Die Struktur des Aucubins, *Tetrahedran Lett.,* 110, 1961.

146. **Birch, A. J., Grimshaw, J., and Juneja, H. R.,** Aucubin, *J. Chem. Soc.,* 5194, 1961.

147. **Butterworth, J. H., Morgan, E. D., and Percy G. R.,** The structure of azadirachtin; the functional groups, *J. C. S. Perkin I,* 2445, 1972.

148. *Aldrich Catalog Handbook of Fine Chemicals,* Catalog 20, 1981—1982, Aldrich Chemical Co., Milwaukee, WI, 1981.

149. **Perkin, Jr., W. H. and Robinson, R.,** XXXIII. — Strychnine, berberine, and allied alkaloids, *J. Chem. Soc.,* 97, 305, 1910.

150. **Karrer, W.,** *Konstitution und Vorkommen der organischen Pflanzenstoffe (exclusive Alkaloide),* Birkhäuser Verlag, Basel, 1958.

151. **Kitagawa, M. and Takani, A.,** Studies on a diaminoacid, canavanin. IV. The constitution of canavanin and canalin, *J. Biochem.,* 23, 181, 1936.

152. **Cadden, J. F.,** Spatial configuration and preparation of canavanine, *Proc. Soc. Exp. Biol. Med.,* 45, 224, 1940.

153. **Hosozawa, S., Kato, N., and Munakata, K.,** Diterpenenoids from *Caryopteris divaricata, Phytochemistry,* 12, 1833, 1973.

154. **Heller, S. R. and Milne, G. W. A.,** *EPA/NIH Mass Spectral Data Base,* U.S. Government Printing Office, Washington, D.C., 1978.

155. **Kato, N. and Munakata, K.,** Crystal and molecular structure of the *p*-bromobenzoate chlorohydrin of clerodendrin A, *J. Chem. Soc. Perkin I,* 69, 1973.

156. **Kato, N., Shibayama, M., and Munakata, K.,** Structure of the diterpene clerodendrin A, *J. Chem. Soc. Perkin I,* 712, 1973.

157. **Kato, N. Shibayama, S., and Munakata, K.,** Structure of the diterpene clerodendrin A., *Chem. Commun.,* 1632, 1971.

158. **Sim, G. A., Hamor, T. A., Paul, I. C., and Robertson, J. M.,** The structure of clerodin, *Proc. Chem. Soc.,* 75, 1961.

159. **Paul, I. C., Sim, G. A., Hamor, T. A., and Robertson, J. M.,** The structure of clerodin: X-ray analysis of clerodin bromo-lactone, *J. Chem. Soc.,* 4133, 1962.

160. **Barton, D. H. R., Cheung, H. T., Cross, A. D., Jackman, L. M., and Martin-Smith, M.,** Diterpenoid bitter principles. Part III. The constitution of clerodin, *J. Chem. Soc.,* 5061, 1961.

161. **Wada, K., Marumo, S., and Munakata, K.,** An insecticidal alkaloid, cocculolidine from *Cocculus trilobus* DC, *Tetrahedron Lett.,* 5179, 1966.

162. **Wada, K., Marumo, S., and Munakata, K.,** An insecticidal alkaloid, cocculolidine from *Cocculus trilobus* DC, *Agric. Biol. Chem.,* 31, 452, 1967.

163. **Bickoff, E. M., Lyman, R. L., Livingston, A. L., and Booth, A. N.,** Characterization of coumestrol, a naturally occurring plant estrogen, *J. Am. Chem. Soc.,* 80, 3969, 1958.

164. **Kupchan, S. M., Hemingway, R. J., and Smith, R. M.,** Tumor inhibitors. XLV. Crotepoxide, a novel cyclohexane diepoxide tumor inhibitor from *Croton macrostachys, J. Org. Chem.,* 34, 3898, 1969.

165. **Lavie, D., Shvo, Y., and Willner, D.,** Interrelationships in the cucurbitacin series, *Chem. Ind. (London),* 951, 1959.

166. **de Koch, W. T., Enslin, P. R., Norton, K. B., Barton, D. H. R., Sklarz, B., and Bothner-By, A. A.,** The constitutions of the cucurbitacins, *J. Chem. Soc.,* 3828, 1963.

167. **Tschesche, R. and Wulff, G.,** Über Saponine der Spirostanolreihe—IX Die Konstitution des Digitonins, *Tetrahedron,* 19, 621, 1963.

168. **Terasaka, M.,** Alkaloids of the root-bark of *Oxira japonica* Thunb. IX. The structure of orixine, *Chem. Pharm. Bull.,* 8, 523, 1960.

169. **Brokke, M. E. and Christensen, B. E.,** Psoralene I: Certain reactions of xanthotoxin, *J. Org. Chem.,* 23, 589, 1958.

170. **Ascher, K. R. S.,** Insect pest control by chemosterilants and antifeedants — Magdegurg 1966 to Milan 1969, *World Rev. Pest Control,* 9, 140, 1971.

171. **Marker, R. E., Tsukamoto, T., and Turner, D. L.,** Sterols. C. Diosgenin, *J. Am. Chem. Soc.,* 62, 2525, 1940.

172. **Hosozawa, S., Kato, N., and Munakata, K.,** Diterpenoids from *Clerodendron calamitosum, Phytochemistry,* 13, 308, 1974.

173. **Matsui, K., Wada, K., and Munakata, K.,** Insect antifeeding substances in *Parabenzoin praecox* and *Piper futokadzura, Agric. Biol. Chem.,* 40, 1045, 1976.

174. **Hughes, G. K., Neill, K. G., and Ritcher, E.,** Alkaloids of the Australian Rutaceae: *Evodia xanthoxyloides* F. Meull., *Aust. J. Sci. Res. Ser. A2,* 401, 1952.

175. **Brownlie, G., Spring, F. S., Stevenson, R., and Strachan, W. S.,** Friedelin and cerin, *Chem. Ind. (London),* 1156, 1955.

176. **Kane, V. V. and Stevenson, R.,** Friedelin and related compounds—V The action of N-bromosuccinimide on friedelin and derivatives, *Tetrahedron* 15, 223, 1961.

177. **Stevenson, R.,** Friedelin and related copounds. VI. Azahomofriedelanes, *J. Org. Chem.,* 28, 188, 1963.

178. **Wada, K. and Munakata, K.,** Naturally occurring insect control chemicals, isoboldine, a feeding inhibitor, and cocculolidine, an insecticide in the leaves of *Cocculus trilobus* DC, *J. Agric. Food Chem.,* 16, 471, 1968.

179. **Pyrek, J. St.,** New pentacyclic diterpene acid trachyloban-19-oic acid from sunflower, *Tetrahedron,* 26, 5029, 1970.

180. **Ourisson, G., Munavalli, S., and Ehret, C.,** *Selected Constants Sesquiterpenoids,* Pergamon Press, New York, 1966.

181. **Virtanen, A. I., Hietala, K., and Wahlroos, O.,** An anti-fungal factor in maize and wheat plants, *Suom. Kemistil.,* 29B, 143, 1956.

182. **Smissman, E. E., LaPidus, J. B., and Beck, S. D.,** Corn plant resistance factor, *J. Org. Chem.,* 22, 220, 1957.

183. **Smissman, E. E., LaPidus, J. B., and Beck, S. D.,** Isolation and synthesis of an insect resistance factor from corn plants, *J. Am. Chem. Soc.,* 79, 4697, 1957.

184. **Brown, R. F. C., Hobbs, J. J., Hughes, G. K., and Ritchie, E.,** The chemical constituents of Australian *Flindersia* species VI. The structure and chemistry of flindersine, *Aust. J. Chem.,* 7, 348, 1954.

185. **Hostettmann, K., Pettei, M. J., Kubo, I., and Nakanishi, K.,** Direct obtention of pure compounds from crude plant extracts by preparative liquid chromatography, *Helv. Chim. Acta,* 60, 670, 1977.

186. **Lichtenstein, E. P. and Casida, J. E.,** Naturally occurring insecticides myristicin, an insecticide and synergist occurring naturally in the edible parts of parsnips, *J. Agric. Food Chem.,* 11, 410, 1963.

187. **McElvain, S. M. and Eisenbraun, E. J.,** The constituents of the volatile oil of catnip. III. The structure of nepetalic acid and related compounds, *J. Am. Chem. Soc.,* 77, 1599, 1955.

188. **Herz, W., Watanabe, H., Miyazaki, M., and Kishida, Y.,** The structures of parthenin and ambrosin, *J. Am. Chem. Soc.,* 84, 2601, 1962.

189. **Seshadri, T. R.,** Biochemistry of natural pigments (exclusive of haeme pigments and carotenoids), *Ann. Rev. Biochem.,* 20, 487, 1951.

190. **Djerassi, C., Brewer, H. W., Clarke, C., and Durham, L. J.,** Alkaloid studies. XXXVIII. Pilocereine — a trimeric cactus alkaloid, *J. Am. Chem. Soc.,* 84, 3210, 1962.

191. **Matsui, K.,** The structure of piperenone, a new insect antifeeding substance from *Piper futokadzura*, *Tetrahedron Lett.*, 1905, 1975.

192. **Matsui, K. and Munakata, K.,** The structure of piperenone, *Agric. Biol. Chem.*, 40, 1113, 1976.

193. **Matsui, K., Fukuyama, K., Tsukihara, K., Tsukihara, T., Katsube, Y., and Munakata, K.,** A structure analysis of a bomoacetate of hydroxypiperenone by the X-ray diffraction method, *Bull. Chem. Soc. Jpn,* 49, 62, 1976.

194. **Barnes, C. S. and Loder, J. W.,** The structure of polygodial: a new sesquiterpene dialdehyde from *Polygonum hydropiper* L., *Aust. J. Chem.*, 15, 322, 1962.

195. Waters Associates, Inc. Application Highlights 3 - Steroids/Insect Molting

196. **Underhill, E. W., Watkin, K. E., and Neish, A. C.,** Biosynthesis of quercetin in buckwheat, *Can. J. Biochem. Physiol.*, 35, 219, 1957.

197. **Henderson, R., McCrindle, R., and Overton, K. H.,** Salannin, *Tetrahedron Lett.*, 3969, 1964.

198. **Henderson, R., McCrindle, R., Melera, A., and Overton, K. H.,** Tetranortriterpenoids—IX The constitution and stereochemistry of salannin, *Tetrahedron*, 24, 1525, 1968.

199. **de Silva, L. B., Stöcklin, W., and Geissman, T. A.,** The isolation of salannin from *Melia dubia*, *Phytochemistry*, 8, 1817, 1969.

200. **Ingham, J. L. and Millar, R. L.,** Sativan: an induced isoflavin from the leaves of *Medicago sativa* L., *Nature (London)*, 242, 125, 1973.

201. **Adams, R. and Govindachari, T. R.,** Senecio alkaloids: the isolation of senecionine from *Senecio cineraria* and some observations on the structure of senecionine, *J. Am. Chem. Soc.*, 71, 1953, 1949.

202. **Wada, K., Enomoto, Y., and Munakata, K.,** Insect feeding inhibitors in plants. Part II. The structures of shiromodiol diacetate, shiromool, and shiromodiol-monoacetate, *Agric. Biol. Chem.*, 34, 946, 1970.

203. **Reichstein, T.,** Besonderheiten der Zucker von herzaktiven Glykosiden, *Angew. Chem.*, 74, 887, 1962.

204. **Brooks, C. J. W. and Draffan, G. H.,** Sesquiterpenoids of *Warburgia* species—II Ugandensolide and ugandensidial (cinnamodial), *Tetrahedron*, 25, 2887, 1969.

205. **Geissman, T. A., Knaack, Jr., W. F., and Knight, J. O.,** Unedoside, a novel iridoid compound, *Tetrahedron Lett.*, 1245, 1966.

206. **Lavie, D., Glotter, E., and Shvo, Y.,** Constituents of *Withania somnifera* Dun. Part IV. The structure of withaferin A, *J. Chem. Soc.*, 7517, 1965.

207. **Kupchan, S. M., Anderson, W. K., Bollinger, P., Doskotch, R. W., Smith, R. M., Renauld, J. A. S., Schnoes, H. K., Burlingame, A. L., and Smith, D. H.,** Tumor inhibitors. XXXIX. Active principles of *Acnistus arborescens*. Isolation and structural and spectral studies of withaferin A and withacnistin, *J. Org. Chem.*, 34, 3858, 1969.

208. **Stermitz, F. R. and Sharifi,** Alkaloids of *Zanthoxylum monophyllum* and *Z. punctatum*, *Phytochemistry*, 16, 2003, 1977.

REFERENCES TO PART B

209. **Begley, M. J. and Grove, J. F.,** Metabolic products of *Phomopsis oblonga*. Part I. 3a,5a,6,7,8,9,9a,9b-Octahydro-7,9b-dimelthylnaphtho(1,2-c)furan-1(3H)-one, *J. Chem. Soc. Perkin Trans.* 1, 861, 1985.

210. **Saleh, M. A., Abdel-Moein, N. M. and Ibrahim, N. A.,** Insect antifeeding azulene derivative from the brown alga *Dictyota dichotoma*, *J. Agric. Food Chem.*, 32, 1432, 1984.

211. **Cardellina, J. H., Raub, M. F., and Van Wagenen, B. C.,** Plant growth regulators and insect control agents from marine organisms, *ACS Symp. Ser.*, 330, 562, 1987.

212. **Dawson, G. W., Griffiths, D. C., Hassanali, A., Pickett, J. A., Plumb, R. T., Pye, B. J., Smart, L. E., and Woodcock, C. M.,** Antifeedants, a new concept for control of barley yellow dwarf virus in winter cereals, *Proc. Br. Crop Prot. Conf. Pests Dis.*, 1001, 1986.

213. **Schmutterer, H., Ascher, K. R. S., and Rembold, H., Eds.** Natural pesticides from the Neem tree, in *Proc. 1st Int. Neem Conf.*, GTZ, Eschborn, W. Germany, 1981.

214. **Schmutterer, H. and Ascher, K. R. S., (Eds.)** Natural pesticides from the Neem tree and other tropical plants, *Proc. 2nd Int. Neem Conf.*, GTZ, Eschborn, W. Germany, 1984.

215. **Schmutterer, H.** Insect growth disrupting and fecundity-reducing ingredients from the Neem and Chinaberry trees, in *Handbook of Natural Pesticides*, Vol. III, Morgan, E. D. and Mandava, N. B., Eds., CRC Press, Boca Raton, FL, 1987, Chap 5.

216. **Jacobson, M., Ed.,** *Focus on Phytochemical Pesticides*, Vol. 1, *The Neem Tree*, CRC Press, Boca Raton, FL, 1989.

217. **Broughton, H. B., Ley, S. V., Slawin, A. M. Z., Williams, D. J., and Morgan, E. D.,** X-ray crystallographic structure determination of detigloyldihydroazadirachtin and reassignment of the structure of the limonoid insect antifeedant azadirachtin, *J. Chem. Soc., Chem. Commun.*, 46, 1986.

218. **Kraus, W., Bokel, M., Klenk, A., and Pöhnl, H.,** The structure of azadirachtin and 22,23-dihydro-23β-methoxyazadirachtin, *Tetrahedron Lett.*, 26, 6435, 1985.

219. **Bilton, J. N., Broughton, H. B., Jones, P. S., Ley, S. V., Lidert, Z., Morgan, E. D., Rzepa, H. S., Sheppard, R. N., Slawin, A. M. Z., and Williams, D. J.,** An X-ray crystallographic, mass spectroscopic and NMR study of the limonoid insect antifeedant azadirachtin and related derivatives, *Tetrahedron,* 43, 2805, 1987.

220. **Kraus, W., Bokel, M., Bruhn, A., Cramer, R., Klaiber, I., Klenk, A., Nagl, G., Pöhnl, H., Sadlo, H., and Vogler, B.,** Structure determination by NMR of azadirachtin and related compounds from *Azadirachta indica* A. Juss (Meliaceae), *Tetrahedron,* 43, 2817, 1987.

221. **Turner, C. J., Tempesta, M. S., Taylor, R. B., Zagorski, M. G., Termini, J. S., Schroeder, D. R. and Nakanishi, K.,** An NMR spectroscopic study of azadirachtin and its trimethyl ether, *Tetrahedron,* 43, 2789, 1987.

222. **Buckingham, J., Ed.** *Dictionary of Organic Compounds,* Chapman and Hall, New York, 1982 and annual supplementary volumes.

223. **Lwande, W., Hassanali, A., Njoroge, P. W., Bentley, M. D., delle Monache, F., and Jondiko, J. I.,** A new 6a-hydroxypterocarpan with insect antifeedent and antifungal properties from the roots of *Tephrosia hildebrandtii* Vatke, *Insect Sci. Its Appl.,* 6, 537, 1985.

224. **Lidert, Z., Taylor, D. A. H., and Thirugnanam, M.,** Insect antifeedant activity of four prieurianin-type liminoids, *J. Nat. Prod.,* 48, 843, 1985.

225. **Kubo, I., and Matsumoto, A.,** Isolation of an insect-antifeedent phloretin 4-0-β-D-glucopyranoside, by rotation locular counter current chromatography and determination of its preferred conformation in solution by nuclear magnetic resonance analysis, *Chem. Pharm. Bull.,* 33, 3817, 1985.

226. **Nakatami, M., Takas, H., Miura, I., and Hase, T.,** Azedarachol, a steroid ester antifeedant from *Melia azedarach* var. *japonica, Phytochemistry,* 24, 1945, 1985.

227. **Hassanali, A., Lwande, W., and Gebreyesus, T.,** Structure activity studies of acridone feeding deterrents, in *Proc. 2nd Int. Neem Conf.,* Schriftenr. GTZ, 161, 75, 1984.

228. **Isman, M. B. and Proksch, P.,** Deterrent and insecticidal chromenes and benzofurans from *Encelia* (Asteraceae), *Phytochemistry,* 24, 1949, 1985.

229. **Dreyer, D. L., Jones, K. C. and Molyneux, R. J.,** Feeding deterrency of some pyrrolizidine, indolizidine and quinolizidine alkaloids towards pea aphids *(Acyrthosiphon pisum)* and evidence for phloem transport of indolizidine alkaloid swainsonine, *J. Chem. Ecol.,* 11, 1045, 1985.

230. **Klocke, J. A., Arisawa, M., Handa, S. S., Kinghorn, A. D., Cordell, G. A., and Farnsworth, N. R.,** Growth inhibitory insecticidal and antifeedant effects of some antileukemic and cytotoxic quassinoids on two species of agricultural pests, *Experientia,* 41, 379, 1985.

231. **Kubo, I. and Matsumoto, T.,** Abyssinin, a potent insect antifeedant from an African medicinal plant *Bersama abyssinica, Tetrahedron Lett.,* 25, 4601, 1984.

232. **Lwande, W., Gebreyesus, T., Chapya, A., McFoy, C., Hassanali, A., and Okech, M.,** 9-Acridone insect antifeedant alkaloids from *Teclea trichocarpa* bark, *Insect Sci. Its Appl.,* 4, 393, 1983.

233. **Bordoloi, M. J., Shukla, V. S., and Sharma, R. R.,** Absolute stereochemistry of the insect antifeedant cadinene from *Eupatorium adenophorum, Tetrahedron Lett.,* 26, 509, 1985.

234. **Corcuera, L. J., Queirolo, C. B., and Argandoma, V. H.,** Effects of 2-β-D-glucosyl-4-hydroxy-7-methoxy-1,4-benzoxazin-3-one on *Schizaphis graminum* (Rondani) (Insecta, Aphididae) feeding on artificial diets, *Experientia.,* 41, 514, 1985.

235. **Matsuda, K. and Senbo, S.,** Chlorogenic acid as a feeding deterrent for the Salicaceae-feeding leaf beetle, *Lochmaeae capreae cribrata* (Coleoptera Chrysomelidae) and other species of leaf beetles, *Appl. Entomol. Zool.,* 21, 411, 1986.

236. **Blaney, W. M., Simmonds, M. S. J., Evans, S. V., and Fellows, L. E.,** The role of the secondary plant compound 2,5-dihydroxymethyl-3,4-dihydroxy pyrrolidine as a feeding inhibitor for insects, *Entomol. Exp. Appl.* 36, 209, 1984.

237. **Vandewalle, M. E.,** Structure of specionin isolated from leaves of the tree *Catalpa speciosa* Warder, Belg. BE 901, 166, *Chem. Abstr.* 103, 33510, 1985.

238. **Zuniga, G. E., Salgada, M. S., and Corcuera, L. J.,** Role of an indole alkaloid in the resistance of barley seedlings to aphids, *Phytochemistry,* 24, 945, 1985.

239. **Miles, D. H., Hankinson, B. L., and Randle, S. A.,** Insect antifeedants from the Peruvian plant *Alchornea triplinervia, ACS Symp. Ser.* 276, 469, 1985.

240. **Reena, C., Afzal, J., Das, K. G., and Sukumazr, K.,** 6,7-Dimethoxyisochroman-3-one: an antifeedant for castor semilooper *Achoea janata* L. *Indian J. Farm Chem.* 1, 63, 1983.

241. **Nawrot, J.,** Principles for grain weevil *(Sitophilus granarius* L. *(Coleoptera: Curculionidae)* control with use of natural chemical compounds affecting the behaviour of beetles, *Pr. Nauk Inst. Ochr. Rosl.* 24, 173, 1983.

242. **Nawrot, J., Bloszyk, E., Grabarczyk, H., Drozdz, B., Daniewski, W. M., and Holub, M.,** Sesquiterpene lactones. Part 28. Further evaluation of feeding deterrency of sesquiterpene lactones to storage pests, *Pr. Nauk Inst. Ochr. Rosl.* 25, 91, 1983.

243. **Nakatani, M., Iwashita, T., Naoki, H., and Hase, T.,** Structure of a limonoid antifeedant from *Trichilia roka, Phytochemistry,* 24, 195, 1985.

244. **Russell, G. B., Shaw, G. J., Christmas, P. E., Yates, M. B., and Sutherland, O. R. W.,** Two 2-aryl benzofurans as insect feeding deterrents from Sainfoin *(Onobrychis viciifolia), Phytochemistry,* 23, 1417, 1984.

245. **Leskinen, V., Polonsky, J., and Bhatnagar, S.,** Antifeedant activity of quassinoids, *J. Chem. Ecol.,* 10, 1497, 1984.

246. **Kubo, I., Matsumoto, T., Tori, M., and Asakawa, Y.,** Structure of Plectrin, an aphid antifeedant diterpene from *Plectranthus barbatus, Chem. Lett.,* 1513, 1984.

247. **Kraus, W. and Bokel, M.,** New tetranortriterpenoids from *Melia azedarach* Linn. (Meliaceae), *Chem. Ber.,* 114, 267, 1981.

248. **Bernays, E. A., Lupi, A., Bettolo, R. M., Mastrofrancesco, C., and Tagliatesta, P.,** Antifeedant nature of the quinone primin and its quinol miconidin from *Miconia* spp., *Experientia,* 40, 1010, 1984.

249. **Bentley, M. D., Leonard, D. E., Reynolds, E. K., Leach, S., Beck, A. B., and Murakoshi, I.,** Lupine alkaloids as larval feeding deterrents for spruce budworm, *Choristoneura fumiferana* (Lepidoptera: Tortricidae), *Ann. Entomol. Soc. Am.,* 77, 398, 1984.

250. **Bentley, M. D., Leonard, D. E., and Bushway, R. J.,** Solanum alkaloids as larval feeding deterrents for spruce budworm, *Choristoneura fumiferana* (Lepidoptera: Tortricidae), *Ann. Entomol. Soc. Am.,* 77, 401, 1984.

251. **Numata, A., Hokimoto, K., Takemura, T., Katsuno, T., and Yamamoto, K.,** Plant constituents biologically active to insects. V. Antifeedants for the larvae of the yellow butterfly, *Eurema hecabe mandarina,* in *Osmunda japonica, Chem. Pharm. Bull.* 32, 2815, 1984.

252. **Bentley, M. D., Leonard, D. E., Stoddard, W. F., and Zalkow, L. H.,** Pyrrolizidine alkaloids as larval feeding deterrents for spruce budworm, *Choristoneura fumiferana* (Lepidoptera: Tortricidae), *Ann. Entomol. Soc. Am.,* 77, 393, 1984.

253. **Schuh, B. A. and Benjamin, D. M.,** Evaluation of commercial resin acids as feeding deterrents against *Neodiprion dubiosus, N. lecontei,* and *N. rugifrons, J. Econ. Entomol.,* 77, 802, 1984.

254. **Della Monache, F., Bettolo, G. B. M., and Bernays, E. A.,** Isolation of insect antifeedant alkaloids from *Maytenus rigida* (Celastraceae), *Z. Angew. Entomol.* 94, 406, 1984.

255. **Champagne, D. E., Arnason, J. T., Philogene, B. J. R., Campbell, I. G., and McLachlan, D. G.,** Photosensitization and feeding deterrence of *Euxoa messoria* (Lepidoptera: Noctuiidae) by α-terthienyl, a naturally occurring thiophene from the Asteraceae, *Experientia,* 40, 577, 1984.

256. **Nawrot, J., Harmatha, J., and Novotny, L.,** On terpenes, Part 277. Insect feeding deterrent activity of bisabolangelone and of some sesquiterpenes of eremophilane type, *Biochem. Syst. Ecol.,* 12, 99, 1984.

257. **Harmatha, J. and Nawrot, J.,** On terpenes. Part 276. Comparison of the feeding deterrent activity of some sesquiterpene lactones and a lignan lactone towards selected insect storage pests, *Biochem. Syst. Ecol.,* 12, 95, 1984.

258. **Corcuera, L. J.,** Effects of indole alkaloids from Gramineae on aphids, *Phytochemistry,* 23, 539, 1984.

259. **Numata, A., Katsuno, T., Yamamoto, K., Nishida, T., Takemura, T., and Seto, K.,** Plant constituents biologically active to insects. IV. Antifeedants for the larvae of the yellow butterfly, *Eurema hecabe mandarina* in *Arachnoides standishii, Chem. Pharm. Bull.,* 32, 325, 1984.

260. **Zuniga, G. E., Argandona, V. H., Niemeyer, H. M., and Corcuera, L. J.,** Hydroxamic acid content of wild and cultivated Graminae, *Phytochemistry,* 22, 2665, 1983.

261. **Ganjian, I., Kubo, I., and Fludzinski, P.,** Insect antifeedant elemanolide lactones from *Veronia arygolalina Phytochemistry,* 22, 2525, 1983.

262. **Altieri, M. A., Lippmann, M., Schmidt, L. L., and Kubo, I.,** Antifeedant effects of nomilin on *Spodoptera frugiperda* (J. E. Smith) (Lepidoptera: Noctuidae) under laboratory and lathehouse conditions, *Prot. Ecol.,* 6, 91, 1984.

263. **Barua, N. C., Barua, P., Goswani, A., Sharma, R. P., and Barua, J. N.,** Synthesis of an insect antifeedant chalcone, *Chem. Ind. (London),* 900, 1983.

264. **Smith, C. M., Kester, K. M., and Fischer, N. H.,** Insect allelochemic effects of sesquiterpene lactones from Melampodium, *Biochem. Syst. Ecol.,* 11, 377, 1983.

265. **Chang, C. C., and Nakanishi, K.,** Specionin, an iridoid insect antifeedant from *Catalpa speciosa, J. Chem. Soc., Chem. Commun.,* 605, 1983.

266. **Numata, A., Takemura, T., Ohbayashi, H., Katsuno, T., Yamamoto, K., Sata, K., and Kobayashi, S.,** Antifeedants for the larvae of the yellow butterfly *Eurema hecabe mandarina* in *Lycoris radiata, Chem. Pharm. Bull.,* 31, 2146, 1983.

267. **Reed, D. K., Kwolek, W. F., and Smith, C. R.,** Investigation of antifeedant and other insecticidal activities of trewiasine towards the striped cucumber beetle and codling moth, *J. Econ. Entomol.,* 76, 641, 1983.

268. **Numata, A., Hokimoto, K., Takemura, T., and Fukui, S.,** Feeding inhibitors for the larvae of the yellow butterfly *Eurema hecabe mandarina* de l'Orza (Lepidoptera: Pieridae) in a flowering fern, *Osmunda japonica* Thunb., *Appl. Entomol. Zool.,* 18, 129, 1983.

269. **Gueskens, R. B. M., Luteijn, J. M., and Schoonhoven, L. M.**, Antifeedant activity of some ajugarin derivatives in three lepidopterous species, *Experientia*, 39, 403, 1983.

270. **Nakanishi, K., Cooper, R., and Nakatani, M.**, Isolation and structures of two insect antifeedants: applications of droplet counter current chromatography, 2-D-H, NMR and a new circular dichroic correlation, *Pr. Nauk Inst. Chem. Org. Fiz. Politech. Wroclaw.*, 1091, 1981.

271. **Koul, O.**, Feeding deterrence induced by plant limonoids in the larvae of *Spodoptera litura* (F.) (Lepidoptera Noctuidae). *Z. Angew. Entomol.*, 95, 166, 1983.

272. **Yano, K.**, Insect antifeeding phenylacetylenes from growing buds of *Artemisia capillaris*, *J. Agric. Food Chem.*, 31, 667, 1983.

273. **Klocke, J. A., and Kubo, I.**, Citrus limonoid by-products as insect control agents, *Entomol. Exp. Appl.* 32, 299, 1982.

274. **Kaethler, F., Pree, D. J., and Brown, A. W.**, Hydrogen cyanide; a feeding deterrent in peach to the oblique-bonded leafroller, *Choristoneura rosaceana* (Lepidoptera: Tortricidae). *Ann. Entomol. Soc. Am.*, 75, 568, 1982.

275. **McLachlan, D., Arnason, J. T., Philogene, B. J., and Champagne, D.**, Antifeedant activity of the polyacetylene, phenylheptatriene (PHT), from the Asteraceae to *Euxoa messoria* (Lepidoptera: Noctuidae), *Experientia*, 38, 1061, 1982.

276. **Azambuja, P. D., Bowers, W. S., Ribeiro, J. M. C., and Garcia, E. S.**, Antifeedant activity of precocenes and analogs on *Rhodnius prolixus*, *Experientia*, 38, 1054, 1982.

277. **Stephenson, A. G.**, Iridoid glycosides in the nectar of *Catalpa speciosa* are unpalatable to nectar thieves, *J. Chem. Ecol.*, 8, 1025, 1982.

278. **Hanson, J. R., Rivett, D. E. A., and Ley, S. V.**, The X-ray structure and absolute configuration of insect antifeedant clerodane diterpenoids from *Teucrium africanum*, *J. Chem. Soc., Perkin Trans.*, 1, 1005, 1982.

279. **Odjo, A., Piart, J., Polonsky, J., and Roth, M.**, Study of the insecticidal effect of two quassinoids on the larvae of *Locusta migratoroides* R and F (Orthoptera: Acridae) *C. R. Seances Acad. Sci. Ser.* 3, 293, 241, 1981.

280. **Liu, H. W., Kusumi, T., and Nakanishi, K.**, A hydroperoxy-chroman with insect antifeedant properties from an African shrub. Characterisation of fully substituted aromatic structures, *J. Chem. Soc. Chem. Commun.*, 1271, 1981.

281. **Liu, H. W., Kubo, I., and Nakanishi, K.**, A southern armyworm antifeedant, 12β-acetoxyharrisonin, from an African shrub *Harrisonia abyssinica*, *Heterocycles*, 17, 67, 1982.

282. **Ascher, K. R. S., Schmutterer, H., Glotter, E., and Kirson, I.**, Withanolides and related ergostane-type steroids as antifeedants for larvae of *Epilachna varivestis* (Coleoptera: Chrysomelidae), *Phytoparasitica*, 9, 197, 1981.

283. **Bernays, E. and De Luca, C.**, Insect antifeedant properties of an iridoid glycoside: ipolamiide, *Experientia*, 37, 1289, 1981.

284. **Ohigashi, H., Wagner, M. R., Matsumura, F., and Benjamin, D. M.**, Chemical basis of differential feeding behaviour of the larch sawfly, *Pristiphora erichsonii* (Hartig), *J. Chem. Ecol.*, 7, 599, 1981.

285. **Nakatani, M., Jones, J. C., and Nakanishi, K.**, Isolation and structures of trichilins, antifeedants against the Southern armyworm, *J. Am. Chem. Soc.*, 103, 1228, 1981.

286. **Kraus, W., and Cramer, R.**, Novel tetranortriterpenoids with insect antifeeding activity from Neem oil, *Liebigs Ann. Chem.*, 181, 1981.

287. **Asakawa, Y., Toyota, M., Takemoto, T., Kubo, I., and Nakanishi, K.**, Insect antifeedant secoaro-madendrone-type sesquiterpenes from *Plagiochila* species, *Phytochemistry*, 19, 2147, 1980.

288. **Alfaro, R. I., Pierce, H. D., Borden, J. H., and Oehlschlager, A. C.**, Insect feeding and oviposition deterrents from western red cedar foliage, *J. Chem. Ecol.*, 7, 39, 1981.

289. **Kraus, W. and Grimminger, W.**, 23-(R,S)-Hydroxytoonacilid and 21-(R,S)-hydroxytoonacilid, two new B-seco-tetranortriterpenoids with feeding inhibitor activity against insects isolated from *Toona ciliata* M. J. Roem. Var *australis* (Meliaceae), *Nouv. J. Chim.*, 4, 651, 1980.

290. **Proksch, P., Proksch, M., Towers, G. H. N., and Rodriguez, E.**, Phototoxic and insecticidal activities of chromenes and benzofurans from Encelia, *J. Nat. Prod.*, 46, 331, 1983.

291. **El-Naggar, S. F., Doskotch, R. W., O'Dell, T. M., and Girard, L.**, Antifeedant diterpenes for the gypsy moth larvae from *Kalmia latifolia:* isolation and characterisation of ten grayanoids, *J. Nat. Prod.*, 43, 617, 1980.

292. **Ascher, K. R. S., Nemny, N. E., Eliyahu, M., Kirson, I., Abraham, A., and Glotter, E.**, Insect antifeedant properties of withanolides and related steroids from Solanaceae, *Experientia*, 36, 998, 1980.

293. **Sutherland, O. R. W., Russell, G. B., Biggs, D. R., and Lane, G. A.**, Insect feeding deterrent activity of phytoalexin isoflavanoids, *Biochem. Syst. Ecol.*, 8, 73, 1980.

294. **Nakatani, M., Okamoto, M., Iwashita, T., Mizukawa, A. K., Naoki, H., and Hase, T.**, Isolation and structure of three seco-limonoids, insect antifeedants from *Trichilia roka* (Meliaceae). *Heterocycles*, 22, 2335, 1984.

295. **Begley, M. J., Crombie, L., Crombie, W. M. L., Gatuma, A. K., and Maradufu, A.,** Germacranolides of *Erlangea cordifolia*. Isolation and structures of cordifolia-54, -55, -P2 and -31 by spectral and X-ray methods, *J. Chem. Soc., Perkin Trans.*, 1, 819, 1984.

296. **Figliuolo, R., Naylor, S., Wang, J., and Langenheim, J. H.,** Unusual nonprotein iminoacid and its relationship to phenolic and nitrogenous compounds in *Copaifera*, *Phytochemistry*, 26, 3255, 1987.

297. **Arnason, J. T., Isman, M. B., Philogene, B. J. R., and Waddell, T. G.,** Mode of action of the sesquiterpene lactone tenulin, from *Helenium amarum* against herbivorous insects, *J. Nat. Prod.*, 50, 690, 1987.

298. **Blust, M. H. and Hopkins, T. L.,** Gustatory responses of a specialist and a generalist grasshopper to terpenoids of *Artemisia ludoviciana*, *Entomol. Exp. Appl.*, 45, 37, 1987.

299. **Yano, K.,** Minor components from growing buds of *Artemisia capillaris* that act as insect antifeedants, *J. Agric. Food Chem.*, 35, 889, 1987.

300. **Jain, D. C.,** Antifeedant active saponin from *Balanites roxburghii* stem bark, *Phytochemistry*, 26, 2223, 1987.

301. **Nakatani, M., Takao, H., Iwashita, T., Naoki, H., and Hase, T.,** The structure of graucin A, a new bitter limonoid from *Evoda grauca* Miq. (Rutaceae). *Bull. Chem. Soc. Jpn.* 60, 2503, 1987.

302. **Chandravadana, M. V.,** Identification of triterpenoid feeding deterrent of red pumpkin beetles *(Aulacophora foveicollis)* from *Momordica charantia*, *J. Chem. Ecol.*, 13, 1689, 1987.

303. **Jones, K. C. and Klocke, J. A.,** Aphid feeding deterrency of ellagitannins, their phenolic hydrolysis products and related phenolic derivatives, *Entomol. Exp. Appl.*, 44, 229, 1987.

304. **Ascher, K. R. S., Eliyahu, M., Glotter, E., Goldman, A., Kirson, I., Abraham, A., Jacobson, M., and Schmutterer, H.,** The antifeedant effect of some new withanolides on three insect species, *Spodoptera littoralis, Epilachna varivestis*, and *Tribolium castaneum*, *Phytoparasitica*, 15, 15, 1987.

305. **Lidert, Z., Wing, K., Polonsky, J., Imakura, Y., Masayoshi, O., Tani, S., Lin, Y. M., Kiyokawa, H., and Lee, K. H.,** Insect antifeedant and growth inhibitory activity of forty six quassinoids on two species of agricultural pests, *J. Nat. Prod.*, 50, 442, 1987.

306. **Miles, D. H., Ly, A. M., Randle, S. A., Hedin, P. A., and Burks, M. L.,** Alkaloidal insect antifeedants from *Virola calophylla* Warb, *J. Agric. Food Chem.*, 35, 794, 1987.

307. **Powell, R. G., Bajaj, R., and McLaughlin, J. L.,** Bioactive stilbenes of *Scirpus maritimus*, *J. Nat. Prod.*, 50, 293, 1987.

308. **Alford, A. R., Cullen, J. A., Storch, R. H., and Bentley, M. D.,** Antifeedant activity of limonin against the Colorado beetle, (Coleoptera: Chrysomelidae), *J. Econ. Entomol.*, 80, 575, 1987.

309. **Hassanali, A., Bentley, M. D., Slawin, A. M. Z., Williams, D. J., Shepherd, R. N., and Chapya, A. W.,** Pedonin, a spirotetranortriterpenoid insect antifeedant from *Harrisonia abyssinica*, *Phytochemistry*, 26, 573, 1987.

310. **Kraus, W., Klenk, A., Bokel, M., and Vogler, B.,** Tetranortriterpenoid lactams with insect antifeeding activity from *Azadirachta indica* A. Juss (Meliaceae). *Liebigs Ann. Chem.*, 337, 1987.

311. **Kraus, W. and Cramer, R.,** Pentanortriterpenoids from *Azadirachta indica* A. Juss (Meliaceae), *Chem. Ber.*, 114, 2375, 1981.

312. **Kraus, W., Cramer, R., Bokel, M., and Sawitzki, G.,** New insect antifeedants from Azadirachta indica and Melia azedarach, in *Proc. 1st Int. Neem Conf.*, GTZ, Eschborn, W. Germany, 1981, 53.

313. **Matsumoto, T. and Sei, T.,** Antifeedant activities of *Ginkgo biloba* L. components against the larvae of, *Pieris rapae crucivora*, *Agric. Biol. Chem.*, 51, 249, 1987.

314. **Van Beek, T. A. and de Groot, Ae.,** Terpenoid antifeedants. Part I. An overview of terpenoid antifeedants of natural origin, *Recl. Trav. Chim. Pay-Bas*, 105, 513, 1986.

315. **Russell, G. B.,** Phytochemical resources for crop protection, *N.Z. J. Technol.*, 2, 127, 1986.

316. **Mikolajczak, K. L., McLaughlin, J. L., and Rupprecht, J. K.,** Control of pests with annonaceous acetogenins, U.S. Pat. Appl. US 860, 357, 1986.

317. **Nawrot, J., Bloszyk, E., Harmatha, J., Novotny, L., and Drozdz, B.,** Action of antifeedants of plant origin on beetles infesting stored products, *Acta Entomol. Bohemoslov.*, 83, 327, 1986.

318. **Rowan, D. D., Hunt, M. B., and Gaynor, D. L.,** Peramine, a novel insect feeding deterrent from rye grass infected with the endophyte *Acremonium loliae*, *J. Chem. Soc., Chem. Commun.*, 935, 1986.

319. **Saxena, B. P., Tikku, K., Atal, C. K., and Koul, O.,** Insect antifertility and antifeedant allelochemics in *Adhatoda vasica*, *Insect Sci. Its Appl.*, 7, 489, 1986.

320. **Chan, T. H., Zhang, Y. J., Sauriol, F., Thomas, A. W., and Strunz, G. M.,** Studies of iridoid chemistry and spruce budworm *(Choristoneura fumiferana)* antifeedants, *Can. J. Chem.*, 65, 1853, 1987.

321. **Shing, T. K. M.,** Oblongolide: synthesis and absolute configuration, *J. Chem. Soc., Chem. Commun.*, 49, 1986.

322. **Blades, D. and Mitchell, B. K.,** Effects of alkaloids on feeding of *Phormia regina*, *Entomol. Exp. Appl.*, 41, 299, 1986.

323. **Arnason, J. T., Philogene, B. J. R., Donskov, N., Muir, A., and Towers, G. H. N.,** Psilotin, an insect feeding deterrent and growth reducer from *Psilotum nudum*, *Biochem. Syst. Ecol.*, 14, 287, 1986.

324. **Dutta, P., Bhattacharyya, P. R., Rabha, L. C., Bordoloi, D., Barua, N. C., Chowdhury, P. K., Sharma, R. P., and Barua, J. N.,** Feeding deterrents for *Philosamia ricini (Samia cynthia* subsp. *ricini)* from *Tithonia diversifolia, Phytoparasitica,* 14, 77, 1986.

325. **Marini Bettolo, G. B., Marta, M., Pomponi, M., and Bernays, E. A.,** Flavan oxygenated pattern and insect feeding deterrence. *Biochem. Sys. Ecol.,* 14, 249, 1986.

326. **Lane, G. A., Biggs, D. R., Russell, G. B., Sutherland, O. R. W., Williams, E. M., Maindonald, J. H., and Donnell, D. J.,** Isoflavonoid feeding deterrents for *Costelytra zealandica.* Structure-activity relationships, *J. Chem. Ecol.,* 11, 1713, 1985.

327. **Strunz, G. M., Giguere, P., and Thomas, A. W.,** Synthesis of pungenin, a foliar constituent of some spruce species, and investigation of its efficacy as a feeding deterrent for spruce budworm (*Choristoneura fumiferana* (Clem.), *J. Chem. Ecol.,* 12, 251, 1986.

328. **Gnanasunderam, C., and Sutherland, O. R. W.,** Hiptagin and other aliphatic nitro esters in *Lotus pedunculatus, Phytochemistry,* 25, 409, 1986.

329. **Simmonds, M. S. J., Blaney, W. M., Delle Monache, F., Marquina, M-Q., and Marini Bettolo, G. B.,** Insect antifeedant properties of anthranoids from the genus *Vismia, J. Chem. Ecol.,* 11, 1593, 1985.

330. **Rees, S., and Harborne, J. B.,** The role of sesquiterpene lactones and phenolics in the chemical defense of the chicory plant, *Phytochemistry,* 24, 2225, 1985.

331. **Belles, X., Camps, F., Coll, J., and Piulachs, M. D.,** Insect antifeedant activity of clerodane diperpenoids against larvae of *Spodoptera littoralis* (Boisd.) (Lepidoptera), *J. Chem. Ecol.,* 11, 1439, 1985.

332. **de Groot, Ae., and von Beek, T. A.,** Terpenoid antifeedants. Part II. The synthesis of drimane and clerodane insect antifeedants, *Recl. Trav. Chim. Pays-Bas,* 106, 1, 1987.

333. **Hubert, T. D., and Wiemer, D. F.,** Ant-repellent terpenoids from *Melampodium divaricatum, Phytochemistry,* 24, 1197, 1985.

334. **Streibl, M., Nawrot, J., and Herout, V.,** Feeding deterrent activity of enantiomeric isoalantones, *Biochem. Syst. Ecol.,* 11, 381, 1983.

335. **Rose, A. F., Jones, K. C., Haddon, W. F., and Dreyer, D. L.,** Grindelane diterpenoid acids from *Grindelia humilis*: feeding deterrency of diterpene acids towards aphids, *Phytochemistry,* 20, 2249, 1981.

336. **Kubo, I., Matsumoto, A., and Shoolery, J. N.,** Structure of deacetylazadirachtinol, application of 2D ^{1}H-^{1}H and ^{1}H-^{13}C shift correlation spectroscopy, *Tetrahedron Lett.,* 25, 4729, 1984.

337. **Klenk, A., Bokel, M., and Kraus, W.,** 3-Tigloylazadirachtol (Tigloyl = 2-methyl-crotonoyl), an insect growth regulating constituent of *Azadirachta indica, J. Chem. Soc., Chem. Commun.,* 523, 1986.

338. **Kubo, I., Matsumoto, A., Matsumoto, T., and Klocke, J. A.,** New insect ecdysis inhibitory limonoid deacetylazadirachtinol isolated from *Azadirachta indica* (Meliaceae) oil, *Tetrahedron,* 42, 489, 1986.

339. **Blaney, W. M., Simmonds, M. S. J., Ley, S. V., and Katz, R. B.,** An electrophysiological and behavioural study of insect antifeedant properties of natural and synthetic drimane-related compounds, *Physiol. Entomol.,* 12, 281, 1987.

340. **Lane, G. A., Sutherland, O. R. W., and Skipp, R. A.,** Isoflavonoids as insect feeding deterrents and antifungal componenets from root of *Lupinus angustifolius, J. Chem. Ecol.,* 13, 771, 1987.

341. **Ley, S. V.,** Synthesis of some insect antifeedants, *Spec. Publ. R. Soc. Chem.,* 53, 307, 1985.

342. **Wisdom, C., Smiley, J. T., and Rodriguez, E.,** Toxicity and deterrency of sesquiterpene lactones and chromenes to the corn earworm (Lepidoptera: Noctindae), *J. Econ. Entomol.* 76, 993, 1983.

343. **Nanayakkara, N. P. D., Klocke, J. A., Compadre, C. M., Hussain, R. A., Pezzuto, J. M., and Kinghorn, A. D.,** Characterization and feeding deterrent effects on the aphid, *Schizaphis graminum,* of some derivatives of the sweet compounds sterioside and rebaudioside A, *J. Nat. Prod.* 50, 434, 1987.

344. **Lindroth, R. L., Hsia, M. T. S., and Scriber, J. M.,** Characterization of phenolic glycosides from quaking aspen, *Biochem. Syst. Ecol.,* 15, 677, 1987.

345. **Cronquist, A.,** The evolution and classification of flowering plants, Thomas Nelson and Sons, London, 1968.

346. **Willis, J. C.,** *A Dictionary of the Flowering Plants and Ferns,* Airy Shaw, H. K., Ed., 8th ed. Cambridge University Press, London, 1973.

REPELLENTS

Dale M. Norris

Though other environmental factors may repel insects, volatile chemicals apparently are the most common type of repellents. Volatile compounds are repellents when they are allomonic, i.e., when they function to the adaptive advantage of the emitting or releasing individual (the source). Such molecules thus elicit an avoidance behavior by the perceiver. If we consider all olfactory chemical communications in the whole of chemical ecology, most, if not all, volatile compounds are repellents to some species or biotypes, but also are attractants to others. This situation, therefore, indicates that we should identify the perceiving species or biotypes when discussing a specific repellency.

Repellents, in contrast to deterrents, may prevent the perceiving organism from even contacting the emitter or releaser. They commonly are the dominant factors among those responsible for a given plant or animal species (including humans) being a nonhost to one or more phytophagous, hematophagous, or carnivorous insects. Repellents thus are major constituents in the natural defense systems of animals or plants. Repellency is an especially common trait in evolved situations where even limited contact between the emitter and perceiver may be detrimental to the emitter.

In this chapter natural molecules which function as repellents to insects are discussed. The intended purpose is to provide a highly representative, if not exhaustive, exposure to repellents.

THE VOLATILE REALM OF CHEMICAL ECOLOGY

Chemical messengers (ligands) which affect insect behavior through their chemosensory system may be conveniently classified as olfactory (i.e., volatile) vs. taste (i.e., contact) substances. According to such classification, repellents especially affect insect behavior through the olfactory mode. However, volatile compounds also affect behavior through contact (i.e., taste) receptors. As an example, the following series of lignin-degradation products[1] have the indicated order of volatile (olfactory) action: syringaldehyde > vanillin > *p*-hydroxybenzaldehyde; their relative effectiveness on taste (contact) receptors is the reverse order. This reversed relative effectiveness as volatile vs. contact messengers is determined largely by the relative volatility of the molecules. Volatility thus influences the arrival of messenger ligands at receptors, but after arrival it does not fundamentally determine their interactions with receptors.

ALLOMONIC VS. KAIROMONIC ACTION

Increasing evidence[2-4] supports the premise that each messenger is, at least, a "double agent". Thus, such a compound is likely to be allomonic to some species or biotypes, but kairomonic to others. The fact that messengers usually are "double agents" creates a very important situation in chemical ecology. Assuming that other environmental factors are similar among given ecological niches, then the distribution of chemical messengers may order species or biotype occurrences and abundances.[4] This would occur because each messenger tells some species or biotypes that they have arrived in a favorable ecological niche and others that they are in hostile (e.g., perhaps life-threatening) territory. Such chemically stimulated movements of biotypes or species to favorable habitats and away from unfavorable niches thus will promote order in such distributions and abundances.[4]

ANIMAL-PRODUCED CHEMICALS AS REPELLENTS TO INSECTS

Animals, and especially insects, produce many diverse chemicals which are repellents to insects.[5-7] However, because such repellents have been thoroughly studied in only a relatively few species, our knowledge of these chemicals is still very incomplete.[7] Such repellent chemicals include hydrocarbons, alcohols, aldehydes, ketones, carboxylic acids, quinones, esters, lactones, phenols, etc.

Esteric repellents of other insects are produced very selectively by species in at least five orders (i.e., Isoptera, Hemiptera, Coleoptera, Lepidoptera and Hymenoptera).[7] Such compounds (Table 1) range from C_5 to C_{26}, but about 70% are C_8 to C_{18}. Nearly 50% are acetates. Even numbered, aliphatic acids predominate in the esters.

Aldehydes which are repellents are released by insects in at least five orders (i.e., Isoptera, Orthoptera, Hemiptera, Coleoptera, and Hymenoptera). A striking array of such compounds is produced (Table 1), but most are aliphatics, and nearly one-half have α,β-conjugated structures. About 25% are terpenoids.

Alcohols which are repellent to some insects are produced in at least six insect orders.[7] Nearly two-thirds are C_{10} to C_{20} compounds (Table 1). More than 50% are primary alcohols. About one-fourth are terpenes.

Ketones which are repellent to some insects are produced in at least six insect orders (i.e., Isoptera, Orthoptera, Hemiptera, Coleoptera, Lepidoptera, and Hymenoptera). Known repellent ketones synthesized by insects are C_4 to C_{22} compounds. Methyl and ethyl ketones are common. Aromatic structures are rare.

Hydrocarbons which are repellent (Table 1) to some insects are released by members in at least seven insect orders. More than 25% of these repellents are terpenes which are methyl branched. Although a majority is saturated, a large variety of alkenes is found. About 25% are terminally unsaturated.

Insect-repellent quinones are produced by insects in four orders. Both 1,4-benzoquinones and 1,4-naphthoquinones are involved.[8] All known repellent phenols (Table 1) are monophenols, but *o, m,* and *p*-substitutions of the hydroxyl group are represented. Lactone insect repellents are produced by some phasmids, cockroaches, beetles, ants, and bees. Animal-produced repellents to insects also include inorganics, sulfides, nitro-compounds, furans, alkaloids, and ethers.

PHYTOCHEMICALS AS REPELLENTS TO INSECTS

Most plant families, genera, or species have characteristic odors, e.g., 5-hydroxy-1,4-naphthoquinone (juglone) in Juglandaceae.[2,4] Such chemical traits usually are allomonic to numerous species of insects. Thus, nonhost decisions by insects regarding plants frequently are based especially on repellent and deterrent chemical constituents.[2,4] Most insect-repellent phytochemicals are products of secondary metabolism.[9]

Though the secondary metabolism of all plants yields a diverse array of products, that of more advanced plants (e.g., angiosperms) is particularly varied.[9] Alkaloids and terpenes dominate numerically as repellents to insects. Quinones, e.g., 1,4-benzoquinones and 1,4-naphthoquinones, also are a very important, but numerically smaller, group of repellent plant chemicals.[2,4,8]

Alkaloids, nitrogenous bases, offer a specific example of the evolution in higher plants of secondary compounds as repellents to many insect species (Table 2). They seem to be especially characteristic of angiosperms. Lower plants (e.g., ferns) have nitrogenous components (e.g., cyanogenetic glycosides) which are repellent to some insects, but these chemicals apparently do not include true alkaloids.[9] Nicotine is a classical example of an alkaloidal

TABLE 1
Representative Animal-Produced Repellents of Insects

Esters	Ref.	Esters	Ref.
n-Butyl butyrate	17, 66	*n*-Hexadecyl acetate	26, 28, 32
n-Butyl hexanoate	17, 42	*trans*-2-Hexenyl acetate	65
Citronellyl acetate	28	*t*-2-Hexenyl butyrate	67
trans-2-Decenyl acetate	7	*n*-Hexeyl acetate	7, 17
n-Decyl acetate	29, 31	*n*-Hexeyl butyrate	17, 42
2,3-Dihydrofarnesyl acetate	28	*n*-Hexyl hexanoate	17, 39
2,6-Dimethyl-5-hepten-1-nonanoate	25	Isobutyl isobutyrate	17
2,6-Dimethyl-5-hepten-1-octanoate	25	Methyl anthranilate	7, 49
n-Docosyl acetate	28	2-Methylbutyl butyrate	17
11-Dodecenyl acetate	72	2-Methylbutyl hexanoate	17
n-Dodecyl acetate	29	2-Methylbutyl isobutyrate	17
n-Eicosyl acetate	28	Methyl 3,4-dihydroxy-benzoate	69
Ethyl decanoate	28	Methyl 2,5-dihydroxy-phenyl acetate	69
Ethyl 3,4-dihydroxy-benzoate	69	Methyl *n*-hexadecanoate	26
Ethyl dodecanoate	28	Methyl *p*-hydroxybenzoate	7, 40, 68
Ethyl hexadecadienoate	28	Methyl 8-hydroxy-quinoline-2-	71
Ethyl hexadecatrienoate	28	carboxylate	7, 49
Ethyl hexadecenoate	28	Methyl 6-methyl salicylate	
Ethyl 9-tetradecenoate	26	Methyl salicylate	7
Farnesyl acetate	7, 26, 28, 29,	*n*-Nonyl acetate	29, 70
	31, 35	*n*-Nonyl formate	70
Farnesyl butyrate	7	9-Octadecenyl acetate	28
Farnesyl hexanoate	7	*n*-Octadecyl acetate	26, 28, 32
Farnesyl octanoate	7	*trans*-2-Octenyl acetate	7, 17
Geranyl decanoate	7	*n*-Octyl acetate	7, 42
Geranylgeranyl acetate	26, 32	*n*-Pentadecyl acetate	7, 31
Geranyl hexanoate	7	2-Phenylethyl isobutyrate	7
Geranyl octanoate	7	2-Phenylethyl 2-methyl butyrate	7
9-Hexadecenyl acetate	7, 30, 32	2-Phenylethyl nonanoate	25
2-Phenylethyl octanoate	25	*n*-Tetradecyl acetate	26, 31, 32
5,13-Tetradecadienyl acetate	22, 72	*n*-Tetradecyl butyrate	18
3,5,13-Tetradecatrienyl acetate	22	*n*-Tridecyl acetate	7
4,6,13-Tetradecatrienyl acetate	22	*n*-Tridecyl butyrate	18
Tetradecenyl acetate	30, 32	*n*-Undecyl acetate	29

Aldehydes	Ref.	Aldehydes	Ref.
Acetaldehyde	37	2-Heptenal	20
Benzaldehyde	7, 21	*trans*-4-Heptenal	20
Butanal	39	Hexadecanal	7
2-Butenal	7	*trans*-2-Hexenal	7, 34, 35, 52
2-*n*-Butyl-2-octenal	53	*p*-Hydroxybenzaldehyde	7, 44
Citral	7, 46	Iridodial	7, 47, 56
Citronellal	7, 48, 57	Isobutanal	17
trans-2-Decenal	20, 35	Isogyrinidal	51
2,3-Dihydro-6-*trans*-farnesal	7, 26	2-Methyl butanal	87
2,6-Dimethyl-5-heptenal	26	2-Methylene butanal	87
Dodecanal	49	2-Methylene butanal dimer	35, 87
Dolichodial	36, 55	2-Methylene pentanal	35
2-Ethylcyclopentane-carbaldehyde	21	2-Methylene propanal	87
5-Ethylcyclopent-1-ene carbaldehyde	21	6-Methyl salicyladehyde	21
Farnesal	58	2-Nonenal	20
Geranylcitronellal	26	Octanal	20, 35, 45
Geranylgeranial	26	*trans*-2-Octenal	35
Gyrinidal	7, 50	4-Oxo-*trans*-2-hexenal	7, 17
Heptanal	20	4-Oxo-*tra s*-2-octenal	88

TABLE 1 (continued)
Representative Animal-Produced Repellents of Insects

Aldehydes	Ref.	Aldehydes	Ref.
2-Pentenal	34	5,8-Tetradecadienal	49
2-Propenal	7	Tetradecanal	28, 49
n-Propanal	38	5-Tetradecenal	49
Salicylaldehyde	7, 43, 54		

Alcohols		Alcohols	
Citronellol	28	11-Hexadecen-1-ol	28
1-Decanol	29	1-Hexanol	17, 20
2-Decen-1-ol	18	trans-2-Hexen-1-ol	17, 23, 87
2,3-Dihydro-6-trans-farnesol	28	Isopiperitenol	19
		2-Methyl-1-butanol	87
2,6-Dimethyl-5-hepten-1-ol	7	3-Methyl-1-butanol	19
		2-Methylene-1-butanol	87
1-Dodecanol	26, 29	4-Methyl-3-heptanol	7
Eicosen-1-ol	28	6-Methyl-5-hepten-2-ol	7
3-Ethyl-2-octanol	89	2-Methyl-3-octanol	7
(5-Ethylcyclopent-1-enyl) methanol	21	2-Nonanol	7
		9-Octadecen-1-ol	28
trans-Farnesol	7	1-Octanol	26
Geraniol	7	3-Octanol	7
Geranylcitronellol	28	2-Octen-1-ol	17
Geranylgeraniol	28	1-Pentadecanol	31
2-Heptadecanol	33	2-Pentadecanol	30
2-Heptanol	7, 87	2-Pentanol	87
3-Heptanol	24	2-Phenylethanol	25
1-Hexadecanol	28	Selin-11-en-4-ol	60
7-Hexadecen-1-ol	28	4,6,13-Tetradecadien-1-ol	22
9-Hexadecen-1-ol	28		
		5,13-Tetradecadien-1-ol	22
		1-Tetradecanol	28
		3,5,13-Tetradecatrien-1-ol	22
		Tetradecenol	32
		1-Tridecanol	31
		2-Tridecanol	30
		1-Undecanol	26, 29
		2-Undecanol	30

Ketones		Ketones	
cis-1-Acetyl-2-methylcyclopentane	46	2-Methyl-5-hepten-2-one	61
		6-Methyl-5-hepten-2-one	7
2-Acetyl-3-methyl-cyclopentene	46	4-Methyl-2-hexanone	55
		4-Methyl-3-hexanone	63
o-Aminoacetophenone	7	6-Methyl-3-octanone	7
2-Butanone	37	2-Nonadecanone	26
3-Decanone	63	2-Nonanone	7
4,6-Dimethyl-4-octen-3-one	63	3-Nonanone	7, 20
		1-Nonen-3-one	20
Gryinidione	51	2-Octanone	7
Gyrinidone	46, 51	3-Octanone	7
2-Heptadecanone	7, 26, 30	2-Pentadecanone	7, 59
2-Heptanone	7, 35	2-Pentanone	7
1,15-Hexadecadien-3-one	89	Romallenone	60
3-Hexadecanone	89	3-Tetradecanone	26

TABLE 1 (continued)
Representative Animal-Produced Repellents of Insects

Esters	Ref.	Esters	Ref.
1-Hexadecen-3-one	59, 89	1-Tetradecen-3-one	59, 89
3-Hexanone	7	2-Tridecanone	7, 59, 62
4-Hexen-4-one	7	2,6,6-Trimethyl-cyclohex-2-ene-1,4-	60
Isophorone	60	dione	
2-Methylcyclopentanone	46	2-Undecanone	7, 62
2-Methyl-4-heptanone	64	Verbenone	60
4-Methyl-3-heptanone	7		

Hydrocarbons		Hydrocarbons	
Bishomofarnesene	83	*n*-Hexacosane	90
Camphene	49, 78	*n*-Hexadecane	77
n-Decane	7, 26, 31	Hexadecene	7, 32, 83
Decene	7	Homofarnesene	83
3,5-Dimethyldodecane	80	Limonene	79, 90
3,4-Dimethyltridecane	80	3-Methyldodecane	80, 81
n-Docosane	90	4-Methyldodecane	32
Docosene	7	5-Methyldodecane	81
Dodecadiene	32	6-Methyldodecane	80
n-Dodecane	7, 26, 31, 32, 74	1-Nonadecene	20
		9-Nonadecene	7, 26, 28, 31, 32, 82, 83, 85
1-Dodecene	7, 26, 29, 31, 32		
n-Eicosane	7, 32, 78, 85	*n*-Nonane	29, 31
Eicosene	7, 32	1-Nonene	20, 76
a-Farnesene	7, 29, 31	*n*-Octadecane	7, 31, 32, 78, 83, 85
trans-Farnesene	7, 73		
n-Heneicosane	7, 26, 32, 78, 85	9-Octadecane	7, 32, 83, 85
		n-Pentacosane	90
Heneicosane	7, 32, 86	Pentadecadiene	32
n-Heptacosane	90	*n*-Pentadecane	48, 75
Heptadecadiene	7, 31, 32, 77, 83—86	1-Pentadecene	20
		7-Pentadecene	7, 26, 31, 32
1-Heptadecene	20	*a*-Phellandrene	49, 90
7-Heptadecene	85	*a*-Pinene	49, 78, 90
cis-8-Heptadecene	7, 26, 29, 32		2-Undecene
3-Methyltridecane	7, 31, 32, 81	β-Pinene	49, 787, 90
5-Methyltridecane	7, 32, 80, 81	Sabinene	49
β-Selinene	60	β-Selinene	60
3-Methylundecane	7, 26, 29, 32	α-Terpinene	49, 78
5-Methylundecane	7, 32	Terpinolene	49, 90
6-Methylundecane	80	*n*-Tetracosane	90
Myrcene	49, 90	Tetracosene	7, 32
3-Methylheptadecane	85	*n*-Tetradecane	7, 31, 32, 75, 78
5-Methylheptadecane	85		
4-Methylheptadecane	85	Tetradecene	7, 32
3-Methylhexadecane	81	Toluene	7
4-Methylhexadecane	85	*n*-Tricosane	90
5-Methylhexadecane	81	4,7-Tridecadiene	31, 49
3-Methylnonadecane	85	*n*-Tridecane	7, 26, 29, 31, 32, 49
3-Methylnonane	7, 32		
3-Methylpentadecane	81	1-Tridecane	20, 30, 76
5-Methylpentadecane	32, 81	4-Tridecene	7, 26, 29, 31, 32, 49
6-Methyltetradecane	80		
Nonadecadiene	7, 32	*n*-Undecane	7, 26, 29, 31, 32, 74, 79
n-Nonadecane	7, 26, 31, 32, 78, 85, 20		

TABLE 1 (continued)
Representative Animal-Produced Repellents of Insects

Esters	Ref.	Esters	Ref.
1-Undecene	7, 76		
7, 31, 32			
p-Quinones		**p-Quinones**	
6-Alkylated naphthoquinones	3	2-Methyl-3-methoxy-1,4-	3
1,4-Benzoquinone	3	benzoquinone	
2,3-Dimethyl-1,4-benzoquinone	3	2-Methyl-1,4-napthoquinone	3
5-Hydroxy-1,4-naphtho-quinone	3	6-Methyl-1,4-naphtho-quinone	3
2-Methoxy-6-*n*-pentyl *p*-benzoquinone	3	1,4-Naphthoquinone	3
		Plumbagin	3
		Toluqinone	3
		2,3,5-Trimethyl benzo-quinone	3
Phenols		**Phenols**	
m-Cresol	3	*m*-Ethylphenol	7
o-Cresol	3	*p*-Ethylphenol	7
p-Cresol	3	Phenol	3
Dimethylphenol	70		

TABLE 2
Representative Plant-Produced Repellents of Insects

Esters	Ref.	Esters	Ref.
Decyl acetate	92	Shiromodiol diacetate	93, 94, 95
Dodecyl acetate	92	Shiromodiol monoacetate	93, 94, 95
(Z,Z)-3,13-Octadecadien-1-ol acetate	96	Undecyl acetate	92
Aldehydes		**Aldehydes**	
Benzaldehyde	98	Polygodial	11
Citral	92, 97	Salicyaldehyde	98
3-Keto-(Z)-9-hexadecenal	98	Ugandensidial	11
3-Ketotetradecanal	98	Warburganal	11
3-Keto-13-tetradecenal	98		

Alcohols	Ref.	Alcohols	Ref.
Benyzyl alcohol	91	*p*-Hydroquinone	103
Carvacrol	97	Linalool	102
Catechol	103	Menthol	98
1,8-Cineol	98	(E,S)-Nerolidol	100
Citronellol	97, 98	Phenol	98
p-Cresol	98	Phylorglucinol	103
m-Cresol	98	Pinitol	99
Eugenol	103	Pyrogallol	103
Farnesol	97, 100	Resorcinol	103
Geraniol	97, 98, 100	*trans*-Verbenol	101
Gossypol	98		
Acids		**Acids**	
Abietic	110	Isobutyric	3, 108, 109
Acetic	108, 109	Isopimaric	110

TABLE 2 (continued)
Representative Plant-Produced Repellents of Insects

Esters	Ref.	Esters	Ref.
Benzoic	3	Isovaleric	108, 109
Butyric	108, 109	12-Methoxyabietic	110
Caproic	108, 109	Propionic	108, 109
Caprylic	3		
Dehydroabietic	110	Sandaracopimaric	110
Formic	3	Tiglic	3
p-Hydroxy benzoic	3	Valeric	3

Nitrogen-containing compounds	Ref.	Nitrogen-containing compounds	Ref.
Acetylcholine	3	Jacobine	104
Actinidine	105	Jacoline	104
N-Atropine	3	Jacozine	104
Berberine	106, 107	Mellein	3
Brucine	107	Nicotine	106
Caffeine	107	Papaverine	106
Chinchonine	106	Quinine	107
Cocaine	106	Sanguinarine	106
HCN	3	Senecionine	104
Histamine	3	Seneciphylline	104
5-Hydroxytryptamine	3	Strychnine	106, 107
Integerrmine	104	a-Tomatine	106

Phenols		Phenols	
Aspirin	3	p-Hydroxybenzaldehyde	3
Catechol	3	Mellein	3
m-Cresol	3	Phenol	3
p-Cresol	3	Salicyaldehyde	3
Guaiacol	3	Syringaldehyde	3
p-Hydroquinone	3		

Quinones	Ref.	Quinones	Ref.
6-Alkylated-1,4-naphthoquinones	3	1,4-Naphthoquinone	12
p-Benzoquinone	4	Plumbagin	3
2,3-Dichloro-1,4-naphthoquinone	12	Toluhydroquinone	92
5-Hydroxy-1,4-naphthoquinone	2	Toluquinone	3
2-Methyl-1,4-naphthoquinone	12	2,3,5-Trimethyl benzoquinone	3

Terpenoids		Terpenoids	
Monoterpenes		Verbenone	101
1,8-Cineol	3	Sesquiterpenes	
Citral	3, 92	Dendrolasin	3
Citronellal	3	β-Ionone	3
Citronellol	3	β-Selinene	3
Geraniol	3	Dimeric sesquiterpenoids	
Limonene	3	6,6'-Dimethoxygossypol	27
Menthol	3	Gossypol	4, 27
Metatabilactone	3	Hemigossypolone	27
Myrcene	3	Hemigossypolone-7-methyl ether	27
β-Phellandrene	3	6-Methoxygossypol	4
α-Pinene	3	Diterpenes	
β-Pinene	3	Caryoptin	11, 95
2-Terpinolene	3	Caryoptin hemiacetal	95

TABLE 2 (continued)
Representative Plant-Produced Repellents of Insects

Esters	Ref.	Esters	Ref.
Caryoptinol	95	Dihydroclerodin-I	95
Clerodendrin-A	11, 95	3-Epicaryoptin	95
Clerodendrin-B	11, 95	Triterpenoids	
Clerodin	11, 95	Azadirachtin	11
Clerodin hemiacetal	95	Meliantriol	11
Dihydrocaryoptin	95		

Flavonoids and other phenylpropanoid derivatives		Flavonoids and other phenylpropanoid derivatives	
Catechin	4	Kaempferol	4
2′,6′-Dihydroxy-4′-methoxy dihydrochalcone	4	Phloretin	4
		Quercetin	4

phytochemical with insect-repellent as well as insecticidal properties. Many of the true alkaloids show such a variety of biological activities, including insect repellency. However, as with most, if not all, phytochemicals, some insects (e.g., *Manduca sexta*) can readily cope with given alkaloids (e.g., nicotine).[10]

Terpenes illustrate the great diversity of repellent molecules which may exist within a class of chemicals. They range from the C_5 hemiterpenes and C_{10} monoterpenes to triterpenoids. Monoterpenoids are especially important as repellents in higher plants. They are dominant components in the "volatile oils" of such plants.[4] Examples are α-pinene and 3-carene in *Pinus silvestris* against *Blastophaga piniperda*.[4] Sesqui- and di-terpenes first show striking diversities among angiosperms.[9] Gossypol in *Gossypium* against *Epicauta* blister beetles is a classic example of the first chemical group. Clerodendrin-A is one of several diterpenoids in the Verbenaceae which have repellent properties to some insect species.[11] Triterpenoidal repellents include azadirachtin, which is extremely active against even the desert locust, *Schistocerca gregaria*.[11] However, this famous repellent and antifeedant lacks significant repellency against North American grasshopper species and certain other insects, e.g., *Oryzaephilus surinamensis*.[11]

COMMON INSECT REPELLENTS IN ANIMALS AND PLANTS

Both plants and animals have evolved chemically based defense systems against a spectrum of stressful agents (biotic and abiotic) in their environments. Given chemicals (e.g., 1,4-benzoquinones) in such defense systems repel not only many insect species, but also many other animals and plants. Thus, organisms apparently have not specifically evolved defenses against insects. Components of such chemical defense systems frequently are quite similar, or even identical, among producing organisms (Table 3). Therefore, common chemical principles seemingly have evolved in the defense systems used by plants and animals.

MOLECULAR STRUCTURE AND REPELLENCY

As previously stated, many volatile chemicals are repellent to certain insect species or biotypes, but are attractive to others.[4] Thus, the correlations between molecular structure and repellency may be poor. However, having presented this initial specific caution about looking for such correlations, some general comments are offered. Numerous molecules which are repellent to many species of insects are relatively oxidized or unsaturated (e.g., *p*-benzoquinone). Reduction of *p*-benzoquinone to its classical redox couple, ρ-hydroqui-

TABLE 3
Representative Repellents Occurring in Both Plants and Animals

Compound	Ref.	Compound	Ref.
Actinidine	3, 48, 82, 105	Isobutyric Acid	3
Benzaldehyde	3, 30, 43	Mellein	3
Benzoquinone	3	Myrcene	3
Cuminaldehyde	3	γ-Pinene	3
Dendrolasin	3	β-Pinene	3
Formic acid	3, 62	Pseudopellefierine	3
Hippodamine	3	Pyrazine	3
Histamine	3	β-Selinene	3
Hydrogen cyanide	3, 43	Solenopsin-A	3
5-Hydroxytryptamine	3	Toluqinone	3
Iridodial	3		

none, destroys the repellency to some insects.[12] Thus, repellency may be more closely allied with molecules which are relatively electrophilic (e.g., oxidizing agents) as compared to neuclophilic (e.g., reducing agents). Further, many molecules which exhibit widespread repellency among species possess α,β-unsaturation and enter readily into 1,4-addition reactions with nucleophilic groups (e.g., sulfhydryls). Reactions involving 1,4-addition frequently show some characteristics of covalent bonding (.e.g, relative irreversibility). Such "relatively irreversible" interactions have sometimes been considered as components of the "general irritability" of organisms as commonly defined in the context of "the general chemical sense" rather than the specific chemical senses.

SENSORY RECEPTION AND ENERGY TRANSDUCTION

Based largely on electrophysiology, it was initially concluded that neurons in certain sensilla on antennae, mouthparts, tarsi, cerci, etc., of insects were specifically sensitive to repellent molecules.[5,6] Historically, electrophysiologists have placed much emphasis on the chemical specificity of given neurons.[5,6] However, as such studies on olfactory messengers have developed beyond a few compounds to involve dozens of chemicals, chemosensory neurons usually have proven to be less specific than was previously thought.[13,14] Most studied neurons now have responded electrophysiologically to a variety of chemical stimulants. These sometimes include both attractants and repellents.

A broader spectrum of individual neuronal sensitivity to attractants and repellents seems compatible with an unifying concept[15] that energy transduction between most, if not all, chemical messengers (e.g., odorants or tastants) and dendritic membranes of sensory neurons involves common functional components, as postulated by Mitchell's chemiosmotic theory of energy coupling. These are (1) a chemiosmotic, membrane-located reversible ATPase; (2) a chemiosmotic, membrane located redox chain; and (3) proton-linked (or hydroxyl ion-linked) solute porter systems. These components of an energy transduction system are all situated in the membrane. Though energy transduction in olfaction and gustation may possibly involve only one basic system,[15] bases for the known specificity of tastants or odors still exist. Specific binding sites for each ligand may easily exist, and the resultant interaction may cause characteristic energy changes in the exposed dendritic membrane. These capabilities of a ligand may provide any necessary stimulatory uniqueness.

PHARMACOLOGICAL ALTERATIONS OF INVOLVED ENERGY TRANSDUCTION

Reagents and drugs may be used to investigate the excitability of cells, and thus whole

TABLE 4

Representative Reactions of Sulfhydryl Reagents With SH Groups in Protein

Reaction	Action
Alkylation	
N-Ethylmaleimide(NEM)	Blocks excitability of olfactory cells in insects
Iodoacetic acid	General effects are similar to those of NEM
Iodoacetamide	General effects are similar to those of NEM
Mercaptide bonding organomercurials	
p-Chloromercuric benzoate (PCMB)	Impairs the function of olfactory and taste receptors in insects
Inorganic Hg^{2+}	
Mercuric chloride	Impairs the function of olfactory and taste receptors in insects
Oxidation	
N-Bromosuccimide	Impairs the function of olfactory and taste receptors in insects

organisms.[16] Such agents usually are chosen to mimic the actions of natural agents (e.g., neurotransmitters or odorants), or to probe parameters of the involved energy transduction. Reagents may mimic actions at the behavioral, physiological, biochemical or biophysical level, or specifically probe energy transduction at the level of the redox chain, ATPase, or solute porter systems.

Many natural neurotransmitters, odorants or tastants interact with neural receptors (e.g., a component of a redox chain) which have a sulfhydryl or disulfide dependency.[15] Repellency especially seems to be associated with such sulfur dependency. Thus, agents which react with sulfur groups (i.e., thiols or disulfides) in amino acids and polypeptides (Table 4) in neural membranes may directly, or indirectly, affect repellency.[16] Some sulfhydryl reagents, e.g., N-ethylmaleimide (NEM), fully mimic naturally occurring repellents (e.g., 1,4-benzoquinone) at the behavioral, electrophysiological, biochemical and biophysical levels.[15,16] These multifaceted repellent effects of sulfhydryl reagents provide strong experimental evidence that membrane proteins play essential roles in the reception and energy transduction involved in repellency.[12]

Reduction of disulfide bonds in nerve-membrane macromolecules may sensitize, or desensitize, an insect's olfactory response.[15,16] Such treatment with the strong reducing agent, dithiothreitol (DTT), sensitized the American cockroach, *Periplaneta americana*, to the potent repellent, 5-hydroxy-1,4-naphthoquinone (juglone).[16] However, such treatment with the weaker disulfide-reducing agent, reduced glutathione, sensitized one *P. americana* biotype but desensitized another to juglone. Thus, pharmacological treatments of SH groups or S-S bridges in chemosensory neural membranes usually alter olfactory responses, but the nature of the alteration, i.e., sensitization vs. densensitization, depends at least on the insect biotype and the reagent. As a result, efforts to alter an insect's sensitivity to repellents through changing the sulfur chemistry in chemosensory neural proteins should test several chemicals including both sulfhydryl reagents and disulfide-reducing agents. Testing several agents over a range of concentrations usually will result in significant pharmacological effects on the repellency.

Inhalation anaesthetics such as diethyl ether variously block insect responses to odors, including repellents. Halothane, chloroform, carbon tetrachloride, or ethyl acetate has similar effects. The mode of action of such anesthetics is apparently to change the fluidity of hydrophobic regions in chemosensory neural membranes.

CONCLUSIONS

Within the "volatile realm" of chemical ecology most, if not all, compounds are repellent to one or more species of insects. A volatile chemical is a repellent when it is allomonic to the perceiving insect. Chemical repellency is frequently a dominant attribute (trait) of organisms that occur in the same environmental area (habitat) as a given insect species, but are nonhosts to that insect. A repellent may have the capability of preventing the perceiving insect from even contacting the emitting organism.

Animals and plants have evolved chemically based defense systems which involve some of the same or similar repellent compounds. Such chemicals are repellent to other parasites and predators besides insects. Thus, organisms do not usually evolve repellents specifically as defense against insects.

Though chemicals which are repellent to one or more insect species include a diversity of molecular structures, repellent compounds frequently are relatively oxidized or unsaturated (i.e., electrophilic) as compared to attractants. Thus, chemically reducing a molecule which is repellent to an insect species (i.e., making it less electrophilic) commonly destroys its repellency to that insect, and may even make it into an attractant for that species or biotype.

Many (perhaps most) chemosensory neurons in an insect's battery of sensilla may respond electrophysiologically to repellent molecules. Thus, the energy transduction between repellents and dendritic membranes apparently involves biophysical principles which are common among the chemical senses. This interpretation does not conflict with the concept that each repellent (volatile ligand) binds uniquely to a receptor site.

Many receptors for repellents involve, directly or indirectly, sulfhydryls or disulfides in membrane proteins. Therefore, some sulfhydryl reagents or disulfide-reducing agents usually may be utilized to alter the sensitivity of an insect to a repellent. Various inhalation anesthetics also usually alter an insect's perception of repellents.

ACKNOWLEDGMENTS

Our research on chemoreception in insects was supported by the College of Agricultural and Life Sciences, University of Wisconsin, Madison, and in part by research grants No. GB-6580, GB-8756, GB-41868, and BNS 74-00953 from the U.S. National Science Foundation. My thanks are extended to the many co-workers who have helped along the way. Special thanks are given to Mary Dietrich and Kim Viney for invaluable technical assistance.

REFERENCES

1. **Meyer, H. J. and Norris, D. M.,** Lignin intermediates and simple phenolics as feeding stimulants for *Scolytus multistriatus, J. Insect Physiol.,* 20, 2015, 1974.
2. **Gilbert, B. L. and Norris, D. M.,** A chemical basis for bark beetle (*Scolytus*) distinction between host and non-host trees, *J. Insect Physiol.,* 14, 1063, 1968.
3. **Rodriquez, E. and Levin, D. A.,** Biochemical parallelisms of repellents and attractants, in *Biochemical Interaction between Plants and Insects,* Wallace, J. W., and Mansell, R. L., Eds., Plenum Press, New York, 1976, 214.
4. **Norris, D. M. and Kogan, M.,** Biochemical and morphological bases of resistance, in *Breeding Plants Resistant to Insects,* Maxwell, F. G., and Jennings, P. R., Eds., John Wiley & Sons, New York, 1980, 22.
5. **Dethier, V.,** *Chemical Insect Attractants and Repellents,* Blakiston, New York, 1947.
6. **Dethier, V.,** Repellents, *Annu. Rev. Entomol.,* 30, 348, 1955.
7. **Blum, M. S.,** *Chemical Defenses of Arthropods,* Academic Press, Orlando, 1981.
8. **Tschinkel, W. R.,** Phenols and quinones from the defensive secretions of the tenebrionid beetle, *Zophobas rugipes, J. Insect Physiol.,* 15, 191, 1969.
9. **Swain, T.,** The effect of plant secondary products on insect plant coevolution, *Proc. Int. Congr. Entomol.,* 15, 249, 1976.

10. **Kogan, M.**, The role of chemical factors in insect/plant relationships, *Proc. Int. Congr. Entomol.*, 15, 211, 1976.

11. **Kubo, I. and Nakanishi, K.**, Insect antifeedants and repellents from African plants, in *Host Plant Resistance to Pests*, Hedin, P. A., Ed., American Chemical Society, Washington, D.C., 1977, 165.

12. **Norris, D. M.**, Anti-feeding compounds, in *Chemistry of Plant Protection*, Vol. 1, Haug, G., and Hoffmann, H. Eds., Springer-Verlag, Berlin, 1986.

13. **O'Connell, R. J.**, The encoding of behaviorally important odorants by insect chemosensory receptor neurons, in *Perception of Behavioral Chemicals*, Norris, D. M., Ed., Elsevier/North-Holland Biomedical Press B.V., Amsterdam, 1981.

14. **Davis, E. E. and Sokolove, P. G.**, Lactic acid-sensitive receptors on the antennae of the mosquito, *Aedes aegypti*, *J. Comp. Physiol.*, 105, 43, 1976.

15. **Norris, D. M.**, Possible unifying principles in energy transduction in the chemical senses, in *Perception of Behavioral Chemicals*, Norris, D. M., Ed., Elsevier/North-Holland Biochemical Press B.V., Amsterdam, 1981.

16. **Ma, W. -C.**, Receptor membrane function in olfaction and gustation: implications from modification by reagents and drugs, in *Perception of Behavioral Chemicals*, Norris, D. M., Ed., Elsevier/North-Holland Biomedical Press B.V., Amsterdam, 1981.

17. **Aldrich, J. R. and Yonke, T. R.**, Natural products of abdominal and metathoracic scent glands of coreoid bugs, *Ann. Entomol. Soc. Am.*, 68, 955, 1975.

18. **MacLeod, J. K., Howe, I., Cable, J., Blake, J. D., Baker, J. T., and Smith, D.**, Volatile scent gland components of some tropical Hemiptera, *J. Insect Physiol.*, 21, 1219, 1975.

19. **Schildknecht, H., Kunzelmann, P., Kraub, D., and Kuhn, C.**, Uber die Chemie der Spinnwebe, I, *Naturwissenschaften*, 59, 98, 1972.

20. **Tschinkel, W. R.**, Unusual occurrence of aldehydes and ketones in the defensive secretion of the tenebrionid beetle, *Eleodes beameri*, *J. Insect Physiol.*, 21, 659, 1975.

21. **Moore, B. P. and Brown, W. V.**, The chemistry of the metasternal gland secretion of the common eucalypt longicorn, *Phoracantha semipunctata* (Coleoptera: Cerambycidae), *Aust. J. Chem.*, 25, 591, 1972.

22. **Trave, R., Garanti, L., Marchesini, A., and Pavan, M.**, Sulla natura chimica del secreto odoroso della larva del Lepidottero *Cossus cossus* L., *Chim. Ind. (Milan)*, 48, 1167, 1966.

23. **Crewe, R. M. and Blum, M. S.**, Alarm pheromones of the attini: their phylogenetic significance, *J. Insect Physiol.*, 18, 31, 1972.

24. **Riley, R. G., Silverstein, R. M., and Moser, J. C.**, Isolation, identification, synthesis and biological activity of volatile compounds from the heads of *Atta* ants, *J. Insect Physiol.*, 20, 1629, 1974.

25. **Lloyd, H. A., Blum, M. S., and Duffield, R. M.**, Chemistry of the male mandibular gland secretion of the ant, *Camponotus clarithorax*, *Insect Biochem.*, 5, 489, 1975.

26. **Bergström, G. and Löfquist, J.**, Chemical basis for odour communication in four species of *Lasius* ants, *J. Insect Physiol.*, 16, 2353, 1970.

27. **Stipanovic, R. D., Bell, A. A., and Lukefahr, M. J.**, Natural insecticides from cotton (*Gossypium*), in *Host Plant Resistance to Pests*, Hedin, P. A., Ed., American Chemical Society, Washington, D.C., 1977, 197.

28. **Kullenberg, B., Bergström, G., and Stallberg-Stenhagen, S.**, Volatile components of the cephalic marking secretion of male bumblebees, *Acta Chem. Scand.*, 24, 1481, 1970.

29. **Bergström, G. and Löfquist, J.**, Odour similarities between the slave-keeping ants *Formica sanguinea* and *Polyergus rufescens* and their slaves *Formica fusca* and *Formica rufibarbis*, *J. Insect Physiol.*, 14, 995, 1968.

30. **Blum, M. S., MacConnell, J. G., Brand, J. M., Duffield, R. M., and Fales, H. M.**, Phenol and benzaldehyde in the defensive secretion of a strongylosomid millipede, *Ann. Entomol. Soc. Am.*, 66, 235, 1973.

31. **Bergström, G. and Löfquist, J.**, *Camponotus ligniperda* Latr. — A model for the composite volatile secretions of Dufour's gland in formicine ants, in *Chemical Releasers in Insects*, Tahori, A. S., Ed., Gordon & Breach, New York, 1971, 195.

32. **Bergström, G. and Löfquist, J.**, Chemical congruence of the complex odoriferous secretions from Dufour's gland in three species of ants of the genus *Formica*, *J. Insect Physiol.*, 19, 887, 1973.

33. **Luby, J. M., Regnier, F. E., Clarke, E. T., Weaver, E. C., and Weaver, N.**, Volatile cephalic substances of the stingless bees, *Trigona mexicana* and *Trigona pectoralis*, *J. Insect Physiol.*, 19, 1111, 1973.

34. **Roth, L. M. and Stahl, W. H.**, Tergal and cercal secretion of *Blatta orientalis* L., *Science*, 123, 798, 1956.

35. **Wallbank, B. E. and Waterhouse, D. E.**, The defensive secretions of *Polyzosteria* and related cockroaches, *J. Insect Physiol.*, 16, 2081, 1970.

36. **Meinwald, J., Chadha, M. S., Hurst, J. J., and Eisner, T.**, Defense mechanisms of arthropods. IX. Anisomorphal, the secretion of a plasmid insect, *Tetrahedron Lett.*, 29, 1962.

37. **Collins, R. P.,** Carbonyl compounds produced by the bedbug, *Cimex lectularius, Am. Entomol. Soc. Am.,* 61, 1338, 1968.

38. **Cmelik, S.,** Volatile aldehydes in the odoriferous secretion of the stink bug, *Libyaspis angolensis, Hoppe-Seyler's Z. Physiol. Chem.,* 350, 1076, 1969.

39. **Baker, J. T. and Kemball, P. A.,** Volatile constituents of the scent gland reservoir of the coreoid, *Pternistria bispina* Stal, *Aust. J. Chem.,* 20, 395, 1967.

40. **Staddon, B. W. and Weatherston, J.,** Constituents of the stink gland of the water bug, *Ilyocoris cimicoides* (L.) (Heteroptera: Naucoridae), *Tetrahedron Lett.,* 4567, 1967.

41. **Murkerji, S. K. and Sharma, H. L.,** Investigations on the offensive odour of hemiptera-bugs, *Tetrahedron Lett.,* 2479, 1966.

42. **Baker, J. T., Blake, J. D., MacLeod, J. K., and Ironside, D. A.,** The volatile constituents of the scent gland reservoir of the fruit-spotting bug, *Amblypelta nitida, Aust. J. Chem.,* 25, 393, 1972.

43. **Eisner, H. E., Eisner, T., and Hurst, J. J.,** Hydrogen cyanide and benzaldehyde produced by millipedes, *Chem. Ind. (London),* 124, 1963.

44. **Chadha, M. S., Joshi, N. K., Mamdapur, V. R., and Sipahimalani, A. T.,** C-21 steroids in the defensive secretions of some Indian water beetles-II, *Tetrahedron,* 2061, 1970.

45. **Wheeler, J. W., Evans, S. L., Blum, M. S., and Torgerson, R. L.,** Cyclopentyl ketones: identification and function in Azteca ants, *Science,* 187, 254, 1975.

46. **Wheeler, J. W., Happ, G. M., Araujo, J., and Pasteels, J. M.,** γ-Dodecalactone from rove beetles, *Tetrahedron Lett.,* 4635, 1972.

47. **Abou-Donia, S. A., Fish, L. J., and Pattenden, G.,** Iridodial from the odoriferous glands of *Staphylinus olens* (Coleoptera: Staphylinidae), *Tetrahedron Lett.,* 4037, 1971.

48. **Bellas, T. E., Brown, W. V., and Moore, B. P.,** The alkaloid actinidine and plausible precursors in defensive secretions of rove beetles, *J. Insect Physiol.,* 20, 277, 1974.

49. **Brand, J. M., Blum, M. S., Fales, H. M., and Pasteels, J. M.,** The chemistry of the defensive secretion of the beetle, *Drusilla canaliculata, J. Insect Physiol.,* 19, 369, 1973.

50. **Meinwald, J., Opheim, K., and Eisner, T.,** Gyrinidal: A sesquiterpenoid aldehyde from the defensive glands of gyrinid beetles, *Proc. Natl. Acad. Sci. U.S.A.,* 69, 1208, 1972.

51. **Miller, J. and Mumma, R.,** Defensive agents of the American water beetles *Agabus seriatus* and *Graphoderus liberus, J. Insect Physiol.,* 19, 917, 1975.

52. **Bevan, C. W. L., Birch, A. J., and Caswell, H.,** An insect repellent from black cocktail ants, *J. Chem. Soc.,* 488, 1961.

53. **Bradshaw, J. W. S., Baker, R., and Howse, P. E.,** Multicomponent alarm pheromones of the weaver ant, *Nature (London),* 258, 230, 1975.

54. **Hefetz, A. and Batra, S. W. T.,** Geranyl acetate and 2-decen-1-ol in the cephalic secretion of the solitary wasp, *Sceliphron caementarium* (Sphecidae: Hymenoptera), *Experientia,* 35, 1138, 1979.

55. **Cavill, G. W. K. and Hinterberger, H.,** The chemistry of ants. IV. Terpenoid constituents of some *Dolichoderus* and *Iridomyrmex* species, *Aust. J. Chem.,* 13, 514, 1960.

56. **Cavill, G. W. K., Ford, D. L., and Locksley, H. D.,** The chemistry of ants. I. Terpenoid constituents of some Australian *Iridomyrmex* species, *Aust. J. Chem.,* 9, 288, 1956.

57. **Chadha, M. S., Eisner, T., Monro, A., and Meinwald, J.,** Defense mechanisms of arthropods. VII. Citronellal and citral in the mandibular gland secretion of the ant *Acanthomyops claviger* (Roger), *J. Insect Physiol.,* 8, 175, 1962.

58. **Bernardi, R., Cardani, C., Ghiringhelli, D., Selva, A., Baggini, A., and Pavan, M.,** On the components of secretion of mandibular glands of the ant *Lasius (Dendrolasius) fuliginosus, Tetrahedron Lett.,* 3893, 1967.

58. **Quennedey, A., Brule, G., Rigaud, J., Dubois, P., and Brossut, R.,** La glande frontale des soldats de *Schedorhinotermes putorius* (Isoptera): analyse chimique et fonctionnement, *Insect Biochem.,* 3, 67, 1973.

60. **Eisner, T., Hendry, L. B., Peakall, D. B., and Meinwald, J.,** 2,5-Dichlorophenol (from ingested herbicide?) in defensive secretion of grasshopper, *Science,* 172, 277, 1971.

61. **Schildknecht, H., Berger, D., Krauss, D., Connert, J., Gehlhaus, J., and Essenbreis, H.,** Defense chemistry of *Stenus comma* (Coleoptera: Staphylinidae), LXI, *J. Chem. Ecol.,* 2, 1, 1976.

62. **Eisner, T., Kluge, A. F., Carrel, J. E., and Meinwald, J.,** Defense mechanisms of arthropods. XXXIV. Formic acid and acyclic ketones in the spray of a caterpillar, *Ann. Entomol. Soc. Am.,* 65, 765, 1972.

63. **Fales, H. M., Blum, M. S., Crewe, R. M., and Brand, J. M.,** Alarm pheromones in the genus *Manica* derived from the mandibular gland, *J. Insect Physiol.,* 18, 1077, 1972.

64. **Trave, R. and Pavan, M.,** Veleni degli insetti. Principi estratti dalla formica *Tapinoma nigerrimum* Nyl., *Chim. Ind. (Milan),* 38, 1015, 1956.

65. **Butenandt, A. and Tam, N.,** Über einen geschlechtsspezifischen duftstoff der wasserwanze *Belostoma indica vitalis (Lethocerus indicus* Lep.), *Hoppe-Seyler's Z. Physiol. Chem.,* 308, 277, 1957.

66. **Waterhouse, D. F. and Gilby, A. R.,** The adult scent glands and scent of nine bugs of the superfamily Coreoidea, *J. Insect Physiol.,* 10, 977, 1964.

67. **Devakul, V. and Maarse, H.**, A secondary compound in the odorous gland liquid of the giant water bug *Lethocerus indicus* (Lep. and Serv.), *Anal. Biochem.*, 7, 269, 1964.
68. **Schildknecht, H. and Weis, K. H.**, Zur kenntnis der pygidialblasensubstanzen vom gelbrandkäfer (*Dytiscus marginalis* L.), *Z. Naturforsch.*, B17b, 448, 1962.
69. **Schildknecht, H.**, The defensive chemistry of land and water beetles, *Angew. Chem., Int. Ed. Engl.*, 9, 1, 1970.
70. **Moore, B. P. and Wallbank, B. E.**, Chemical composition of the defensive secretion in carabid beetles and its importance as a taxonomic character, *Proc. R. Entomol. Soc. London, Ser. B.*, 37, 62, 1968.
71. **Schildknecht, H. and Birringer, H.**, Die steroid des schlammschwimmers *Ilybius fenestratus*, III, *Z. Naturforsch.*, 24B, 1529, 1969.
72. **Marchesini, A., Garanti, L., and Pavan, M.**, Sulla natura chimica del secreto delle glandole mandibolari della larva di *Zeuzera pyrina* L. (Lep. Cossidae), *Ric. Sci.*, 39, 874, 1969.
73. **Bowers, W. S., Nault, L. R., Webb, R. E., and Dutky, S. R.**, Aphid alarm pheromone: isolation, identification, synthesis, *Science*, 177, 1121, 1972.
74. **Gilby, A. R. and Waterhouse, D. F.**, The identification of *trans*-4-ketohex-2-enal by its proton magnetic resonance spectrum, *Aust. J. Chem.*, 17, 1311, 1964.
75. **Calam, D. H. and Youdeowei, A.**, Identification and functions of secretion from the posterior scent gland of fifth instar larva of the bug *Dysdercus intermedius*, *J. Insect Physiol.*, 14, 1147, 1968.
76. **Hurst, J. J., Meinwald, J., and Eisner, T.**, Defense mechanisms of arthropods. XII. Glucose and hydrocarbons in the quinone-containing secretion of *Eleodes longicollis*, *Ann. Entomol. Soc. Am.*, 57, 44, 1964.
77. **Keville, R. and Kannowski, P. B.**, Sexual excitation by pheromones of the confused flour beetle, *J. Insect Physiol.*, 21, 81, 1975.
78. **Hayashi, N., Komae, H., and Hiyama, H.**, Monoterpene hydrocarbons from ants, *Z. Naturforsch.*, 28C, 626, 1973.
79. **Quilico, A., Grünanger, P., and Pavan, M.**, Sul componente odoroso del veleno del formicide *Myrmicaria natalensis* Fred., *Proc. Int. Congr. Entomol., 11th*, 3, 66, 1962.
80. **Regnier, F. E., Nieh, M., and Hölldobler, B.**, The volatile Dufour's gland components of the harvester ants, *Pogonomyrmex rugosus* and *P. barbatus*, *J. Insect Physiol.*, 19, 981, 1973.
81. **Brophy, J. J., Cavill, G. W. K., and Shannon, J. S.**, Venom and Dufour's gland secretions in an Australian species of *Camponotus*, *J. Insect Physiol.*, 19, 791, 1973.
82. **Wheeler, J. W., Olagbemiro, T., Nash, A., and Blum, M. S.**, Actinidine from the defensive secretions of dolichoderine ants, *J. Chem. Ecol.*, 3, 241, 1977.
83. **Morgan, E. D. and Wadhams, L. J.**, Chemical constituents of Dufour's gland in the ant, *Myrmica rubra*, *J. Insect Physiol.*, 18, 1125, 1972.
84. **Cavill, G. W. K. and Williams, P. J.**, Constituents of Dufour's gland in *Myrmecia gulosa*, *J. Insect Physiol.*, 13, 1097, 1967.
85. **Cavill, G. W. K. and Houghton, E.**, Hydrocarbon constituents of the Argentine ant, *Iridomyrmex humilis*, *Aust. J. Chem.*, 26, 1131, 1973.
86. **Wheeler, J. W., Blum, M. S., Daly, H. V., Kislow, C. J., and Brand, J. M.**, Chemistry of mandibular gland secretions of small carpenter bees (*Ceratina* spp.), *Ann. Entomol. Soc. Am.*, 70, 635, 1977.
87. **Waterhouse, D. F. and Wallbank, B. E.**, 2-Methylene butanal and related compounds in the defensive scent of *Platyzosteria* cockroaches (Blattidae: Polyzosteriinae), *J. Insect Physiol.*, 13, 1657, 1967.
88. **Games, D. E. and Staddon, B. W.**, Composition of scents from the larva of the milkweed bug *Oncopeltus fasciatus*, *J. Insect Physiol.*, 19, 1527, 1973.
89. **Prestwich, G. D.**, Chemical analysis of soldier defensive secretions of several species of East African termites, in *Pheromones and Defensive Secretions in Social Insects*, Noirot, C., Howse, P. E., LeMasne, G., Eds., University of Dijon Press, Dijon, France, 1975, 149.
90. **Moore, B. P.**, Studies on the chemical composition and function of the cephalic gland secretion in Australian termites, *J. Insect Physiol.*, 14, 33, 1968.
91. **Hedin, P. A. Jenkins, J. N., and Maxwell, F. G.**, Behavioral and developmental factors affecting host plant resistance to insects, in *Host Plant Resistance to Pests*, Hedin, P. A., Ed., American Chemical Society, Washington, D.C., 1977, 231.
92. **Dettner, K. and Schwinger, G.**, Defensive secretions of three Oxytelinae rove beetles (Coleoptera: Staphylinidae), *J. Chem. Ecol.*, 8, 1411, 1982.
93. **Munakata, K.**, Insect feeding deterrents in plants, in *Chemical Control of Insect Behavior: Theory and Application*, Shorey, H. H., and McKelvey, Jr., J. J., Eds., John Wiley & Sons, New York, 1977, 93.
94. **Munakata, K.**, Insect antifeedants in plants, in *Control of Insect Behavior by Natural Products*, Wood, D. L., Silverstein, R. M., and Nakajima, M., Eds., Academic Press, Orlando, 1970, 179.
95. **Munakata, K.**, Insect antifeedants of *Spodoptera litura* in plants, in *Host Plant Resistance to Pests*, Hedin, P. A., Ed., American Chemical Society, Washington, D.C., 1977, 185.

96. **Gentry, C. R., Bierl-Leonhardt, B. A., McLaughlin, J. R., and Plimmer, J. R.,** Air permeation tests with (Z,Z)-3,13-octadecadien-1-ol acetate for reduction in trap catch of peachtree and lesser peachtree borer moths, *J. Chem. Ecol.,* 7, 575, 1981.

97. **Saxena, K. N. and Basit, A.,** Inhibition of oviposition by volatiles of certain plants and chemicals in the leafhopper *Amrasca devastans* (Distant), *J. Chem. Ecol.,* 8, 329, 1982.

98. **Prestwich, G. D. and Collins, M. S.,** Chemical defense secretions of the termite soldiers of *Acorhinotermes* and *Rhinotermes* (Isoptera, Rhinotermitinae): ketones, vinyl ketones, and β-ketoaldehydes derived from fatty acids, *J. Chem. Ecol.,* 8, 147, 1982.

99. **Reese, J. C., Chan, B. G., and Waiss, Jr., A. C.,** Effects of cotton condensed tannins, maysin (corn) and pinitol (soybean) on *Heliothis zea* growth and development, *J. Chem. Ecol.,* 8, 1429, 1982.

100. **Doskotch, R. W., Cheng, H. -Y., Odell, T. M., and Girard, L.,** Nerolidol: An antifeeding sesquiterpene alcohol for gypsy moth larvae from *Melaleuca leucadendron, J. Chem. Ecol.,* 6, 845, 1980.

101. **Bedard, W. D., Wood, D. L., Tilden, P. E., Lindahl, Jr., K. Q., Silverstein, R. M., and Rodin, J. O.,** Field responses of the western pine beetle and one of its predators to host- and beetle-produced compounds, *J. Chem. Ecol.,* 6, 625, 1980.

102. **Chapman, R. F., Bernays, E. A., and Simpson, S. J.,** Attraction and repulsion of the aphid, *Cavariella aegopodii,* by plant odors, *J. Chem. Ecol.,* 7, 881, 1981.

103. **Todd, G. W., Getchum, A., and Cress, D. C.,** Resistance in barley to greenbug, *Schizaphis graminum* L. Toxicity of the phenolic and related compounds and related substances, *Ann. Entomol. Soc. Am.,* 64, 718, 1971.

104. **Rothschild, M.,** Secondary plant substances and warning colouration in insects, in *Insect-Plant Relationships,* van Enden, H. F., Ed., Symp. R. Entomol. Soc. Lond., 1972.

105. **Bellas, T. E., Brown, W. V., and Moore, B. P.,** The alkaloid actinidine and plausible precursors in defensive secretions of rove beetles, *J. Insect Physiol.,* 20, 277, 1974.

106. **Buhr, H., Toball, R., and Schreiber, K.,** Die wirkung von einigen pflanzenlichen Sonderstoffen, insbesondere des Kartoffelkäfers (*Leptinotarsa decemlineata* Say), *Entomol. Exp. Appl.,* 1, 209, 1958.

107. **Ma, W. -C.,** Some properties of gustation in the larvae of *Pieris brassicae, Entomol. Exp. Appl.,* 12, 584, 1969.

108. **Hwang, Y. -S., Kramer, W. L., and Mulla, M. S.,** Oviposition attractants and repellents of mosquitoes. Isolation and indentification of oviposition repellents for *Culex* mosquitoes, *J. Chem. Ecol.,* 6, 71, 1980.

109. **Kramer, W. L., Hwang, Y. -S., and Mulla, M. S.,** Oviposition repellents of mosquitoes: Negative responses elicited by lower aliphatic carboxylic acids, *J. Chem. Ecol.,* 6, 415, 1980.

110. **Ohigashi, H., Wagner, M. R., Matsumura, F., and Benjamin, D. M.,** Chemical basis of differential feeding behavior of the larch sawfly, *Pristiphora erichsonii* (Hartig), *J. Chem. Ecol.,* 7, 599, 1981.

OVIPOSITION STIMULANTS AND DETERRENTS

J. A. A. Renwick

INTRODUCTION

The importance of understanding the relationships between plants and insects is well recognized by agricultural entomologists, especially when public pressure requires more emphasis on natural means of pest control.[1] The factors that determine whether a plant is utilized by a specific insect are numerous and complex. However, the involvement of chemical constituents of the plants is well documented.[1-4] Plant compounds mediate the processes of orientation to host plants, oviposition, feeding and subsequent growth, and development of the insects.

Oviposition is a critical step in the life cycle of phytophagous insects. The success of an insect's progeny depends on the judicious selection of a suitable source of food by the adult female.[5] Although the role of plant chemicals in the acceptance or rejection of a plant has been amply demonstrated, few of the chemicals responsible for the insects' behavioral responses have been identified. Due to the lack of specific examples of chemicals that mediate host selection for oviposition, a review of this group of allelochemicals must be extremely limited.

The purpose of this chapter is to discuss the factors that play a role in the choice of plants for oviposition and to compile a list of known stimulants and deterrents. Since numerous studies are reported where chemical and physical factors are not adequately separated, this list will be limited to examples where chemical mediation of oviposition is clearly shown, and preferably where some progress has been made in characterizing the chemical compounds involved.

GENERAL FACTORS AFFECTING OVIPOSITION BY INSECTS

Assessment of the role played by stimulants and deterrents in oviposition by insects depends on a sound knowledge of all the physical as well as chemical stimuli that affect landing on and acceptance of a plant. A complex array of conditions must be met before oviposition can occur. The physiological condition of the ovipositing female must be just right, and a threshold of motivation must be reached.[6-7] Among the physical factors influencing oviposition, visual stimuli are often critical. Color, size, or shape of the plant may be important.[8-9] Tactile stimuli often affect acceptance or rejection of a plant. Thus, texture of the contact surface is also important. Bioassays for stimulant or deterrent activity must provide an environment and substrates for oviposition that meet all these requirements.

The behavioral steps leading to oviposition are often quite complex. Orientation to a host plant may be guided by physical and olfactory stimuli. Then, once the plant has been reached, contact stimuli come into play. Some specialized insects such as onion flies and cabbage flies exhibit a typical series of behavioral responses after contact with a host plant. Leaf and stem runs as well as probing of the soil appear to be necessary to assess the chemistry, structure, and environment of the plant before eggs are laid.[10-12]

CHEMICAL STIMULI MEDIATING OVIPOSITION

Chemical constituents of plants are likely to be involved in all the behavioral responses of insects that eventually result in oviposition. Attractants and repellents determine whether positive or negative orientation occurs, and may provide the ultimate cue for landing or

$$CH_2OH \quad CH_3 \qquad\qquad\qquad H$$
$$O\ O-CH(CH_2)_6CH(CH_2)_6CONCH_2CH_2SO_3H$$

(1) Rhagoletis cerasi
oviposition deterrent

contact. Stimulants or deterrents may be present at the leaf surface. Many butterflies exhibit "drumming" behavior after landing on a leaf.[13] This rapid movement of the foretarsi over the leaf surface is thought to release chemicals that provide a more positive signal for acceptance of the plant.[14] Other insects are influenced by chemicals released from trichomes or glands when these structures at the leaf surface are disrupted.[15-16] A combination of volatile and nonvolatile plant constituents may be necessary for ultimate acceptance of the plant as a suitable host.

The role of stimulants and deterrents in mediating oviposition is often difficult to demonstrate. Both may be present in a plant so that their effects are masked. A balance between positive and negative cues that may be tipped in either direction has been suggested,[7] and the stimultaneous occurrence of stimulants and deterrents in a plant has recently been clearly shown.[17] Other complications that make interpretation of bioassay results difficult include learning effects, where the insects associate definite visual characteristics with a preferred plant.[9,18] Also, physical interference due to biologically inactive material in extracts can lead to misleading results. For example, oils or waxes in an extract may coat the test substrate in a way that prevents perception of active stimuli by the insects. Interpretation of results can also be complicated when physiological and sensory effects of plant chemicals are not clearly separated. In some cases, stimulation of oviposition may be confused with stimulation of oogenesis.[19-20]

The possible application of oviposition deterrents in the protection of crop plants from insect pests has often been suggested, and the argument in favor of this concept has been strengthened by the discovery of oviposition-deterring pheromones or "epideictic pheromones".[21] These are chemicals released by the insect during or after oviposition to discourage oviposition by conspecific females on the same limited food source. This subject has been adequately reviewed, and new examples are continually being demonstrated.[22] However, work on the chemistry of these pheromones has been slow, and only one complete identification (Structure 1) has been reported, for the European cherry fruit fly (Rhagoletis cerasi).[23] Field tests with crude pheromone extracts have shown the potential of these chemicals for pest management.[24] The promise of deterrents in discouraging oviposition by agriculturally important pests has prompted many studies on the effects of known chemicals sprayed on the plants.[25-29] But usually these tests have little bearing on the natural forces that determine whether a plant is accepted for oviposition.

PERCEPTION OF STIMULANTS AND DETERRENTS

The chemical stimuli involved in mediating oviposition may be perceived by various chemoreceptors on different parts of an insect's body.[30] In general, antennal receptors are responsible for perception of volatile stimuli (olfaction), and contact receptors are involved for nonvolatile cues. However, the separation is not always clear. It may be difficult to distinguish between stimulants and attractants and between deterrents and repellents. In most cases, olfactory cues mediate behavior before contact with a plant, whereas contact chemoreception comes into play after landing. But the possibility of continued olfactory input after reaching the plant surface must be considered. The subject of contact chemoreception has recently been thoroughly reviewed.[31] Tarsal receptors are generally involved in the perception of oviposition stimuli by all the phytophagous insects studied.

OVIPOSITION STIMULANTS AND DETERRENTS ISOLATED FROM PLANTS

Many literature reports on the isolation of oviposition stimulants from plants do not clearly distinguish between attractants and stimulants. Thus, a list of known stimulants is difficult to compile without inadvertently including some compounds or extracts that might be considered attractants. Examples of plant constituents that stimulate oviposition by insects are listed in Table 1. Cases where volatile compounds, referred to as oviposition stimulants, appear to be primarily involved in orientation to the plant have been omitted. Other plant volatiles that are clearly active as close range stimuli, and may still be active after contact with the plant, are included.

The list of oviposition deterrents in Table 2 includes only one case of volatile plant constituents that might be considered repellents.[32] Most of the deterrents have been isolated from nonhost plants and may be responsible in part for rejection of these plants by the insect studied. However, some deterrents have also been isolated from host plants which have been damaged by previous feeding or oviposition.[33-36] These compounds appear to function as markers, in the same way as oviposition deterring pheromones, to discourage egg deposition on previously occupied food sources. Despite the relatively large number of cases where deterrent activity of host or nonhost plant tissue has been demonstrated, very few of the chemicals responsible for this activity have been identified.

CONCLUSIONS

The potential use of oviposition deterrents to protect crop plants from insects at a stage before any feeding damage can occur is particularly appealing to agricultural entomologists. In addition, a knowledge of the specific plant chemicals that stimulate or deter oviposition can be extremely valuable to plant breeders in the design of resistant varieties. Yet the chemistry of this group of allelochemicals is only beginning to be unravelled. Identification of the active constituents is likely to lead to the elucidation of structure-activity relationships, reception mechanisms and behavioral aspects of plant-insect interactions that could play an important role in our approach to pest management in the future.

TABLE 1
Plant Constituents That Stimulate Oviposition by Insects

Insect	Plant	Active material	Ref.
Coleoptera			
Rice weevil, *Sitophilus zeamais*	Rice	Ferrulates, diglycerides, sterols	37
Bruchid, *Callosobruchus chinensis*	Soybean	Saponin fraction	38
Bean bruchid, *Acanthoscelides obtectus*	Bean	Nonpolar and polar extracts	39, 40, 41
Diptera			
Beet fly, *Pegomya betae*	Beet	Mixture of alkanes	42
Cabbage root fly, *Delia radicum*	Cabbage	Sinigrin, allylisothiocyanate, β-phenyl-ethylamine, carbon disulfide	43

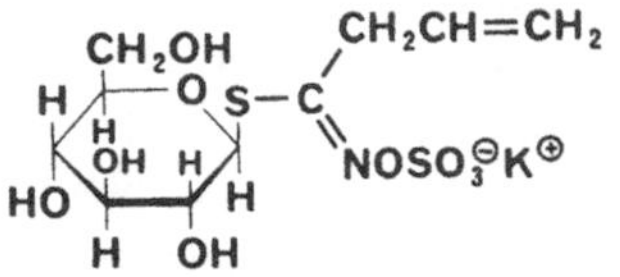

(2) Sinigrin

		Glucosinolates	44
		Isothiocyanates with glucosinolates	45
		Sinigrin and less polar compounds	46
Onion maggot, *Hylemya antigua*	Onion	n-Propyl disulfide, n-propyl mercaptan	47
Carrot fly *Psila rosae*	Carrot	Methyl isoeugenol	48

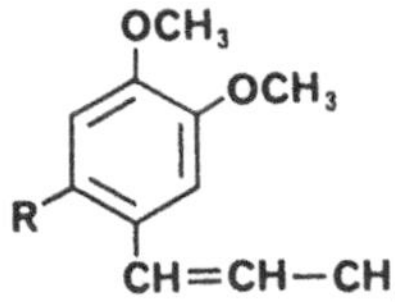

(3) R = H, Methylisoeugenol
(4) R = OCH₃, Asarone

		trans-Asarone, trans-methyl isoeu-genol, bergaptene, osthol, xantho-toxin, falcarindiol	49, 50

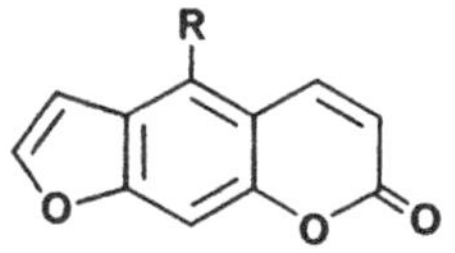

(5) R = OCH₃, Bergaptene
(6) R = H, Xanthotoxin
or 8-methoxypsoralene

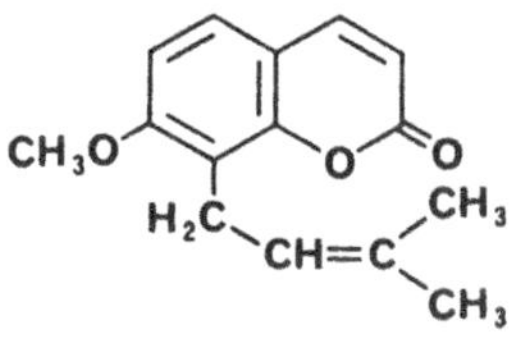

(7) Osthole

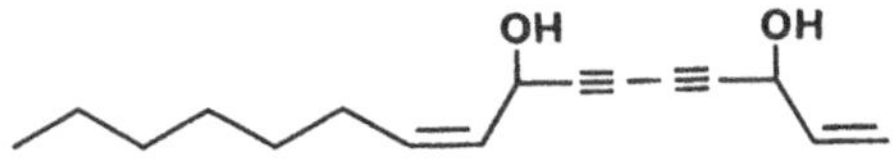

(8) Falcarindiol

TABLE 1 (continued)
Plant Constituents That Stimulate Oviposition by Insects

Insect	Plant	Active material	Ref.
Sorghum shootfly *Atherigona soccata*	Sorghum	Acetone extract	51
Lepidoptera			
Diamondback moth, *Plutella xylostella*	Mustard	Allylisothiocyanate	52
Small white cabbage butterly, *Pieris rapae*	Cabbage	Sinigrin (see Structure 2)	18
		Unknown polar compounds	53
Large white cabbage butterfly, *Pieris brassicae*	Cabbage	Mustard oil glycosides	54
Swallowtail butterfly, *Papilio polyxenes*	Carrot	Luteolin 7-O-[6″-malonyl-glucoside] *trans*-chlorogenic acid	55

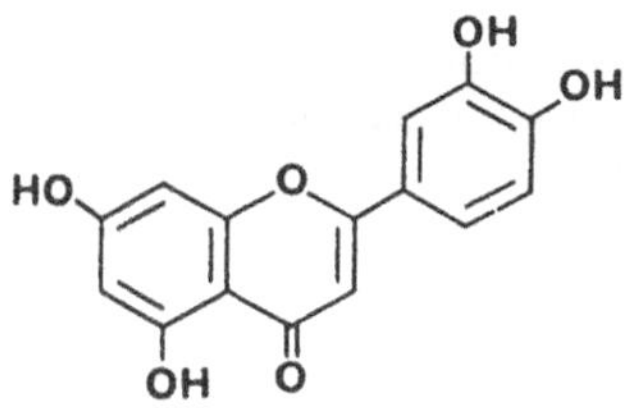

(9) Luteolin

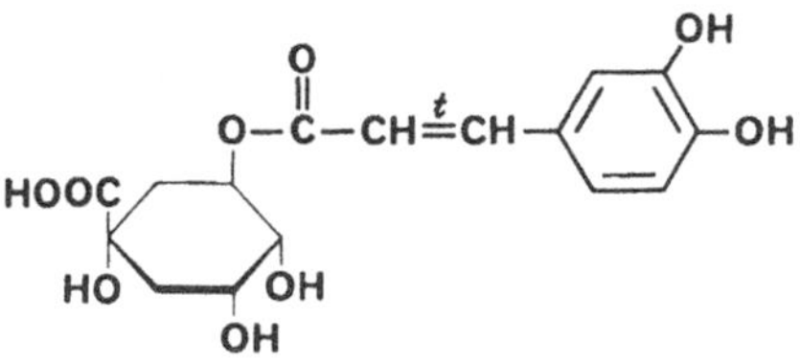

(10) *trans*-Chlorogenic acid

Citrus swallowtail, *Papilio protenor*	Orange	Naringin, hesperidin, (with other compounds)	56

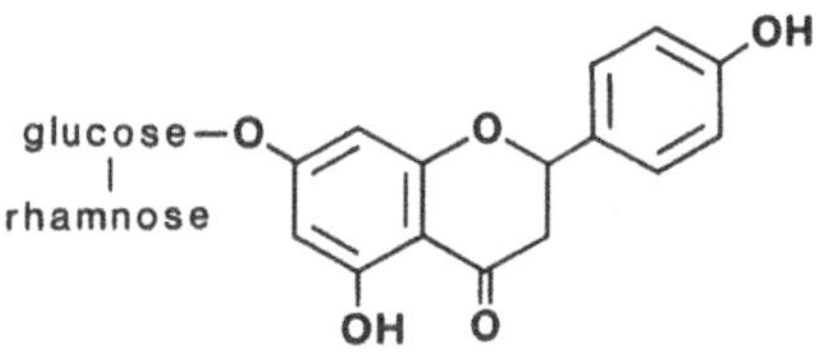

(11) Naringin

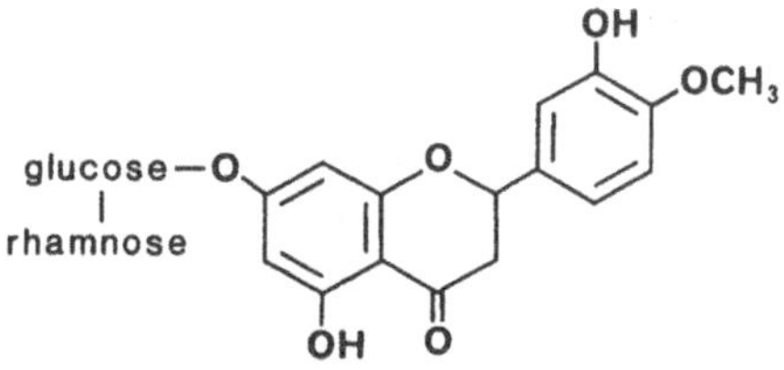

(12) Hesperidin

Tobacco budworm, *Heliothis virescens*	Tobacco	α- and β-4,8,13-duvatrien-1-ols and 1,3-diols	57, 58

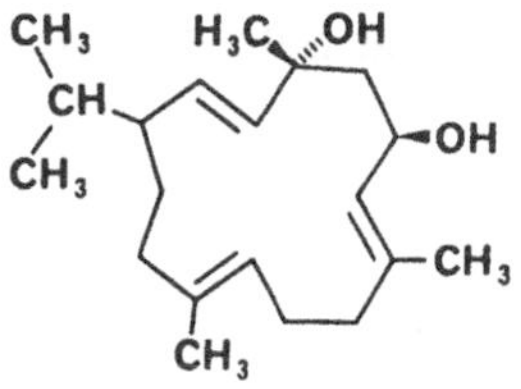

(13) Duvatriene-1,3-diol

Heliothis subflexa	Ground-cherry	Methanol extract	59

TABLE 1 (continued)
Plant Constituents That Stimulate Oviposition by Insects

Insect	Plant	Active material	Ref.
Tomato pinworm moth, *Keiferia lycopersicella*	Tomato	Polar compounds from leaf surface	60
Spruce budworm, *Choristoneura fumiferana*	Balsam fir	Nonpolar extract	61
Leek moth, *Acrolepia assectella*	Leek	Methanolic extracts	62
Codling moth, *Laspeyresia pomonella*	Apple	α-Farnesene	63

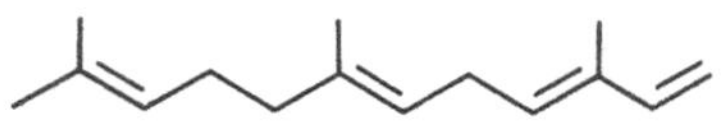

(14) α-Farnesene

Papilio xuthus	Citrus	Vicenin-2, narirutin, rutin, hesperidin, 5-hydroxy-*N,W*-methyltryptamine, adenosine	64

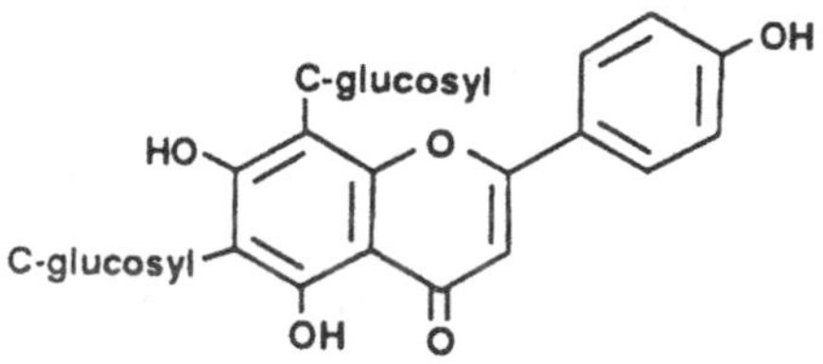

(15) Vicenin-2

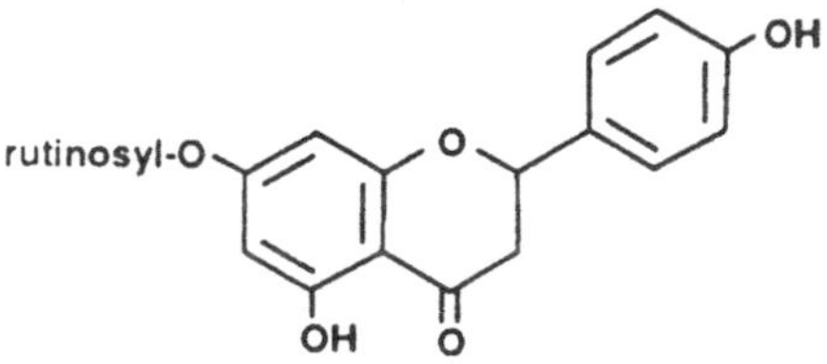

(16) Narirutin

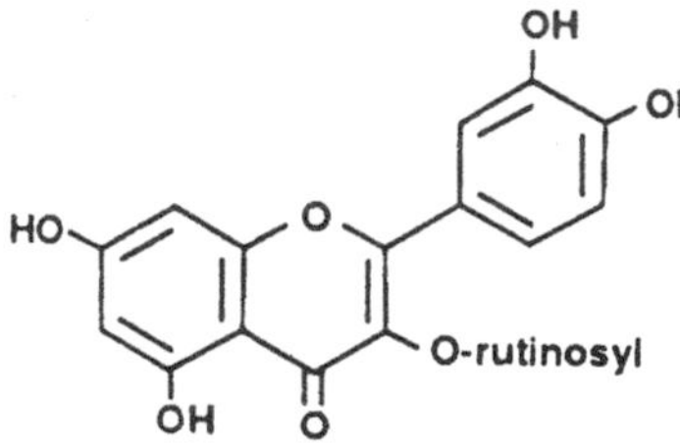

(17) Rutin

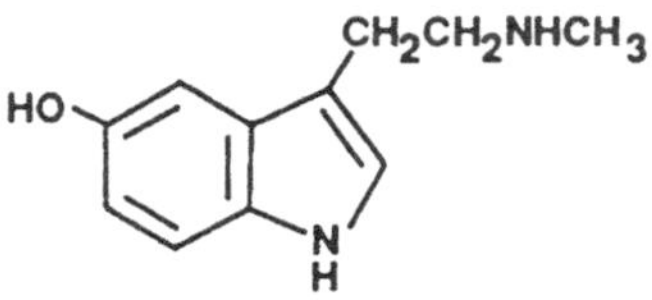

(18) 5-Hydroxy-N ω-methyltryptamine

TABLE 2
Plant Constituents That Deter Oviposition by Insects

Insect	Plant	Active material	Ref.
Coleoptera			
Bruchid beetle, *Callosobruchus chinensis*	Various	Petroleum ether extracts	29
Diptera			
Onion root maggot, *Hylemya antiqua*	Western red cedar	Leaf oil	65
	Bean seeds	Nonpolar extracts	66
Olive fruit fly, *Dacus oleae*	Olive fruit	Lipophilic compounds in juice	36

TABLE 2 (continued)
Plant Constituents That Deter Oviposition by Insects

Insect	Plant	Active material	Ref.
Cabbage root fly, *Delia radicum*	Cabbage	Fraction of polar extract	46
Carrot fly, *Psila rosae*	Cabbage	Polar extracts	46
Homoptera			
Sweetpotato whitefly, *Bemisia tabaci*	Neem	Aqueous extracts	67
Leafhopper, *Amarasca devastans*	Various	Citral and terpene alcohols	32
Lepidoptera			
Codling moth, *Cydia pomonella*	Various medicinal plants	Aqueous extracts	28
Diamondback moth, *Plutella xylostella*	Hyssop, rosemary, sage, thyme, white clover	Alcohol extracts	68
Beet armyworm, *Spodoptera exigua*	Pigweed	Ethanol or aqueous extract	69
Southern armyworm, *Spodoptera eridania*	Pigweed	Aqueous extract	69
Tobacco budworm, *Heliothis virescens*	Various non host plants	Aqueous extracts	70
	Elderberry	Polar and nonpolar extracts	71
Cabbage butterflies, *Pieris brassicae, P. rapae, P. napi*	Various nonhost plants	Polar or nonpolar extracts	72
Cabbage butterfly, *Pieris rapae*	Host plants	Hexane extracts	73
	Nonhost plants	Aqueous extracts	73
	Erysimum cheiranthoides *Capsella bursa-pastoris*	Polar extracts	17
Cabbage looper, *Trichoplusia ni*	Various host and non-host plants	Macerated leaf tissues	34
	Cabbage	Ether-soluble constituents	74
Sugar beet moth, *Scrobipalpa ocellatella*	Chestnut leaves	Aqueous extracts	75
Large white butterfly, *Pieris brassicae*	Siberian wallflower	A strophanthidine glycoside	76

REFERENCES

1. **Städler, E.**, Attractants, arrestants, feeding and oviposition stimulants in insect-plant relationships: application for pest control, in *Natural Products for Innovative Pest Management*, Whitehead, D. L. and Bowers, W. S., Eds., Pergamon Press, New York, 1983, 243.
2. **Kogan, M.**, The role of chemical factors in insect/plant relationships, in *Proc. 15th Int. Congr. Entomol.*, Washington, D.C., 1977, 211.
3. **Fraenkel, G. S.**, The *raison d'etre* of secondary plant substances, *Science*, New York, 129, 1466, 1959.
4. **Schoonhoven, L. M.**, Chemical mediators between plants and phytophagous insects, in *Semiochemicals: Their role in Pest Control*, Nordlund, D. A., Jones, R. L. and Lewis, W. J., Eds., John Wiley & Sons, New York, 1981, 31.
5. **Renwick, J. A. A.**, Non-preference mechanisms: plant characteristics influencing insect behavior, in *Plant Resistance to Insects*, Hedin, P. A., Ed., Am. Chem. Soc. Symp. Ser., 208, 1983, 199.
6. **Dethier, V. G.**, Mechanisms of host plant recognition, *Entomol. Exp. Appl.*, 31, 49, 1982.
7. **Miller, J. R. and Strickler, K. L.**, Finding and accepting host plants, in *Chemical Ecology of Insects*, Bell, W. J. and Cardé, R. T., Eds., Chapman and Hall Ltd., London, 1984, 127.
8. **Prokopy, R. J. and Owens, E. D.**, Visual detection of plants by herbivorous insects, *Annu. Rev. Entomol.*, 28, 337.
9. **Rausher, M. D.**, Search image for leaf shape in a butterfly, *Science*, New York, 200, 1071, 1978.

10. **Harris, M. O. and Miller, J. R.,** Foliar form influences ovipositional behavior of the onion fly, *Physiol. Entomol.,* 9, 145, 1984.
11. **Zohren, E.,** Laboruntersuchungen zur Massenzucht, Lebenweise, Eiablage und Eiablageverhalten der Kohlfliege *Chortophila brassicae* (Diptera: Anthomyiidae), *Z. Angew. Entomol.,* 62, 139, 1968.
12. **Schöni, R. and Städler, E.,** Personal communication, 1987.
13. **Ilse, D.,** New observations on responses to colors in egg-laying butterflies, *Nature, (London),* 140, 544, 1937.
14. **Fox, R. M.,** Forelegs of butterflies. I. Introduction: chemoreception, *J. Res. Lepid.,* 5, 1, 1966.
15. **Duffey, S. S.,** Plant glandular trichomes: their partial role in defense against insects, in *Insects and the Plant Surface,* Juniper, B., and Southwood, R., Eds., Edward Arnold, London, 1986, 151.
16. **Gregory, P., Avé, D. A., Bouthyette, P. J., and Tingey, W. M.,** Insect-defensive chemistry of potato glandular trichomes, in *Insects and the Plant Surface,* Juniper, B., and Southwood, R., Eds., Edward Arnold, London, 1986, 173.
17. **Renwick, J. A. A. and Radke, C. D.,** Chemical stimulants and deterrents regulating acceptance or rejection of crucifers by cabbage butterflies, *J. Chem. Ecol.,* 13, 1771, 1987.
18. **Traynier, R. M. M.,** Visual learning in assays of sinigrin solution as an oviposition releaser for the cabbage butterfly, *Pieris rapae, Entomol. Exp. Appl.,* 40, 25, 1986.
19. **Hedin, P. A., Maxwell, F. G., and Jenkins, J. N.,** Insect plant attractants, feeding stimulants, repellents, deterrents and other related factors affecting insect behavior, in *Proc. Summer Inst. Biol. Control Plant Insects Dis.,* Maxwell, F. G., and Harris, F. A., Eds., University Press, Jackson, Mississippi, 1974, 494.
20. **Robert, P. C.,** Les relations plantes — insectes phytophages chez les femelles pondeuses: le rôle des stimulus chimiques et physiques. Une mise au point bibliographique, *Agronomie,* 6, 127, 1986.
21. **Prokopy, R. J.,** Evidence for a marking pheromone deterring repeated oviposition in apple maggot flies, *Environ. Entomol.,* 1, 326, 1972.
22. **Prokopy, R. J.,** Epideictic pheromones that influence spacing patterns of phytophagous insects, in *Semiochemicals: Their Role in Pest Control,* Nordlund, D. A., Jones, R. L., and Lewis, W. J., Eds., John Wiley & Sons, New York, 1981, 181.
23. **Hurter, J., Boller, E. F., Städler, E., Blattmann, B., Buser, H. R., Bosshard, N. U., Damm, L., Kozlowski, M. W., Schöni, R., Raschdorf, F., Dahinden, R., Schlumpf, E., Fritz, H., Richter, W. J., and Schreiber, J.,** Oviposition-deterring pheromone in *Rhagoletis cerasi* L.: Purification and determination of the chemical constitution, *Experientia,* 43, 157, 1987.
24. **Katsoyannos, B. I. and Boller, E. F.,** First field application of oviposition-deterring marking pheromone of European cherry fruit fly, *Environ. Entomol.,* 5, 151, 1976.
25. **Tabashnik, B. E.,** Deterrence of diamondback moth (Lepidoptera: Plutellidae) oviposition by plant compounds, *Environ. Entomol.,* 14, 575, 1985.
26. **Tabashnik, B. E.,** Plant secondary compounds as oviposition deterrents for cabbage butterfly, *Pieris rapae* (Lepidoptera: Pieridae), *J. Chem. Ecol.,* 13, 309, 1987.
27. **Flint, H. M., Smith, R. L., Pomonis, J. G., Forey, D. E., and Horn, B. R.,** Phenylacetaldehyde: oviposition inhibitor for the pink bollworm, *J. Econ. Entomol.,* 70, 547, 1977.
28. **Abivardi, C. and Benz, G.,** Oviposition-deterring activity of sixteen extracts of medicinal plants, extensively used in modern medicine, against *Cydia pomonella* L. (Lepidoptera: Tortricidae), *Mitt. Schweiz., Entomol. Ges.,* 59, 31, 1986.
29. **Abo El-Ghar, G. E. S. and El-Sheikh, A. E.,** Effectiveness of some plant extracts as surface protectants of cowpea seeds against the pulse beetle, *Callosobruchus chinensis, Phytoparasitica,* 15, 109, 1987.
30. **Städler, E.,** Sensory aspects of insect plant interactions, *Proc. 15th Int. Congr. Entomol.,* Washington, D.C., 1977, 228.
31. **Städler, E.,** Contact chemoreception, in *Chemical Ecology of Insects,* Bell, W. J., and Cardé, R. T., Eds., Chapman & Hall, London, 1984, 3.
32. **Saxena, K. N. and Basit, A.,** Inhibition of oviposition by volatiles of certain plants and chemicals in the leafhopper *Amarasca devastans* (Distant), *J. Chem. Ecol.,* 8, 329, 1982.
33. **Renwick, J. A. A. and Radke, C. D.,** An oviposition deterrent associated with frass from feeding larvae of the cabbage looper, *Trichoplusia ni* (Lepidoptera: Noctuidae), *Environ. Entomol.,* 9, 318, 1980.
34. **Renwick, J. A. A. and Radke, C. D.,** Host plant constituents as oviposition deterrents for the cabbage looper, *Trichoplusia ni, Entomol. Exp. Appl.,* 30, 201, 1981.
35. **Cirio, U.,** Reperti sul meccanismo stimolo-risposta nell'ovideposizione del *Dacus oleae* Gmelin (Diptera; Trypetidae), *Redia,* 52, 577, 1971.
36. **Girolami, V., Vianello, A., Strapazzon, A., Ragazzi, E., and Veronese, G.,** Ovipositional deterrents in *Dacus oleae, Entomol. Exp. Appl.,* 29, 177, 1981.
37. **Maeshima, K., Hayashi, N., Murakami, T., Takahashi, F., and Komae, H.,** Identification of chemical oviposition stimulants from rice grain for *Sitophilus zeamais* Motschulsky (Coleoptera, Curculionidae), *J. Chem. Ecol.,* 11, 1, 1985.

38. **Applebaum, S. W., Gestetner, B., and Birk, Y.**, Physiological aspects of host specificity in the Bruchidae. IV. Developmental incompatibility of soybeans for *Callosobruchus, J. Insect Physiol.*, 11, 611, 1965.

39. **Pouzat, J.**, Le comportement de ponté de la bruche du haricot en présence d'extrait de plante-hôte. Mise en évidence d'interaction gustatives et tactiles, *C. R. Acad. Sci. (Paris) D.*, 282, 1971, 1976.

40. **Pouzat, J.**, Effet des stimulations provenant de la plante-hôte, le haricot (*Phaseolus vulgaris* L.) sur le comportement de ponté de la Bruche du Haricot (*Acanthoscelides obtectus* Say), *Colloq. Int. C.N.R.S.*, 265, 115, 1977.

41. **Jermy, T. and Szentesi, Á.**, The role of inhibitory stimuli in the choice of oviposition site by phytophagous insects, *Entomol. Exp. Appl.*, 24, 258, 1978.

42. **Röttger, U.**, Blattwachskomponenten als Schlüsselreize der Eiablage von *Pegomya betae* Curt. (Diptera, Anthomyidae), *Mitt. Dtsch. Ges. Allg. Angew. Entomol.*, 1, 22, 1978.

43. **Traynier, R. M. M.**, Stimulation of oviposition by the cabbage root fly *Erioischia brassicae, Entomol. Exp. Appl.*, 10, 401, 1967.

44. **Städler, E.**, Chemoreception of host plant chemicals by ovipositing females of *Delia (Hylemya) brassicae, Entomol. Exp. Appl.*, 24, 511, 1978.

45. **Finch, S.**, Volatile plant chemicals and their effect on host plant finding by the cabbage root fly *(Delia brassicae), Entomol. Exp. Appl.*, 24, 350, 1978.

46. **Schöni, R., Städler, E., Renwick, J. A. A., and Radke, C. D.**, Host and non-host plant chemicals influencing the oviposition behavior of several herbivorous insects, in *Insects-Plants*, Labeyrie, V., Fabres, G., and Lachaise, D., Eds., Dr. W. Junk Publishers, Dordrecht, 1987, 31.

47. **Matsumoto, Y. and Thorsteinson, A. J.**, Effect of organic sulfur compounds on oviposition in onion maggot, *Hylemya antiqua* Meigen (Diptera: Anthomyiidae), *Appl. Entomol. Zool.*, 3, 5, 1968.

48. **Berüter, J. and Städler, E.**, An oviposition stimulant for the carrot rust fly from carrot leaves, *Z. Naturforsch.*, 26b, 339, 1971.

49. **Guerin, P. M., Städler, E., and Buser, H. R.**, Identification of host plant attractants for the carrot fly, *Psila rosae, J. Chem. Ecol.*, 9, 843, 1983.

50. **Städler, E. and Buser, H. R.**, Defense chemicals in leaf surface wax synergistically stimulate oviposition by a phytophagous insect, *Experientia*, 40, 1157, 1984.

51. **Unnithan, G. C., Saxena, K. N., Bentley, M. D., and Hassanali, A.**, Role of sorghum extract in eliciting oviposition on a non-host by the sorghum shootfly, *Atherigona soccata* Rondani (Diptera: Muscidae), *Environ. Entomol.*, 16, 967, 1987.

52. **Gupta, P. D. and Thorsteinson, A. J.**, Food plant relationships of the diamondback moth (*Plutella maculipennis* (Curt)) II. Sensory regulation of oviposition of the adult female, *Entomol. Exp. Appl.*, 3, 305, 1960.

53. **Renwick, J. A. A. and Radke, C. D.**, Chemical recognition of host plants for oviposition by the cabbage butterfly, *Pieris rapae* (Lepidoptera: Pieridae), *Environ. Entomol.*, 12, 446, 1983.

54. **Ma, W. C. and Schoonhover, L. M.**, Tarsal contact chemosensory hairs of the large white butterfly *Pieris brassicae* and their possible role in oviposition behavior, *Entomol. Exp. Appl.*, 16, 343, 1973.

55. **Feeny, P. P., Sachdev, K., Rosenberry, L., and Carter, M.**, *Phytochemistry*, 27, 3439, 1988.

56. **Honda, K.**, Flavanone glycosides as oviposition stimulants in a papilionid butterfly, *Papilio protenor, J. Chem. Ecol.*, 12, 1999, 1986.

57. **Jackson, D. M., Severson, R. F., Johnson, A. W., Chaplin, J. F., and Stephenson, M. G.**, Ovipositional response of tobacco budworm moths (Lepidoptera: Noctuidae) to cuticular chemical isolate from green tobacco leaves, *Environ. Entomol.*, 13, 1023, 1984.

58. **Jackson, D. M., Severson, R. F., Johnson, A. W., and Herzog, G. A.**, Effects of cuticular duvane diterpenes from green tobacco leaves on tobacco budworm (Lepidoptera: Noctuidae) oviposition, *J. Chem. Ecol.*, 12, 1349, 1986.

59. **Mitchell, E. R. and Heath, R. R.**, *Heliothis subflexa* (GN) (Lepidoptera: Noctuidae): demonstration of oviposition stimulant from ground cherry using novel bioassay, *J. Chem. Ecol.*, 13, 1849, 1987.

60. **Burton, R. L. and Schuster, D. J.**, Oviposition stimulant for tomato pinworms from surfaces of tomato plants, *Ann. Entomol. Soc. Am.*, 74, 512, 1981.

61. **Städler, E.**, Host plant stimuli affecting oviposition behavior of the eastern spruce budworm, *Entomol. Exp. Appl.*, 17, 176, 1974.

62. **Thibout, E., Auger, J., and Lecomte, C.**, Host plant chemicals responsible for attraction and oviposition in *Acrolepiopsis assectella*, in *Proc. 5th Int. Symp. Insect-Plant Relationships*, Visser, J. H., and Minks, A. K., Eds., Pudoc, Wageningen, 1982, 107.

63. **Wearing, C. H. and Hutchins, R. F. N.**, α-Farnesene, a naturally occurring oviposition stimulant for the codling moth, *Laspeyresia pomonella, J. Insect Physiol.*, 19, 1251, 1973.

64. **Nishida, R., Ohsugi, T., Kokubo, S., and Fukami, H.**, Oviposition stimulants of a citrus-feeding swallowtail butterfly, *Papilio xuthus* L., *Experientia*, 43, 342, 1987.

65. **Alfaro, R. I., Pierce, H. D., Jr., Borden, J. H., and Oehlschlager, A. C.**, Insect feeding and oviposition deterrents from western red cedar foliage, *J. Chem. Ecol.*, 7, 39, 1981.

66. **Wiens, M. N., Rahe, J. E., Vernon, R. S., and McLean, J. A.,** Oviposition deterrents for *Hylemya antiqua* in hydrated seeds of *Phaseolus vulgaris, Environ. Entomol.,* 7, 165, 1978.

67. **Coudriet, D. L., Prabhaker, N., and Meyerdirk, D. E.,** Sweet potato whitefly (Homoptera: Aleyrodidae): effects of neem-seed extract on oviposition and immature stages, *Environ. Entomol.,* 14, 776, 1985.

68. **Dover, J. W.,** The responses of some Lepidoptera to labiate herb and white clover extracts, *Entomol. Exp. Appl.,* 39, 177, 1985.

69. **Mitchell, E. R. and Heath, R. R.,** Influence of *Amaranthus hybridus* L. allelochemics on oviposition behavior of *Spodoptera exigua* and *S. eridania* (Lepidoptera: Noctuidae), *J. Chem. Ecol.,* 11, 609, 1985.

70. **Tingle, F. C. and Mitchell, E. R.,** Aqueous extracts from indigenous plants as oviposition deterrents for *Heliothis virescens* (F.), *J. Chem. Ecol.,* 10, 101, 1984.

71. **Tingle, F. C. and Mitchell, E. R.,** Behavior of *Heliothis virescens* (F.) in presence of oviposition deterrents from elderberry, *J. Chem. Ecol.,* 12, 1523, 1986.

72. **Lundgren, L.,** Natural plant chemicals acting as oviposition deterrents on cabbage butterflies (*Pieris brassicae* [L.], *P. rapae* [L.] and *P. napi* [L.]), *Zoologica Scripta,* 4, 253, 1975.

73. **Renwick, J. A. A. and Radke, C. D.,** Constituents of host- and non-host plants deterring oviposition by the cabbage butterfly, *Pieris rapae, Entomol. Exp. Appl.,* 39, 21, 1985.

74. **Renwick, J. A. A. and Radke, C. D.,** Activity of cabbage extracts in deterring oviposition by the cabbage looper, *Trichoplusia ni,* in *Proc. 5th Int. Symp. Insect-Plant Relationships,* Visser, J. H. and Minks, A. K., Eds., Pudoc, Wageningen, 1982, 139.

75. **Robert, P. C. and Blaisinger, P.,** Role of non-host plant chemicals in the reproduction of an oligophagous insect: the sugar beet moth *Scrobipalpa ocellatella* (Lepidoptera: Gelechiidae), *Entomol. Exp. Appl.,* 24, 432, 1978.

76. **Rothschild, M., Alborn, H., Stenhagen, G., and Schoonhoven, L. M.,** A strophanthidine glycoside in the Siberian wallflower (*Cheiranthus* x *allionii* Horte Bois): a naturally occurring contact deterrent for the large white butterfly (*Pieris brassicae* L.), *Phytochemistry,* 27, 101, 1988.

PLANT STIMULANTS AND ATTRACTANTS (KAIROMONES)*

Susan K. Waage and Paul A. Hedin

INTRODUCTION

For the purpose of this discussion, an attractant is defined as a stimulus to which an insect responds by orienting movements toward the source. A stimulant is descriptive of a sequence of events: (1) initiation of feeding, (2) maintenance of feeding, and (3) cessation of feeding. There has been a gradual evolution in the terminology of behavioral chemicals. In the 1960s, the terms feeding stimulant, attractant, repellent, and deterrent were used widely. The compounds responsible for these activities were known as secondary plant constituents, compounds that were not required for primary growth and development of the plant. However, it was recognized that they played several roles in plant protection and propagation.

Whittaker[1] suggested that the secondary plant constituents (non-nutritional chemicals affecting behavior and development) be called *"allelochemics"* because of their effects on other organisms. He also proposed that factors giving the host plant an advantage be called "allomones", and factors giving the insect an adaptive advantage (i.e., attractants and stimulants) be called "kairomones". The term "kairomone" has also had a more restrictive meaning, that of one or more chemicals in a host insect (sequestered from a plant on which the insect feeds) attracting a second insect. However, the term "kairomone" has been used widely for plant compounds that attract or initiate feeding by an insect.

Some plant compounds may be toxic (allomones) to some insects but kairomones to others. In evolutionary time, apparently some insects develop mechanisms by adaptation to detoxify compounds in plants on which they must feed, while others do not. In nearly all instances, these plants are recognized as quite host specific for the given insect. Some structure-activity studies have shown that seemingly minor differences such as stereochemistry as well as group functionality can convert kairomones to allomones, and vice versa.

The focus of this chapter is to provide in-depth information about a few selected examples of stimulants and attractants. However, the following review articles are cited to provide a fairly exhaustive listing of categories of behavioral responses and the related allelochemics.[2-6]

FEEDING STIMULANTS

The boll weevil, *Anthonomus grandis* Boheman, oviposits in the flower bud (square) of cotton, *Gossypium hirsutum* L. The larvae feed on the square, resulting in a decreased yield of cotton. It was observed[7] that aqueous extracts of the bud stimulated feeding by the boll weevil. In subsequent isolational studies,[8,9] each of a series of successive solvent extracts of increasing polarity was found to elicit feeding activity. Further fractionation indicated that several classes of compounds were involved, but when the pure components were isolated, they showed little or no activity. Only a partial regeneration of activity was found after recombination of fractions.

In a study[10] of 286 compounds including known cotton constituents, common metabolites, and compounds inducing primary mammalian sensations of taste and odor, 52 were found to have substantial activity, and among these 14 had been previously reported in cotton. Large amounts of flavonoids were contained in the alcoholic and aqueous extracts,

* In partial fulfillment, Ph.D. requirement, Department of Chemistry, Mississippi State University, Mississippi State, MS 39762.

which were highly active.[8,9] Although these extracts retained most of their activity after removal of the flavonoids, this class did exhibit appreciable activity. Of these, quercetin (Structure 1), quercetin-7-glucoside (quercimeritrin) (Structure 2), and quercetin-3′-glucoside (Structure 3) showed moderate activity. These were the first active compounds isolated from cotton bud extracts.

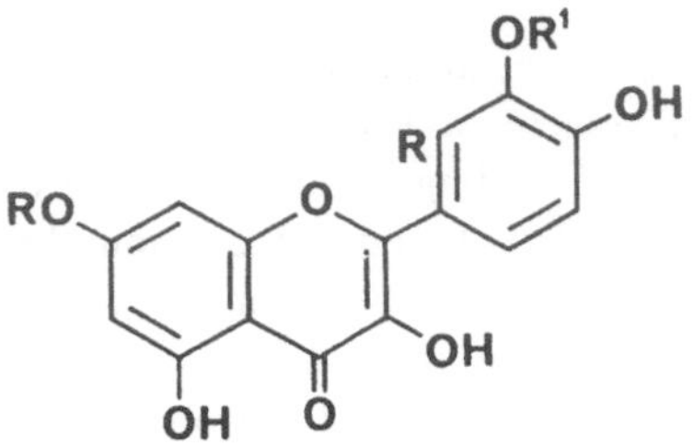

(1) R = R^1 = H
(2) R = glucose, R^1 = H
(3) R = H, R^1 = glucose

The isolation of the flavonoid components from the cotton bud was described by Struck et al.[9] Successive 24 h extractions (soxhlet) of freeze-dried powder were made with petroleum ether, benzene, chloroform, ethanol, methanol, and water. The methanol fraction in which flavonoids had been detected was subjected to preparative paper chromatography with 22% isopropanol in water. The slowest migrating fraction contained the flavonoid monoglycosides. The compounds were separated by polyamide thin-layer and paper chromatography and crystallized from water.

The assay used to test the compounds for feeding stimulant activity was described by Hedin et al.[8] A solution of the compound to be tested was applied to a 37 mm square of Whatman no. 1 filter paper, and the paper was wrapped around a cylindrical agar-water plug and fastened with a staple. Each plug was placed in a 100 mm Petri dish with 10 weevils and with another plug that was wrapped with paper that was not impregnated. The Petri dishes were placed in an incubator for 4 h, after which the papers were unwrapped and the number of punctures in each paper was counted. After several replications of one test a positive or negative score was calculated by subtracting the mean number of punctures in the blank from the mean number of punctures in the test plugs. An index (T/S) was obtained by multiplying the score by 100 and then dividing by the response obtained to an aqueous extract of freeze-dried cotton bud powder. On this scale, quercetin had a T/S value of 44 (for a 100 μg sample), and quercetin-7-glucoside had a T/S value of 52.

Quercetin, 3,5,7,3′,4′-pentahydroxyflavone ($C_{15}H_{10}O_7$, molecular weight 302.33), is the most common flavonoid, occurring in many plants in the free form and as several glycosides.[11,12] It is crystallized as yellow needles from dilute alcohol in the form of the dihydrate ($C_{15}H_{10}O_7 \cdot 2H_2O$). It becomes anhydrous at 95° to 97°C, and decomposes at 314°C. Its ultraviolet spectrum shows λ_{max} at 255 and 370 nm with shoulders at 269 and 301 nm.[13] An alcoholic solution of quercetin has a very bitter taste for humans.[12]

Quercimeritrin, 3,3′,4′,5,7-pentahydroxyflavone-7-D-glucoside ($C_{21}H_{20}O_{12}$, molecular weight 463.37) is crystallized from aqueous pyridine as the trihydrate ($C_{21}H_{20}O_{12} \cdot 3H_2O$) in the form of yellow plates. It loses the water of crystallization at 100°C and melts at 247° to 249°C.

Another flavonol glycoside, kaempferol-3-O-xylosylgalactoside (Structure 4), has been found to stimulate feeding in the monophagous flea beetle, *Phyllotreta armoraciae*.[14] This compound, along

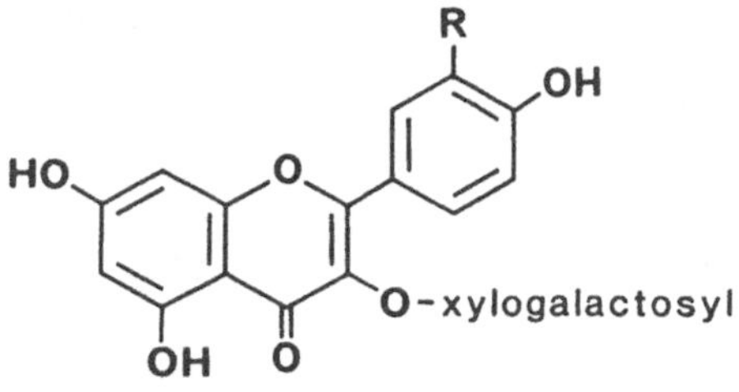

(4) R = H
(5) R = OH

with quercetin-3-O-xylosylgalactoside (Structure 5) was isolated from aqueous methanol extracts of the freeze-dried leaves of horseradish, *Armoracia rusticana*. This was the first allelochemic which is not a glucosinolate to stimulate feeding by a crucifer feeding insect.

OVIPOSITION STIMULANTS

The eastern spruce budworm, *Choristoneura fumiferana*, a forest pest of eastern North America, feeds on the foliage of various genera of coniferous trees. The balsam fir and the native spruce are its principal hosts. The feeding response of the spruce budworm has been found to be influenced only by nonspecific phagostimulants, such as sugars and the amino acid L-proline.[15] However, the oviposition behavior of the female seems to be affected by several factors, including light, gravity, and density of the needles on the twigs. The female spruce budworm distinguishes between the different host species with the following order of decreasing preference: white spruce > balsam fir > red and black spruce. Oviposition is influenced by chemical stimuli from the host needles.

By using electrophysiological techniques, the effects of several different pinenes on the olfactory chemoreceptors on the antennae were observed.[15] Significant activity was shown by (+)-α-pinene (Structure 6) and (−)-β-pinene (Structure 7).

(6)    (7)

Pinene ($C_{10}H_{16}$, molecular weight 136.23) is a common constituent of many volatile oils. (+)-α-Pinene[16] is a liquid with a boiling point of 155 to 156°C and an optical rotation at 20°C (solvent not specified) of +51.14° (for mass spectral data see Table 1).

Higgs et al.[19] reported that two mono-oxygenated monoterpenes caused an oviposition response in the house longhorn beetle, *Hylotrupes bajulus*. The frass of wood-boring larvae which had been reared on blocks of untreated *Pinus sylvestis* wood was extracted, and the fraction which gave the major electroantennographic response was analyzed. The major components were (−)-verbenone (Structure 8), *p*-cymen-9-ol (Structure 9), and myrtenol (Structure 10).

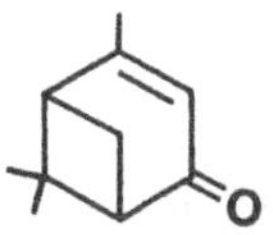

(8)

(9)

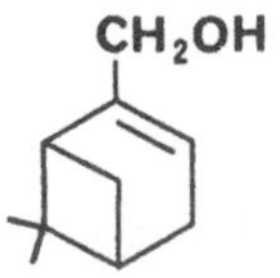

CH$_2$OH

(10)

TABLE 1
Selected Mass Spectral Data for Some Insect Attractants

Compound	Structure	Fragmentation[a]	Ref.
α-Pinene	6	136,M$^+$(8), 93(100), 92(30), 77(22), 41(23), 39(24)	17
β-Pinene	7	136,M$^+$(7), 93 (100), 69(47), 41(64), 39(33), 27(31)	18
α-Farnesene (E,E isomer)	11	204,M$^+$(1), 189(1), 123(18), 119(23), 107(27), 93(65), 79(30), 69(52), 55(47), 41(65)	26
β-Selinene	15	204,M$^+$(26), 121(88), 107(59), 80(100), 41(99), 29(56)	30
(+)-Limonene	17	136,M$^+$(20), 93(47), 68(100), 67(37), 41(29), 39(36)	39
Caryophyllene	18	204,M$^+$(2), 91(98), 79(100), 77(68), 53(73), 39(83)	40
α-Bisabolol	21	222,M$^+$(0.5), 204(12), 135(15), 119(29), 109(30), 93(26), 81(27), 69(100), 43(30), 41(34)	32
Bisabolene oxide	22	220,M$^+$(40), 138(100), 117(35), 115(40), 109(56), 96(59), 95(50), 67(35), 41(52)	33
Tetrahydro-β-bisabolol	—	226,M$^+$(1), 208(9), 124(44), 123(44), 113(100), 110(22), 95(53), 81(31), 55(29), 43(28), 41(38)	33
Phenylacetaldehyde	27	120,M$^+$(26), 92(23), 91(100), 90(2), 89(41), 65(13)	44

[a]   Fragmentations with molecular ion (M$^+$) and relative abundancies in parentheses.

In a simple choice test, the number of eggs oviposited on host wood vs. host wood plus each test compound was observed. (−)-Verbenone and *p*-cymen-8-ol were the most active constituents.

In 1973, Wearing et al.[20] noted that one or more volatile compounds produced by mature apples of the cultivar "Sturmer Pippin" stimulated oviposition and flight activity of female codling moths, *Laspeyresia pomonella* L., and in another study,[21] it was found that volatiles which were present in a chloroform extract of the apples were attractive to the larvae. Sutherland and Hutchins[22] isolated the active compound and identified it as an acyclic sesquiterpene, α-farnesene (Structure 11). A report in 1973[23] confirmed that two stereo-isomers, (E,E)-α- and (Z,E)-α-farnesene, both present in apples, stimulated oviposition by female codling moths.

CH$_3$–C=CH–CH$_2$–CH$_2$–C=CH–CH$_2$–CH=C–CH=CH$_2$ Struc 11
            | | |
           CH$_3$ CH$_3$ CH$_3$

α-Farnesene; 3,7,11-trimethyl-1,3,6,10-dodecatetraene (C$_{15}$H$_{24}$, molecular weight 204.34), is a thin oil with a boiling point of 204°C.[24] It is insoluble in water, but is miscible with hydrocarbon solvents.[25] (For mass spectral data see Table 1.)

The procedure for the isolation of α-farnesene from apples was described by Sutherland and Hutchins.[22] Three portions (300 ml) of chloroform were used to wash six small apples, and the extracts were combined and evaporated under reduced pressure at less than 40°C. Light petroleum ether was used to redissolve the residue, which was then chromatographed on Florisil. Rechromatography of the highly active hydrocarbon fraction on silicic acid impregnated with 20% $AgNO_3$ separated the saturated from the unsaturated hydrocarbons. Further chromatography of the unsaturated fraction on silicic acid impregnated with 20% $AgNO_3$ concentrated all of the activity in the polyunsaturated fraction. A pure fraction was obtained by column chromatography using 25% ether in benzene. It was shown by gas-liquid chromatography that the fraction had one major component whose retention time was the same as that of α-farnesene. When the extracted fraction was chromatographed with the pure substance, the peaks coincided.

Two methods were used to test for activity.[23] A solution was prepared by dissolving 4 μg of α-farnesene in 0.1 ml of chloroform. The solution was then pipetted into a ball of cotton wool 3 cm in diameter and weighing 0.1 ± 0.03 g. A control was prepared in the same way, but using pure chloroform. After the solvent had evaporated, the balls of cotton wool were placed in Petri dishes with lids.

In one test, each cotton wool ball was placed in a separate Petri dish (9.5 cm in diameter) with 4 female moths. After 30 min the number of eggs laid was counted. In the other bioassay, one test and one control cotton wool ball were placed on opposite sides of the same Petri dish (14 cm in diameter) with 8 female moths. After 30 min the number of eggs laid in each half of the dish was counted, and the distribution of the eggs was plotted. Ten to twenty replicates were done for each test. The experiments were done in the dark at a temperature of 24 ± 1°C and a relative humidity of 80 to 90%. They were begun at 4:45 p.m. ± 15 min.

In the first bioassay, for 10 replicates the total number of eggs laid in test dishes containing natural α-farnesene was 149, and the total number of eggs laid in control dishes was 30. In the second bioassay, for 10 replicates the total number of eggs laid on the test side was 143 and on the control side was 34.

HOST HABITAT STIMULANTS

Compounds that stimulate host seeking responses in parasites are defined as host habitat stimulants.[27] These compounds generally are sequestered from the host plant.

A compound that stimulates the host-seeking response in the parasite, *Microplitis croceipes*, was isolated from the feces of the host, the corn earworm, *Heliothis zea*.[28] 13-Methylhentriacotane (Structure 12) was identified as the first chemical that mediates the host-parasite relation.

$$CH_3(CH_2)_{11}\overset{\displaystyle CH_3}{\overset{|}{CH}}(CH_2)_{17}CH_3$$

(12)

Dry feces (100 g) of the corn earworm were homogenized with 500 ml of hexane and 10 g of sodium sulfate. The residue was extracted in a soxhlet apparatus for 2 h using the decanted liquid. The active material was isolated from the hexane extract by column chromatography on silica gel (2.5 × 40 cm). The active fraction was concentrated and chromatographed on a charcoal column (4 × 35 cm). The column was eluted successively with 500 ml of hexane and 2000 ml of 2% benzene in hexane. The active material was in the portion from 1500 to 1700 ml of the second solvent. Following rechromatography on the charcoal column, the active compound was purified by preparative gas chromatography (g.c.). The final product yield was 1 mg.[28]

The g.c. retention index (Kovats method) was calculated to be 3125 using a column (3 mm I.D. by 150 cm) containing 3% OV-I on 80 to 100 mesh Var Aport 30 at 260° and helium carrier gas (15 ml/min).[28]

13-Methylhentriacotane was synthesized by Equation 1. The g.c. retention index (Kovats method) of the synthesized compound was 3132.[28]

$$
\begin{array}{c}
\text{2-tetradecanone} \\
+ \\
\text{octadecyl bromide}
\end{array}
\xrightarrow[\text{2. hydrogenation}]{\text{1. Wittig reaction}}
\text{13-methylhentriacotane}
\qquad (1)
$$

For bioassay, the sample was applied to a 0.25 cm² area on a piece of filter paper which was placed in a Petri dish with a parasite. A positive response indicated an intense search of the area surrounding the sample and rubbing the surface with the antennae. After 3 approaches, a score of 3, 2, or 1 was given, depending on whether the 1st, 2nd, or 3rd approach resulted in a positive response.[28]

When under attack by predators, aphids produce a secretion from specialized structures called cornicles. The predator may release its prey when it comes in contact with droplets of the secretion. In addition, an alarm pheromone is present in the droplets, which is a warning to other aphids which can escape by walking, falling, or jumping away. Germacrene A (Structure 13) has been identified in the alarm pheromone of the spotted alfalfa aphid, *Therioaphis maculata*.[29]

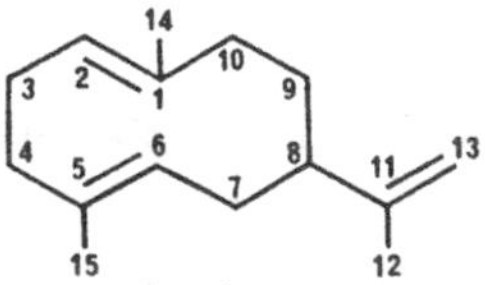

(13)

Aphids were reared in the field on susceptible alfalfa plants of the variety *caliverde* and were collected and placed in methanol at room temperature for 20 d. The methanol extract was filtered and concentrated and extracted in a separatory funnel with hexane. The hexane extract was washed with water and dried over Na_2SO_4. The hydrocarbons were separated by chromatography over florisil (50 g) with petroleum ether and purified by column chromatography over silica gel (3 g) with petroleum ether.[29]

Germacrene A was characterized by spectral data (Tables 1, 2, 3) and by the preparation of two derivatives. The mass spectrum showed the molecular ion at 204 mass units, which was consistent with a molecular formula of $C_{15}H_{24}$. From the ORD curve (C = 1.0, CCl_4), the optical rotation (at 25°C) was found to be −26.8°.[29]

11,13-Dihydrogermacrene A (Structure 14) was prepared by the reduction of 2.5 mg of the pure pheromone with palladium-charcoal (11 mg) in 2.5 ml of methanol. Hydrogen gas was bubbled through the mixture, which was stirred for 30 min at room temperature. The product had a molecular ion peak at 206 in the mass spectrum, corresponding to a molecular formula of $C_{15}H_{26}$. The infrared spectrum had major peaks at 1650, 1375, 1385, and 860 cm^{-1}.[29]

A sample (6.0 mg) of the pheromone was dissolved in 5 ml of hexane and stirred with silic acid at room temperature for 8 h. Analysis of the product by spectroscopic methods indicated β-selinene (Structure 15). The major infrared peaks were at 3080, 1780, 1650, and 890 cm^{-1} (Tables 1 and 2).[29]

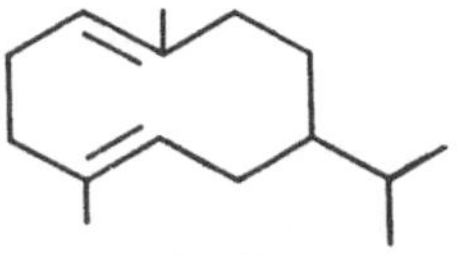

(14) 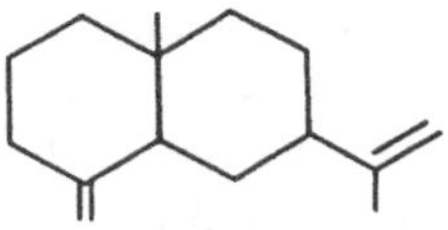   (15)

TABLE 2
Selected ^{1}H NMR Data for Some Insect Attractants

Compound	Structure	Spectral shifts[a]	Assignment[b]	Ref.
Germacrene A (in CCl$_4$)	13	1.52[6H,bs, half width 5 Hz]	Not assigned	29
		1.73 [3H, d, J = 1.5 Hz]	Vinyl methyl	
		4.59 [1H, m]	Terminal methylene	
		4.66 [1H, m]	Terminal methylene	
		4.75—5.25 [2H, m]	Not assigned	
β-Selinene (in CCl$_4$)	15	0.78 (3H, s)	Tertiary methyl	29
		1.78 (3H, s)	Vinyl methyl	
		4.43—4.70 (4H,bs)	4 terminal methylene protons	
α-Pinene (in CDCl$_3$)	19	0.84 (s, 3H)	C–H 9 or 10	38
		1.27 (s, 3H)	9 or 10	
		1.65 (s, 3H)	8	
		5.17 (m, 1H)	3	
α-Bisabolol (in CCl$_4$)	21	1.15 (t,3H,J = 10.5 Hz)	RR′CH–C<u>H</u>$_3$	
		1.48 (s, 1H)	–O<u>H</u>	
		1.62 (d, 9H)	Vinyl methyls	
		1.77—2.15 (m,8H)	Methylenes	
		2.22 (bs, 1H)	R′R″CHR‴	
		2.34 (bm,2H)	R′R″C=CH–C<u>H</u>$_2$R‴	
		5.10 (s,1H)	RC<u>H</u>=C(CH$_3$)$_2$	
		5.35 (s,1H)	R′CH$_3$C=C<u>H</u>-R″	
Tetrahydro-α-bisabolol (in CCl$_4$)		0.82 (s,1H)	–O<u>H</u>	
		0.94 (d,6H,J = 5.0)	RCH(C<u>H</u>$_3$)$_2$	
		0.96 (d,3H,J = 5.0)	R′R″CHC<u>H</u>$_3$	
		1.15—1.50 (bs,12H)	Methylenes	
		1.50—1090 (bs,3H)	R′R″C<u>H</u>C–OH–CH$_3$R‴ and R′R″CH$_2$C–OH–C<u>H</u>$_3$R″	
Bisabolene oxide (in CCl$_4$)	22	0.90 (d,3H)	C<u>H</u>$_3$CHR′R″	
		1.20 (bs,2H)	R′R″R‴C<u>H</u>$_2$R⁗(C-10)	
		1.55 (s,9H)	Vinyl methyls	
		1.35—2.20 (m,7H)	Methylenes and methinyl	
		3.11 (m,1H)	R′R″C=C<u>H</u>R‴	
		4.92 (bs,1H)	R′R″C=C<u>H</u>R‴	
		5.07 (bs,1H)	R′CH$_3$C=C<u>H</u>R″	
Ipsenol (in CDCl$_3$)	28	7.40 (dd,1H,J$_{trans}$18Hz,J$_{cis}$12Hz)	–C–C<u>H</u>=CH$_2$	50
		4.9—5.35 (m,4H)	Terminal methylenes	
		3.80 (m,1H)	C<u>H</u>–OH	
		2.47(dd,1H,J$_{gem}$14Hz,J$_{vic}$4Hz)	=C–C<u>H</u>$_2$–CHOH	
		2.18(dd,1H,J$_{gem}$14Hz,J$_{vic}$9Hz)		
		1.75 (m,1H)	–C<u>H</u>(CH$_3$)$_2$	
		1.60 (s,1H)	–O<u>H</u>	
		1.35 (m,2 noneq. H)	–CH–C<u>H</u>$_2$–CHOH	
		0.92 (d,3H,J7Hz), 0.90 (d,3H)	–CH(C<u>H</u>$_3$)$_2$	
Ipsdienol (in CCl$_4$)	29	6.28(dd,1H,J$_{trans}$ 18Hz,J$_{cis}$11Hz)	CH$_2$=CH–C=CH$_2$	
		4.88—5.2 (m,5H)	Olefinic H	

		4.32 (m,1H)	–CH$_2$–C<u>H</u>–CH
			        |
			       OH	

		2.29 (d,2H)		
		1.69 (s,3H), 1.61 (s,3H)	(CH$_3$)$_2$C=C	
		1.66 (s,1H)	–O<u>H</u>	
Verbenol (in CCl$_4$)	30	6.27 (broadened)	=C<u>H</u>	
		4.29 (broadened)	–C<u>H</u>–OH	
		1.8—2.5 (m,4H)	Not assigned	
		1.7 (s,3H), 1.32 (s,3H)	=C–(C<u>H</u>$_3$)$_2$	
		1.23 (s,1H)	–O<u>H</u>	
		1.03 (s,3H)	–CH$_3$	

[a] bs, broad singlet; s, singlet; d, doublet; dd, doublet of doublets; t, triplet; m, multiplet.

[b] As with RR′ CH–C<u>H</u>$_3$, C<u>H</u>$_3$ identifies the protons for which the shift data is given.

TABLE 3
Selected Infrared Spectral Data for Some Insect Attractants

Compound	Structure	Frequency (cm^{-1})	Assignment	Ref.
Germacrene A	13	3030 cm^{-1}	Terminal methylene	29
		1780, 1650	Trisubstituted double bond	
		1385, 1375	Isopropyl doublet	
		880	Terminal methylene	
		850	Trisubstituted double bond	
Ipsenol	28	2.96 μm	O–H	50
		3.23	Conjugated double bonds	
		7.21, 7.29	Geminal dimethyl	
		10.1, 11.1	Vinyl	
		11.26 (shoulder)	Terminal methylene	
Ipsdienol	29	3.02 μm	O–H	50
		6.26	Conjugated double bonds	
		9.80	C–OH	
		10.08, 11.10	Vinyl group	
Verbenol	30	2.98 μm	O–H	
		6.04	C=C	
		7.27 (notched)	–C(CH$_3$)$_2$	
		9.65, 9.90	C–OH	

PLANT ATTRACTANTS

Several mono- and sesquiterpenes are responsible for the attractancy of the cotton plant, *Gossypium hirsutum*, to the boll weevil, *Anthonomus grandis* Boheman. Minyard et al.[31] reported that (+)-α-pinene (Structure 16), (+)-limonene (Structure 17), (−)-β-caryophyllene (Structure 18), (+)-β-bisabolol (Structure 19), and caryophyllene oxide (Structure 20) were active components of the plant attractant complex; and α-bisabolol (Structure 21)[32] and bisabolene oxide (Structure 22)[33] were later identified as active constituents.

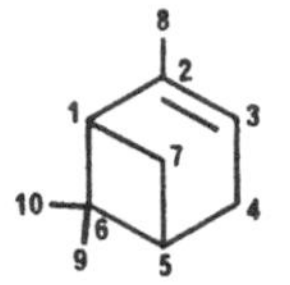

(16)

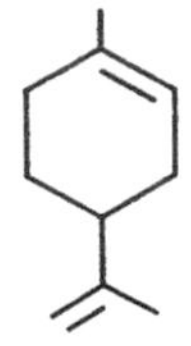

(17)

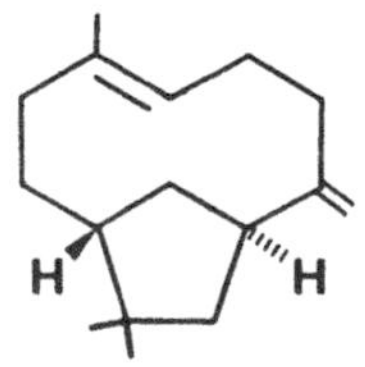

(18)

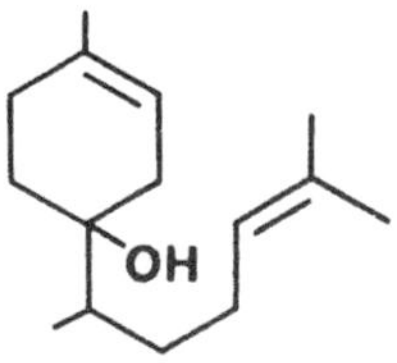

(19)

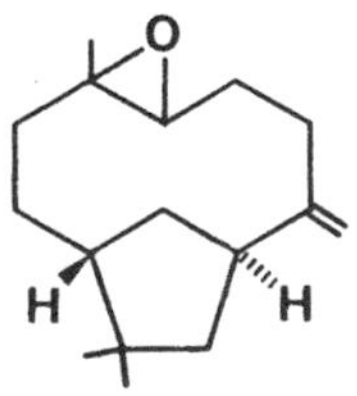

(20)

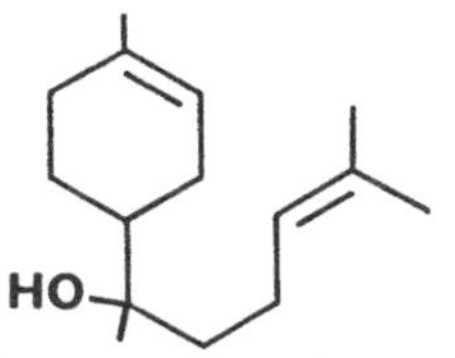

(21)

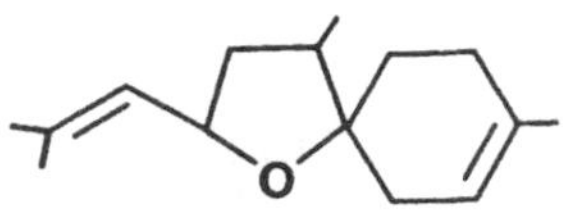

(22)

TABLE 4
Attractiveness of Hydrocarbons Isolated from Cotton Square
Oil to the Boll Weevil[31]

Compound	Concentration (ppm)	Attractiveness (T/S)
(+)-α-Pinene	1	1.13
	3	0.54
	10	0.46
	30	0.42
	100	0.35
(+)-Limonene	0.3	0.86
	1	0.57
	3	0.66
	10	0.66
	30	0.60
	100	0.51
(−)-β-Caryophyllene	0.1	0.17
	1	0.36
	10	0.70
	100	0.36

The attractants were isolated according to the procedure of Minyard et al.[34] Column chromatography of crude cotton square oil on silica gel yielded three fractions. The hydrocarbon fraction, which was eluted from the column with pentane, was further separated by gas-liquid chromatography to give a fraction containing terpenes and one containing sesquiterpenes. A fraction containing midpolar compounds was eluted from the column using 2-chloropropane. These compounds were separated by the Girard T procedure and by thin-layer and column chromatography on silica gel G. The polar compounds were eluted with methanol. Solvent partitioning between glycol and carbon tetrachloride separated the alcohols from the nonalcohols. The alcohols were purified by vacuum distillation and gas-liquid chromatography, and the nonalcohols were purified by vacuum distillation.

The bioassay technique was described by Hardee et al.[35] Aqueous solutions or suspensions of the samples were made, with the concentrations approximately equal to plant levels.[31] The best available attractant mixture from cotton buds was used as a standard. The results were reported as the ratio of net test to net standard values (T/S). This net value was the difference between the number of weevils responding to the sample and a water check. The hydrocarbon fraction did not show much activity (T/S = −0.04), but several individual components had moderate to high activity (Table 4).

(+)-α-Pinene (Tables 1 and 2) was previously mentioned as an oviposition stimulant for the spruce budworm. Another monoterpene, (+)-limonene (Table 1), is a common constituent of various essential oils, including orange and lemon. It has a boiling point of 175.5 to 176.5°C, and an optical rotation (at 19.5°C) of +123.8°.[36] Caryophyllene is a sesquiterpene that occurs in many essential oils (Table 1). The optical rotation (at 15°C) is −5.2°.

The attractancy of the midpolar (T/S = 0.78) and polar (T/S = 1.06) fractions was attributed to several oxygenated sesquiterpenes. (+)-β-Bisabolol was the major constituent (5.6% by weight) of the polar fraction. Mixtures of (+)-β-bisabolol and caryophyllene oxide were quite attractive to weevils (Table 5). Two other compounds were moderately active (Table 6) and were later identified as α-bisabolol[32] and bisabolene oxide.[33]

α-Bisabolol was identified by its spectral properties and by its conversion to tetrahydro-α-bisabolol (Tables 1, 2). Bisabolene oxide was characterized by spectroscopic analysis and by hydrogenation to give tetrahydro-β-bisabolol (Tables 1, 2).

TABLE 5
Attractiveness of (+)-β-Bisabolol and Caryophyllene Oxide to the Boll Weevil[31]

Test substance	Concentration (ppm)	Attractiveness (T/S)
(+)-β-Bisabolol	1	0.38
	3	0.46
	10	0.54
	100	0.37
	1000	0.12
Caryophyllene oxide	1	0.41
	3	1.03
	10	0.62
	30	0.54
	100	0.11
(+)-β-Bisabolol, caryophyllene oxide	10.1	0.60
	10.10	0.80
	100.10	0.54
	100.100	0.34

TABLE 6
Attractiveness of α-Bisabolol and Bisabolene Oxide to the Boll Weevil[31]

Compound	Concentration (ppm)	Attractiveness (T/S)
α-Bisabolol	1	0.37
	2	0.64
	6	0.44
	10	0.30
	20	0.08
Bisabolene oxide	1	0.47
	2	0.43
	6	0.58
	10	0.49
	20	0.23
	100	0.14

Pine plantations in New Zealand and Tasmania have suffered much damage by the wood wasp, *Sirex noctilio*. Damaged trees (*Pinus radiata*) have been found to be more attractive to *S. noctilio*.[41] High activity was seen in felled trees, girdled trees, and trees under induced attack by *S. noctilio*.

The change in composition of the volatiles with time produced by felled *P. radiata* was studied, and EAG methods were used to test the activity of the substances on the antennae of female *S. noctilio*. The amplitude of the signals produced by wasp antennae are related to the amount of test material [(amplitude of EAG signal) α log (amount of test material)], and different compounds elicit different responses.

Traps which were used to collect the volatiles were rinsed with ethyl ether. The solution was distilled under reduced pressure (20 mm Hg) using a Vigreux column to remove most of the monoterpenes, and micropreparative gas chromatography was used to fractionate the rest of the material. The active components were characterized by gas chromatography and by coupled gas chromatography-mass spectroscopy.

The less volatile fraction contained several components which were present in only trace amounts immediately after felling but increased to nearly 1% of the total volatiles by the 23rd day. A high antennal response from *S. noctilio* was elicited by camphor (Structure 23), pinocamphone (Structure 24), isopinocamphone (Structure 25), and *trans*-pinocarveol (Structure 26).

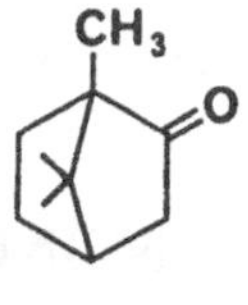

(23) 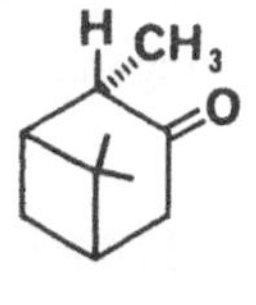   (24)

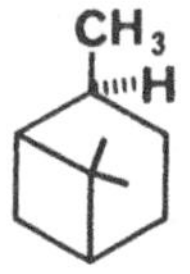

(25) 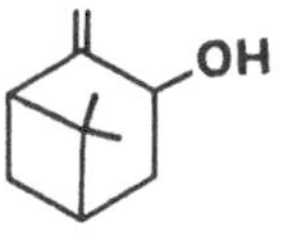   (26)

In separate studies, phenylacetaldehyde (Structure 27) was found to be an attractant for the codling moth, *Laspeyresia pomonella*,[42] and also for the corn earworm, *Heliothis zea*.[43]

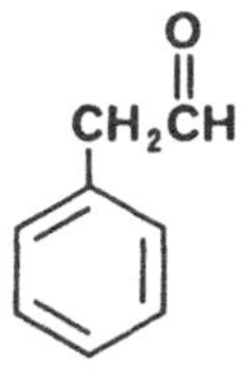

(27)

The bladder flower, *Araujia sericofera*, has been of interest because of its ability to attract and trap many species of moths. In a study[42] to determine the attractant component, the ether extract of bladder flowers was steam-distilled to separate the volatiles from the nonvolatile portion. Column chromatography of the volatile fraction, followed by microdistillation and gas chromatography, gave a pure distillate. The mass spectrum showed a molecular ion at m/e 120 (C_8H_8O), and it had the same fragmentation pattern as phenylacetaldehyde (Table 1).

The distillate was tested for activity by using Jackson traps with cotton wicks.[42] Traps which were baited with pure phenylacetaldehyde caught a mean of 14 ± 2.6 moths as compared to 1.0 ± 1.0 moths collected in the control traps. In another test, blacklight traps baited with 10 mg of phenylacetaldehyde trapped more macrolepidoptera (2238 moths) than unbaited traps (1697 moths).

Phenylacetaldehyde has also been isolated from the silk of sweet corn, *Zea mays* L.[43] The volatiles were collected by placing the corn silk in a glass chamber through which air was drawn by vacuum. The volatiles were collected in a tube packed with Porapak-Q. The adsorbent was extracted with ether, and the ether extract was concentrated and analyzed by gas chromatography-mass spectroscopy. Of the 9 volatiles that were identified, phenylacetaldehyde was found to be the attractive component.

In the field, Jackson traps baited with phenylacetaldehyde captured the corn earworm, the European corn borer, the soybean looper, the tarnished plant bug, *Cisseps fulvicollis*, and the forage looper. Because phenylacetaldehyde is a broad range attractant it may be useful in the control of harmful insect populations.

Caryophyllene (Structure 18) was previously mentioned as a component of cotton bud oil that is attractive to the boll weevil.[31] In another study,[45] it was reported to be also an attractant for the green lacewing, *Chrysopa carnea*. The green lacewing feeds primarily on mites, aphids, and other small insects and insect eggs.[46] The discovery of caryophyllene as an attractant for the green lacewing may be very important in the biological control of insects that inhabit cotton fields.

HOST HABITAT PHEROMONES

Some insects have the capability to sequester plant compounds, and/or to biochemically alter them to thence be employed as pheromones.[47] The biochemical alteration in several instances involves the process of allylic oxidation for the conversion of terpene hydrocarbons to alcohols. Examples follow.

Three components of the aggregation pheromone of the spruce bark beetle, *Ips typographus*, have also been found to be attractants for the bark beetle predator, *Thanasimus formicarius*.[48,49] The compounds ipsenol (Structure 28), ipsdienol (Structure 29), and verbenol (Structure 30) were isolated from the hindgut of the bark beetle and were characterized by comparison of their gas chromatographic retention times with those of authentic samples.

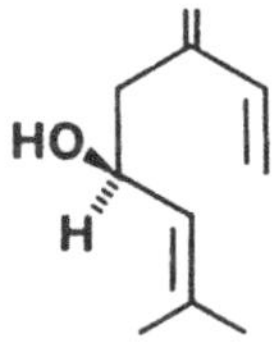

(28) 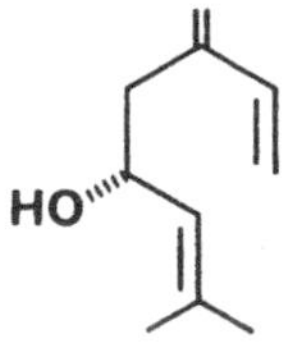   (29) 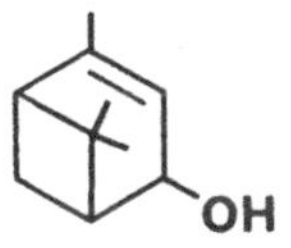   (30)

Ipsenol, 2-methyl-6-methylene-7-octen-4-ol (Tables 1 to 3), has a molecular ion peak at m/e 154 (1.1% of base) in its mass spectrum, which corresponds to a molecular formula of $C_{10}H_{18}O$. The base peak is at m/e 68. The ultraviolet spectrum in hexane shows a peak at 226 nm (ϵ = 20,000). The synthesis of ipsenol is illustrated by reaction 2.[50]

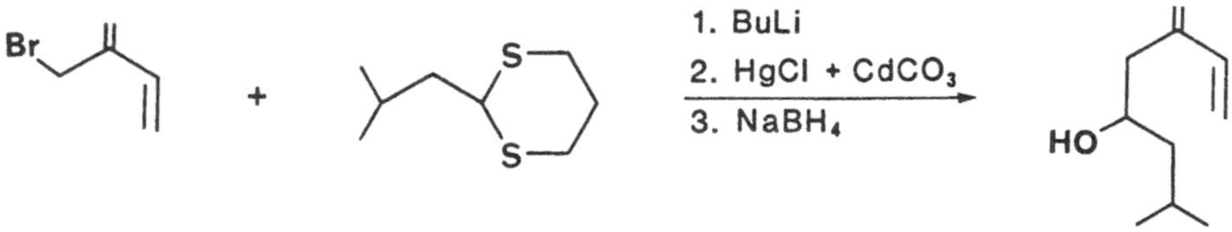

1. BuLi
2. HgCl + CdCO$_3$
3. NaBH$_4$

(2)

Ipsdienol, 2-methyl-6-methylene-2,7-octadien-4-ol (Tables 1 to 3), is synthesized as shown in reaction 3.[50]

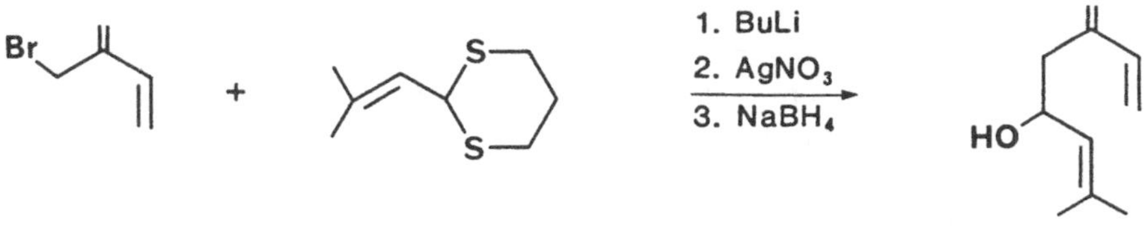

1. BuLi
2. AgNO$_3$
3. NaBH$_4$

(3)

Mass spectral analysis of verbenol, 2,6,6-trimethylbicyclo [3.1.1]-2-hepten-4-ol (Tables 1 to 3), reveals the molecular ion at m/e 152, corresponding to a molecular formula of $C_{10}H_{16}O$. The base peak is at m/e 43, and other major fragmentation peaks at m/e 119, 134, and 137. In the ultraviolet spectrum (in hexane), there is a peak at 215 nm (ϵ = 4000). Verbenol can be synthesized by treating verbenone with sodium borohydride (reaction 4).[50]

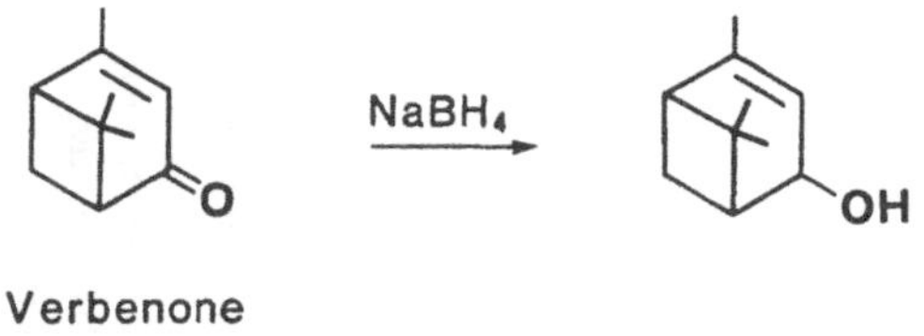

Verbenone

$$(4)$$

The terpenoid compounds (+)-*cis*-2-isopropenyl-1-methylcyclobutaneethanol, (Z)-3,3-dimethyl-δ^1,β-cyclohexaneethanol, and β-caryophyllene (Structure 18) were isolated from frass of the female boll weevil. In laboratory bioassays, a mixture of these compounds attracted primarily males,[51] whereas the male pheromone, grandlure, attracted primarily females.[52] β-Caryophyllene was previously mentioned as a component of cotton bud oil that is attractive to the boll weevil.[31]

Another compound that has exhibited activity is tricosane ($C_{23}H_{48}$), a straight-chain hydrocarbon isolated from scales of the corn earworm, *Heliothis zea*, and attractant for the egg parasite, *Trichogramma evanescens* Westwood.[53] Tricosane is included in this category because high molecular weight hydrocarbons can be found in the waxes of plants as well as insects and other animals.

A vacuum assembly was used to collect moth scales in an insect rearing lab. The concentrated hexane extract was fractionated on a silica gel column using 100 ml volumes of hexane, ether/hexane (1:9), ether/hexane (1:3), ether/hexane (1:1), ether, acetone, and water. Active fractions were concentrated, then analyzed by gas chromatography-mass spectroscopy.

Fractions were tested in a greenhouse by spraying them on pea seedlings to which host eggs had been applied. Ten parasites were released, and after 1 to 2 h the eggs were collected and dissected. A control consisted of pea seedlings sprayed only with solvent.

In the field, host eggs were placed on plants in a plot, which were sprayed alternatively with the test fraction or the solvent control. Plants which were not sprayed at all were used as checks. Then, newly emerged *T. evanescens* were released between rows.

The hexane fraction which was the most active was analyzed by gas-liquid chromatography on 5% SE-30 (210°C, 70 ml helium/min). The first peak had a retention time of 13.2 min, thus a Kovats retention index of 2300, corresponding to a C_{23} hydrocarbon, tricosane. Subsequent GC-MS showed a molecular ion (M^+ = 324, 1.7%) consistent with the formula $C_{23}H_{48}$. In tests on the pure compound, tricosane was found to be the most active component of the *H. zea* moth scale.

Finally, some additional insect stimulants and attractants are listed in Table 7. These are representative of the variety of chemicals involved in the insect-host relationship. Increased knowledge of the chemicals that control insect behavior may lead to a better understanding of insect-plant co-evolution.

TABLE 7
Additional Compounds Affecting Insect Behavior

Compound	Insect		Kind of activity	Plant	Ref.
	Scientific name	Common name			
Sinigrin	Pieris rapae L.	Imported cabbageworm	Feeding stimulant	Cruciferae	54
Carvone, methyl chavicol, coriandrol	Papilo polyxenes asterius Stoll.	Black swallowtail	Feeding stimulant, attractant	Umbelliferae	55
Guanine, monophosphate	Musca domestica L.	Housefly	Feeding stimulant	Yeast	56
Gossypol, quercetin, β-sitosterol, quercetin-7-glucoside, quercetin-3'-glucoside, cyanidin-3-glucoside	Anthonomus grandis Boh.	Boll weevil	Feeding stimulant	Cotton	10
Sinigrin, sinalbin, glucocheirolin	Plutella maculipennis (Curt)	Diamondback moth	Feeding stimulant	Cruciferae	57
Phaseolunatin, lotaustralin	Epilachna varivestis Mulsant	Mexican bean beetle	Feeding stimulant	Phaseolus	58
Amygdalin	Malacosoma americana (Fab.)	Eastern tent caterpillar	Feeding stimulant, attractant	Rosaceae	54
Allyl sulfide	Hylemia antiqua (Meig.)	Onion maggot	Feeding stimulant, attractant	Onion	59
Salicin	Phyllodecta vitallinae	Willow beetle	Feeding stimulant	Willow var.	60
Cucurbitacins	Diabrotica undecimpunctata howardi (Barb.)	Spotted cucumber beetle	Feeding stimulant, attractant	Curcurbitaceae	61, 62
Calotropin	Poekilocerus bufonius (Klug)	Grasshopper	Feeding stimulant	Milkweed, Ascelepius syrica (L.)	63
7-α-L-rhamnosyl-6-methoxyluteolin	Agasicles sp. (Nov.)	Chrysomelid beetle	Feeding stimulant	Alligatorweed, Alternanthera phylloxeroides, Amaranthaceae	64
p-hydroxyacetophenone, o-hydroxybenzyl alcohol, p-hydroxybenzaldehyde, (+)-catechin-5-α-D-xylopyranoside, lupeyl cerotate	Scolytus multistriatus (Marsham)	Smaller European elm bark beetle	Feeding stimulant	Elm, Ulmus americana (L.)	65, 66
Anethole, anisic aldehyde	Papilio polyxenes asterius (Stoll.)	Black swallowtail	Feeding stimulant	Anise, coriander, celery, angelica, citrus	67
Hypericin	Calliphora brunsvicensis		Feeding stimulant	Hypericum sp. (St. John's wort)	68

Compound(s)	Insect	Scientific name	Action	Host/Source	Ref.
Allyl isothiocyanate	Imported cabbageworm	Pieris rapae L.	Attractant	Cruciferae	69
2-hexenol, 3-hexenol, citral, terpinyl acetate, linalyl acetate, linalool	Silkworm	Bombyx mori L.	Attractant	Mulberry, Morus alba L.	70, 71, 72, 73, 74, 75
Oryzanone	Rice stem borer	Chilo suppressalis (Walker)	Attractant	Rice, Oryza sativa L.	76
Coumarin	Vegetable weevil	Listroderes costirustris oblignus King	Attractant	Sweet clover, Melilotus officinalis L.	77
Amyl salicylate	Tobacco hornworm	Protoparce sexta (Johan.)	Attractant	Jimson weed, Datura stromonium	78
Methyl eugenol	Oriental fruit fly	Dacus dorsalis (Hendel.)	Attractant	Fruits	79
α-Pinene	Douglas-fir beetle	Dendroctonus pseudotsugae (Hopkins)	Attractant	Douglas fir, Pseudotsuga menziezei Mirb. Franco	80
1,3-Diolein	House fly	Musca domestica	Attractant	Mushroom, Amarita muscaria L.	81
Iridodiol, matatabiol, 5-hydroxy-matatabiether, 7-hydroxydehydro-matatabiether, allomatatabiol.	Lacewings	Chrysopa septempunctata (Wesmael)	Attractant	Matatabi, Actinidia polygama (Miq.)	82
Benzoic acid, α-terpineol	Pine beetle	Blastophagus piniperda (L.)	Attractant	Pinus densiflora, Pinus silvestris	83
Ethanol	Ambrosia beetle	Gnatotrichus sulcatus (LeConte)	Attractant	Western hemlock, Tsuga heterophylla (R.) Sargent	84
Eugenol, anethole, α-Pinene	Pales weevil	Hylobius pales (Herbst.)	Attractant	Loblolly pine stem	85
α-Pinene	Western pine beetle	Dendroctonus brevicomis (LeConte)	Attractant	Pinus ponderosa (Laws)	86
Syringaldehyde, vanillin	Smaller European elm bark beetle	Scolytus multistriatus (Marsham)	Attractant	Decaying hardwood	87
Geraniol, citronellol	Japanese beetle	Popilia japonica	Attractant		88
α-pinene, β-pinene, limonene	Checkered beetle	Enoclerus sphegeus (Fab.) Enoclerus undatulus (Say)	Attractant	Douglas fir, Ponderosa pine, Grand fir	89
Terpinyl acetate, anethole, terpene alcohols	Codling moth Oriental fruit	Laspeyresia pomonella L. Grapholitha (= Laspeyresia) molesta Busck	Attractant		90, 91
Vanillic acid, p-hydroxybenzoic acid, p-coumaric acid, protocatechuic acid, ferulic acid	Termites	Kalotermes flavicollis (Fab.) Zootermopsis nevadensis (Hagen) Heterotermes indicola (Wasmann)	Attractant	Wood infected Basidomycetes	92

TABLE 7 (continued)
Additional Compounds Affecting Insect Behavior

Compound	Insect Scientific name	Insect Common name	Kind of activity	Plant	Ref.
β-carotene, niacin, Vitamin D₂, cholesterol, diethylstilbesterol, DL-aspartic acid, L-proline, histidine, pangamic acid	*Reticuliternes* *Bruchophagus roddi* (Gus.)	Alfalfa seed chalcid	Attractant	Alfalfa	93
α-Pinene	*Hylotrupes bajulus* (Gyll.) and *H. ater* Payk. (Coleoptera: Scolytidae)	Bark beetles	Attractant	*Pinus* sp. and others	94
Turpentine, smoke	*Cerambycid* sp.	Wood boring beetles	Attractant	Wood	95
Geraniol	*Apis mellifera* (L.)	Honey bees	Attractant	General	96
Allyl isothiocyanate	*Plutella xylostella* (Curt)	Diamondback moth	Oviposition stimulant	*Brassica nigra* L. and other Cruciferae	97
Sinigrin, β-phenyl-ethylamine, allyl isothiocyanate	*Hylemya brassicae* (Meigen)	Cabbage maggot	Oviposition stimulant	Swede, *Brassicae napus* L.	98
n-propyldisulfide, methyl disulfide, n-propyl mercaptan, n-propyl alcohol	*Hylemya antiqua* (Meigen)	Onion maggot	Oviposition stimulant	Onion	99
α-Pinene, β-Pinene, limonene, eugenol	*Schistocerca gregaria* (Forskal)	A desert locust	Oviposition stimulant	*Commiphora*	100
Methyl isoeugenol, *trans*-1,2-di-methoxy-4-propenylbenzene	*Psila rosae* (F.)	Carrot rust fly	Oviposition stimulant	Carrot	101

REFERENCES

1. **Whittaker, R. H.,** The biochemical ecology of higher plants, in *Chemical Ecology,* Sondheimer, E. and Simeone, J. B., Eds., Academic Press, Orlando, 1970, 43.
2. **Hedin, P. A., Maxwell, F. G., and Jenkins, J. N.,** Insect plant attractants, feeding stimulants, repellents, deterrents, and other related factors affecting insect behavior, in *Proc. Inst. Biol. Control Plant Insects Dis.,* Maxwell, F. G., Ed., University Press, 1974, 494.
3. **Hedin, P. A.,** *Host Plant Resistance to Pests,* Am. Chem. Soc. Symp. Ser., No. 62, Washington, D.C., 1977, 1.
4. **Rosenthal, G. A. and Janzen, D. H.,** *Herbivores: Their Interaction with Secondary Plant Metabolites,* Academic Press, Orlando, 1979, 1.
5. **Wallace, J. W. and Mansell, R. L.,** *Recent Advances in Phytochemistry,* Vol. 10, 1976.
6. **Nordlund, D. A., Jones, R. L., and Lewis, W. J.,** *Semiochemicals: Their Role in Pest Control,* John Wiley & Sons, New York, 1981, 1.
7. **Keller, J. C., Maxwell, F. G., Jenkins, J. N.,** Cotton extracts as arrestants and feeding stimulants for the boll weevil, *J. Econ. Entomol.,* 59, 800, 1962.
8. **Hedin, P. A., Thompson, A. C., and Minyard, J. P.,** Constituents of the cotton bud. Factors that stimulant feeding by the boll weevil, *J. Econ. Entomol.,* 59, 181, 1966.
9. **Struck, R. F., Frye, J., Shealy, Y. F., Hedin, P. A., Thompson, A. C., and Minyard, J. P.,** Constituents of the cotton bud. IX. Further studies on a polar feeding stimulant complex, *J. Econ. Entomol.,* 61, 270, 1968.
10. **Hedin, P. A., Miles, L. R., Thompson, A. C., and Minyard, J. P.,** Constituents of the cotton bud. Formulation of a boll weevil feeding stimulant mixture, *J. Agric. Food Chem.,* 16, 513, 1968.
11. **Gripenberg, J.,** Flavones, in *The Chemistry of Flavonoid Compounds,* Geissman, T. A., Ed., MacMillan, New York, 1962, 423.
12. **Stecher, P. G., Windholz, M., and Leahy, D. S.,** Eds., *The Merck Index,* 8th ed., Merck, Rahway, N.J., 1968, 899.
13. **Mabry, T. J., Markham, K. R., and Thomas, M. B.,** *The Systematic Identification of Flavonoids,* Springer-Verlag, Berlin, 1970, 65.
14. **Nielsen, J. K., Larsen, L. M., and Sorensen, H.,** Host plant selection of the horseradish flea beetle *Phyllotreta armorciae* (Coleoptera: Chrysomelidae): identification of two flavonol glycosides stimulating feeding in combination with glucosinolates, *Entomol. Exp. Appl.,* 26, 40, 1979.
15. **Stadler, E.,** Host plant stimuli affecting oviposition behavior of the eastern spruce budworm, *Entomol. Exp. Appl.,* 17, 176, 1974.
16. **Stecher, P. G., Windholz, M., and Leahy, D. S.,** Eds., *The Merck Index,* 8th ed., Merck, Rahway, N.J., 1968, 835.
17. **Stenhagen, E., Abrahamson, S., and McLafferty, F. W.,** Eds., *Atlas of Mass Spectral Data,* Vol. 1, John Wiley & Sons, New York, 1969, 705.
18. **Stenhagen, E., Abrahamson, S., and McLafferty, F. W.,** Eds., *Atlas of Mass Spectral Data,* Vol. 1, John Wiley & Sons, New York, 1969, 709.
19. **Higgs, M. D. and Evans, D. A.,** Chemical mediators in the oviposition behaviour of the house longhorn beetle, *Hylotrupes bajulus, Experientia,* 34, 46, 1978.
20. **Wearing, C. H., Connor, P. J., and Ambler, K. D.,** The olfactory stimulation of oviposition and flight activity of the codling moth, *Laspeyresia pomonella* L., using apples in an automated olfactometer, *N.Z. J. Sci.,* 16, 697, 1973.
21. **Sutherland, O. R. W.,** The attraction of newly-hatched codling moth (*Laspeyrisia pomonella*) larvae to apple, *Entomol. Exp. Appl.,* 15, 481, 1972.
22. **Sutherland, O. R. W. and Hutchins, R. F. N.,** α-Farnesene, a natural attractant for codling moth larvae, *Nature,* 239, 170, 1972.
23. **Wearing, C. H. and Hutchins, R. F. N.,** α-Farnesene, a naturally occurring oviposition stimulant for the codling moth, *Laspeyresia pomonella, J. Insect Physiol.,* 19, 1251, 1973.
24. **Devon, T. K. and Scott, A. I.,** *Handbook of Naturally Occurring Compounds,* Vol. II, Academic Press, Orlando, 1972, 73.
25. **Stecher, P. G., Windholz, M., and Leahy, D. S.,** Eds., *The Merck Index,* 8th ed., Merck, Rahway, N.J., 1968, 448.
26. **Stenhagen, E., Abrahamsson, S., and McLafferty, F. W.,** Eds., *Registry of Mass Spectral Data,* Vol. 2, John Wiley & Sons, New York, 1974, 1023.
27. **Vinson, S. B.,** Host selection by insect parasitoids, *Ann. Rev. Entomol.,* 21, 109, 1976.
28. **Jones, R. L., Lewis, W. J., Bowman, M. C., Beroza, M., and Bierl, B. A.,** Host-seeking stimulant for parasite of corn earworm: isolation, identification, and synthesis, *Science,* 173, 842, 1972.

29. **Nishino, C., Bowers, W. S., Montgomery, M. E., Nault, L. R., and Nielson, M. W.,** Alarm pheromone of the spotted alfalfa aphid, *Therioaphis maculata* Buckton (Homoptera: Aphididae), *J. Chem. Ecol.,* 3, 349, 1977.

30. **Stenhagen, E., Abrahamsson, S., and McLafferty, F. W., Eds.,** *Atlas of Mass Spectra Data,* Vol. 2, John Wiley & Sons, New York, 1969, 1450.

31. **Minyard, J. P., Hardee, D. D., Gueldner, R. C., Thompson, A. C., Wiygul, G., and Hedin, P. A.,** Constituents of the cotton bud. Compounds attractive to the boll weevil, *J. Agric. Food Chem.,* 17, 1093, 1969.

32. **Hedin, P. A., Thompson, A. C., Gueldner, R. C., and Minyard, J. P.,** Isolation of α-bisabolol from the cotton bud, *Phytochemistry,* 10, 1693, 1971.

33. **Hedin, P. A., Thompson, A. C., Gueldner, R. C., Ruth, J. M.,** Isolation of bisabolene oxide from the cotton bud, *Phytochemistry,* 11, 2118, 1972.

34. **Minyard, J. P., Tumlinson, J. H., Thompson, A. C., and Hedin, P. A.,** Constituents of the Cotton Bud. The Carbonyl Compounds, *J. Agric. Food Chem.,* 15, 517, 1967.

35. **Hardee, D. D., Mitchell, E. B., Huddleston, P. M., and Davich, T. B.,** A laboratory technique for bioassay of plant attractants for the boll weevil, *J. Econ. Entomol.,* 59, 240, 1966.

36. **Stecher, P. G., Windholz, M., and Leahy, D. S., Eds.,** *The Merck Index,* 8th ed., Merck, Rahway, N.J., 1968, 619.

37. **Stecher, P. G., Windholz, M., and Leahy, D. S., Eds.,** *The Merck Index,* 8th ed., Merck, Rahway, N.J., 1968, 215.

38. **Bhacca, N. S., Johnson, L. F., Shoolery, J. N., Eds.,** *NMR Spectra Catalog,* Varian Associates, Palo Alto, Calif., 1963.

39. **Stenhagen, E., Abrahamsson, S., McLafferty, F. W., Eds.,** *Atlas of Mass Spectral Data,* Vol. 1, John Wiley & Sons, New York, 1969, 704.

40. **Stenhagen, E., Abrahamsson, S., McLafferty, F. W., Eds.,** *Atlas of Mass Spectral Data,* Vol. 1, John Wiley & Sons, New York, 1969, 1453.

41. **Simpson, R. F. and McQuilkin, R. M.,** Identification of volatiles from felled *Pinus radiata* and the electroantennograms they elicit from *Sirex noctilio, Entomol. Exp. Appl.,* 19, 205, 1976.

42. **Cantelo, W. W. and Jacobson, M.,** Phenylacetaldehyde attracts moths to bladder flower and to blacklight traps, *Environ. Entomol.,* 8, 444, 1979.

43. **Cantelo, W. W. and Jacobson, M.,** Corn silk volatiles attract many pest species of moths, *J. Environ. Sci. Health,* A14, 695, 1979.

44. **Stenhagen, E., Abrahamsson, S., and McLafferty, F. W., Eds.,** *Atlas of Mass Spectral Data,* Vol. 1, John Wiley & Sons, New York, 1969, 477.

45. **Flint, H. M., Salter, S. S., Walters, S.,** Caryophyllene: an attractant for the green lacewing, *Environ. Entomol.,* 8, 1123, 1979.

46. **Sweetman, H. R.,** *The Principles of Biological Control,* William C. Brown Company, Dubuque, Iowa, 1958, chap. 10.

47. **Vinson, S. B.,** Habitat location, in *Semiochemicals: Their Role in Pest Control,* Nordlund, D. A., Jones, R. L., and Lewis, W. J., Eds., John Wiley & Sons, New York, 1981, 57.

48. **Bakke, A. and Kvamme, T.,** Kairomone response in *Thanasimus* predators to pheromone components of *Ips typographus, J. Chem. Ecol.,* 7, 305, 1981.

49. **Vité, J. P., Bakke, A., Renwick, J. A. A.,** Pheromones in *Ips* (Coleoptera: Scolytidae): occurrence and production, *Can. Entomol.,* 104, 1967, 1972.

50. **Reece, C. A., Rodin, J. O., Brownlee, R. G., Duncan, W. G., and Silverstein, R. M.,** Synthesis of the principal components of the sex attractant from male *Ips confusus* frass: 2-methyl-6-methylene-7-octen-4-ol, 2-methyl-6-methylene-2,7-octadien-4-ol, and (+)-*cis*-verbenol, *Tetrahedron,* 24, 4249, 1968.

51. **Hedin, P. A., McKibben, G. H., Mitchell, E. B., and Johnson, W. L.,** Identification and field evaluation of the compounds comprising the sex pheromone of the female boll weevil, *J. Chem. Ecol.,* 5, 617, 1979.

52. **Tumlinson, J. H., Hardee, D. D., Gueldner, R. C., Thompson, A. C., Hedin, P. A., and Minyard, J. P.,** Sex pheromone produced by the male boll weevil: Isolation, identification, and synthesis, *Science,* 166, 1010, 1969.

53. **Jones, R. L., Lewis, W. J., Beroza, M., Bierl, B. A., Sparks, A. N.,** Host-seeking stimulants (kairomones) for the egg parasite, *Trichogramma evanescens, Environ. Entomol.,* 2, 593, 1973.

54. **Verschaeffelt, E.,** The cause determining the selection of food in some herbivorous insects, *Proc. Acad. Sci. Amsterdam,* 13, 536, 1910.

55. **Dethier, V. G.,** Chemical factors determining the choice of food plants by *Papilio* larvae, *Am. Nat.,* 75, 61, 1941.

56. **Robbins, W. E., Thompson, M. J., Yamamoto, R. T., and Shortino, T. J.,** Feeding stimulants for the female house fly, *Musca domestica* L., *Science,* 147, 628, 1965.

57. **Thorsteinson, A. J.,** The chemotactic responses that determine host specificity in an oligophagous insect (*Plutella maculipennis* [Curt.]), *Can. J. Zool.,* 31, 52, 1953.

58. **Klingenburg, M. and Bucher, J.,** Biological oxidations, *Annu. Rev. Biochem.,* 29, 669, 1960.
59. **Peterson, A.,** Some chemicals attractive to adults of the onion maggot (*Hylemia antiqua* Meig.) and the seed corn maggot (*Hylemia cilicrura* Rond.), *J. Econ. Entomol.,* 17, 87, 1924.
60. **Kearns, H. G. H.,** *Annu. Rep. Hortic. Stn.,* Long Ashton, 199, 1931.
61. **Chambliss, O. Y. and Jones, C. M.,** Curcurbitacins: Specific insect attractants in Curcurbitaceae, *Science,* 153, 1392, 1966.
62. **DaCosta, C. P. and Jones, C. M.,** Cucumber beetle resistance and mite susceptibility controlled by the bitter gene in *Cucumis sativus L., Science,* 172, 1145, 1971.
63. **Euw, J., Fishelson, L., Parsons, J., Rechstein, T., and Rothschild, M.,** Cardenolides (heart poisons) in a grasshopper feeding on milkweeds, *Nature,* 214, 35, 1967.
64. **Zielske, A., Simons, J. and Silverstein, R.,** A flavone feeding stimulant in alligatorweed, *Phytochemistry,* 11, 393, 1972.
65. **Baker, J. and Norris, D.,** Behavioral responses of the smaller European elm bark beetle, *Scolytus multistriatus,* to extracts of non-host tree tissues, *Entomol. Exp. Appl.,* 11, 464, 1968.
66. **Doskotch, R., Chatterji, S., and Peacock, J.,** Elm bark derived feeding stimulants for the smaller European elm bark beetle, *Science,* 167, 380, 1970.
67. **Erlich, P. R. and Raven, P. H.,** Butterflies and plants, *Sci. Am.,* 216, 104, 1967.
68. **Rees, C. J. C.,** A study of the mechanism and functions of chemoreception especially in some phytophagous insects in *Calliphora,* Ph.D. dissertation, Oxford University, England, 1966.
69. **Hovanitz, W. and Chang, V. C. S.,** Comparison of the selective effect of two mustard oils and their glucosides to *Pieris* larvae, *J. Res. Lepid.,* 2, 281, 1963.
70. **Guenther, E.,** *The Essential Oils,* Vol. 2, D. van Nostrand, New York, 1979.
71. **Watanabe, T.,** Substances in mulberry leaves which attract silkworm larvae *(Bombyx mori), Nature,* 182, 325, 1958.
72. **Hamamura, Y.,** Food selection by silkworm larvae, *Nature,* 183, 1746, 1959.
73. **Hamamura, Y., Hayashiya, K., Naito, K., Matsuura, K., and Nishida, J.,** Food selection by silkworm larvae, *Nature,* 194, 754, 1962.
74. **Horie, Y.,** Effects of various fractions of mulberry leaves on the feeding of the silkworm, *Bombyx mori* L., *J. Sericult. Sci. Japan,* 31, 258, 1962.
75. **Ito, T.,** Nutrition of the silkworm, *Bombyx mori.* IV. Effects of ascorbic acid, *Bull. Sericult. Exp. Stn. (Tokyo),* 17, 119, 1961.
76. **Munakata, K.,** Insect antifeedants in plants, in *Control of Insect Behavior by Natural Products,* Wood, L., Silverstein, R. M., and Nakajima, N., Eds., Academic Press, Orlando, FL, 179, 1970.
77. **Matsumoto, Y. A.,** A dual effect of coumarin, olfactory attraction, and feeding inhibition on the vegetable weevil adult in relation to the uneatability of sweet clover leaves, *Jpn. J. Appl. Entomol. Zool.,* 6, 141, 1962.
78. **Morgan, A. C. and Lyon, S. C.,** Notes on amysalicylate as an attractant to the tobacco hornworm moth, *J. Econ. Entomol.,* 21, 189, 1928.
79. **Howlett, F. M.,** Chemical reactions of fruit flies, *Bull. Entomol. Res.,* 6, 297, 1915.
80. **Heikkenen, H. J. and Hruitfiord, B. F.,** *Dendroctonus pseudotsugae:* a hypothesis regarding its primary attractant, *Science,* 150, 1457, 1965.
81. **Nuto, T. and Sugawara, R.,** The housefly attractants in mushrooms. I. Extraction and activities of the attractive components in *Amanita muscaria* (L.), *Agric. Biol. Chem. (Japan),* 29, 949, 1965.
82. **Sakan, T., Issac, S., and Hyeon, G.,** The chemistry of attractants for Chrysopidae from *Actinidia polygama* (Mig.), in *Control of Insect Behavior by Natural Products,* Wood, D. L., Silverstein, R. M., and Nakajima, N., Eds., Academic Press, Orlando, FL, 237, 1970.
83. **Kangas, E., Pertunnen, V., Oksanen, H., and Rinne, M.,** Orientation of *Blastophagus pineperda* L. to its breeding material. Attractant effect of α-terpineol isolated from pine rind, *Ann. Entomol. Fenn.,* 31, 61, 1965.
84. **Cade, S., Hruitfiord, B., and Gara, R.,** Identification of primary attractant for *Gnatotrichus sulcatus* isolated from western hemlock logs, *J. Econ. Entomol.,* 63, 1010, 1970.
85. **Thomas, H. and Hertell, G.,** Responses of the boll weevil to natural and synthetic host attractants, *J. Econ. Entomol.,* 62, 383, 1969.
86. **Vité, J. and Pitman, G.,** Insect and host odors in the aggregation of the western pine beetle, *Can. Entomol.,* 101, 113, 1969.
87. **Meyer, H. J. and Norris, P. M.,** Vanillin and syring aldehyde as attractants for *Scolytus multistriatus,* *Ann. Entomol. Soc. Am.,* 60, 858, 1967.
88. **de Wilde, J.,** Vergeten Hoofdstukken Uit de Phytopharmacie, *Ghent Landbouw Hogeschoal Mededelingen,* 22, 335, 1957.
89. **Harwood, W. G. and Rudinsky, J. A.,** The flight and olfactory behavior of checkered beetles (Coleoptera: Cleridae) predatory on the Douglas fir beetle, *Tech. Bull. 95,* Agric. Exp. Stn., Corvallis, Oregon, 1966.

90. **Eyer, J. R. and Rhodes, H.**, Preliminary notes on the chemistry of codling moth baits, *J. Econ. Entomol.*, 24, 702, 1931.
91. **Dethier, V. G.**, *Chemical Insect Attractants and Repellents*, Blakiston, Philadelphia, 1947.
92. **von Becker, G.**, Termiten-anlockende Wirkung einiger bei Basidiomyceten-Angriff in Holz entstehender Verbindungen, *Holzforschung*, 18, 168, 1964.
93. **Kamon, J. A. and Frank, W. D.**, Olfactory response of the alfalfa seed chalcid, *Bruchophagus voddi* Guss., to chemicals found in alfalfa, *University of Wyoming Agric. Exp. Stn. Bull. 413*, 1964.
94. **Pertunnen, V.**, Reactions of two bark beetle species, *Hylurgops palliatus* Gyll. and *Hyastes ater* Payk. (Col., Scolytidae) to the terpene α-pinene, *Ann. Entomol. Fennici*, 23, 101, 1957.
95. **Gardiner, L. M.**, Collecting wood-boring beetle adults by turpentine and smoke, *Canada Dept. Agric. Forest Biol., Bimonthly Prog. Rep.*, 13, 1, 1957.
96. **Free, J. B.**, The attractiveness of geraniol to foraging honey bees, *J. Apic. Res.*, 1, 52, 1962.
97. **Gupta, P. D., Thorsteinson, A. J.**, Food plant relationships of the diamond-back moth (*Plutella maculipennis* [Curt.]) II. Sensory regulation of the adult female, *Entomol. Exp. Appl.*, 3, 305, 1969.
98. **Traynier, R. M. M.**, Chemostimulation of oviposition by the cabbage rootfly, *Erioischia brassicae* (Bouche), *Nature*, 207, 201, 1965.
99. **Matsumoto, Y., Thorsteinson, A. J.**, Effect of organic sulfur compounds on oviposition in the onion maggot, *Hylemya antiqua*, *Appl. Entomol. Zool.*, 3, 5, 1958.
100. **Carlisle, D. B., Ellis, P. E., Betts, E.**, The influence of aromatic shrubs on sexual maturation in the desert locust *Schistocerca gregaria*, *J. Insect Physiol.*, 11, 1541, 1965.
101. **Berentes, J., Staedler, E.**, An oviposition stimulator for the carrot rust fly from carrot leaves, *Z. Naturforsch.*, B., 26, 339, 1971.

DETOXIFICATION OF PLANT SUBSTANCES BY INSECTS

Patrick F. Dowd

THE PARAMETERS INVOLVED

One of the factors that limits the ability of insects to feed successfully on plants is the presence of toxins. These secondary metabolites are presumably produced by the plant in addition to primary metabolites (nutrients), and may play an evolutionary-derived intentional role in insect or other pest control by plants. Since few plants are immune to the deprivations of at least one species of insect, herbivorous insects must possess some means of dealing with toxins that are encountered in feeding. While selective feeding on plant parts low in toxins certainly assists some insects in successfully developing on plants, many insects possess general or specific means of detoxifying plant toxins they ingest, and the detoxification of these potentially toxic ingested substances will be the subject of the following discussion. The reader is directed to the Chapter entitled Insect Growth Regulators, Part B for discussion on the relative toxicity of plant substances.

The method(s) of detoxification varies among the different insect species, as would be expected from the varying host range and physiology of the insect species. A particular species of insect may have many means of detoxifying a toxin, and this may vary according to the plant and its chemical composition that is being fed upon at a particular instance. Hence, although the following discussion may deal with only one or a few means of detoxification in an insect towards a particular toxin, this does not necessarily mean that other modes still awaiting discovery are not involved, or even that the method of detoxification under discussion is the most important one to the insect species involved.

In this chapter the methods of detoxification to be considered will be examined with a few brief examples, followed by a comprehensive discussion of the examples of detoxification of plant toxins where the structures are known, arranged into different classes. This classification will follow the one that was set up in a previous review of the subject.[1] The following section will be concerned with the factors that influence detoxification of plant substances, namely, strategies of detoxification by different insect species, modifiers of detoxification methods (primarily enzymes), the sources of the ability (e.g. enzyme location when metabolism is involved), and examples of situations when toxins (or their metabolites) have been used to the benefit of the herbivorous insect. Finally, postulated trends and needs for future research will be discussed.

METHODS OF DETOXIFICATION

As has been found from studies on insecticide efficacy towards insects, there are different means whereby an insect may detoxify a plant toxin. These means are the same as, or analogous to, those reported for insecticides. However, since the mode of entry of plant toxins is presumably oral, slight differences in orientation do exist. Modes of detoxification will be considered to fall into three categories: resistant target site, limited penetration, and biochemical conversion.

RESISTANT TARGET SITE

Although a resistant target site is a fairly common means of insecticide resistance, its involvement in plant chemical detoxification has seen limited study (see Berenbaum for discussion[2]). One of the problems in studying this mechanism is that the mode of action of most plant toxins is unknown at the molecular level. Hence, examples of this phenomenon are often limited to those plant chemicals that have been used commercially as insecticides or drugs, or appear to have that potential.

LIMITED PENETRATION

Limited penetration is another method by which insects may effectively detoxify plant chemicals. Penetration may be limited by different means. The plant material may pass so rapidly through the gut that toxins do not have a chance to penetrate in significant quantities. The gut itself may be resistant to penetration by toxins, or the Malpighian tubules may excrete the substance before it penetrates to the target site. In many instances the penetration is limited or directed by binding (as to the peritrophic membrane) areas that act as sinks for molecules with similar polarities (such as the fat body or the hemolymph), and sequestering the chemicals (either unchanged or after some metabolism) both in general (once again the fat body), or specific areas in the insects. Physical processes involved in sequestration in insects may include diffusional, solvational, adsorptive, phasic, and entrapment (see Duffey for discussion[3]).

METABOLISM

Probably the most extensively studied means of detoxification is that of biochemical conversion, or metabolism. This method may be expressed from as wide a range of factors as changes in the gut pH, which limit protein binding by tannins,[4] to multienzyme complexes, which are presumably inducible by plant compounds (i.e. unspecific monooxygenases). The classes of enzymes involved in metabolism vary, and additional research should show increasing revelation of their nature in the near future. For the purposes of the following discussion, all specific referral to enzyme classes will follow the most recent conventions established by the Commission on Enzyme Nomenclature by the International Union of Biochemistry.[5] For more information on the enzymes involved in detoxification in insects, the reader is directed to the discussion by Ahmad et al.[6]

OXIDOREDUCTASES

Unspecific monooxygenases (E.C. 1.14.14.1), also known as mixed-function oxidases or polysubstrate monooxygenases, have probably received the most attention with respect to plant chemical metabolism in insects. They appear to serve a common function in xenobiotic metabolism in a variety of organisms, including insects. Actually, an enzyme complex is involved since the system includes NADPH-ferrihemoprotein reductase (E.C. 1.6.2.4), which is a flavoprotein enzyme that catalyses the reduction of the unspecific monooxygenase through the oxidation of NADPH.[5] The unsubstituted monooxygenase is a heme-thiolate enzyme (P-450) which requires oxygen, and catalyzes such reactions as hydroxylation, epoxidation, N-oxidation, sulfoxidation, N-, S-, and O-dealkylation, desulfuration, and deamination.[5] Reduction of azo, nitro, and N-oxide groups may also be catalyzed.[5] Substrates not only include xenobiotics (including insecticides), but also fatty acids, prostaglandins, steroids, and vitamins.[5] Whether all of these reactions are catalyzed in insects remains to be seen. More specific monooxygenases are involved in nutrient biosynthesis.[5] Although the involvement of the unspecific monooxygenases in plant chemical metabolism has apparently been demonstrated in many cases (see below), there still exists a need for more conclusive research in the areas of induction and substrate specificity, in spite of recent advances.[1,7] These phenomena will be more thoroughly discussed in a later section.

Other oxidoreductases may also be involved in plant chemical metabolic detoxification by insects. One example will be discussed in the section on cardenolide detoxification.

TRANSFERASES

Glutathione transferases (E.C. 2.5.1.18), formerly referred to as various S-X (X =

various combinations of alkyl or aryl)-transferases, also catalyze a fairly wide range of reactions.[5] These enzymes conjugate the substrate with glutathione and at the same time remove another group and add it to a proton.[5] The resulting glutathione conjugate may be an aliphatic, aromatic, or heterocyclic compound, while the group complexed to the proton may be a halide, nitrogen, or sulfur group.[5] Additional reactions catalyzed include additions to aliphatic epoxides and arene oxides by glutathione, disulfide interchanges, some isomerization reactions, and the reduction of polyolnitrate to its component parts.[5] Once again, not all of these reactions have been demonstrated in insects, although the involvement of this enzyme in plant toxin metabolism seems certain. Other transferases are also involved in plant chemical detoxification. One of these is thiosulfate sulfurtransferase (E.C. 2.8.1.1), which was formerly known as rhodanese.[5] This enzyme adds a thiosulfate group to cyanide to form a sulfite and a thiocyanate, although other sulfur groups may also be used in the reaction.[5] Although examples of catalysis by this enzyme exist (see following discussion on cyanogenic compound detoxification), its involvement in plant cyanogenic metabolism may be more limited than previously suspected, since its presence in insects does not correlate well with the cyanogenic compounds present in the host plants.[8,9] Some of the cyanogenic compound metabolism may in fact be performed by glutathione transferase, although this has not yet been investigated. Other transferases that are probably involved include aryl sulfotransferase (E.C. 2.8.2.1) and phenol β-glucosyltransferase (2.4.1.35), both of which conjugate phenolic compounds.[5] For more information on the activity of these enzymes in insects, the reader is directed to a recent discussion by Wilkinson.[10]

HYDROLASES

Although the substrate 1-naphthyl acetate is commonly used to indicate carboxyl ester hydrolase (E.C. 3.1.1) activity, studies are limited on the metabolism of plant substrates that contain carboxyl or other esters by insects. This is unfortunate, since ester linkages are part of many secondary plant compounds, especially those found in the Compositae.[11] Proteolytic enzymes may be responsible for an unknown amount of *O, S,* and *N* ester hydrolysis, since they may be relatively nonspecific.[5] However, isomer-specific forms of epoxide hydrolases (E.C. 3.3.2.3) which convert epoxides to glycols and act on a variety of epoxides and arene oxides[5] appear to be associated with plant-feeding insects.[12] Some other hydrolytic enzymes which appear to be involved in plant chemical detoxification (although some activation may also occur) are the glucosidases (α-glucosidase, E.C. 3.2.1.20, and β-glucosidase, E.C. 3.2.1.21).

Other enzymes may also be involved in plant chemical detoxification in insects, whether they are specifically designed as such or merely serve as incidental metabolizers when actually they have another primary function. Correlation studies are needed which demonstrate the interaction of specific enzymes (isolated through purification) and detoxification of specific plant compounds which are appropriate substrates.

In reading the following section on detoxification of plant compounds by insects, several factors should be considered. In many cases, the type of metabolism is inferred from the structure that results (i.e. in the case of apparently oxidized compounds), but other enzyme types may be capable of performing the same reaction. Conjugated products may be relatively unstable and the resulting conjugate group may be lost during the separation or purification process, with the result that (for example) conjugative metabolism may appear to be oxidative. This is even more of a problem with epoxides that are formed through metabolism, since epoxide groups are often highly reactive. The toxicity of metabolites is seldom tested. In addition, most studies do not examine all routes of possible detoxification that are described in the previous section, so the relative importance of the investigated route may be less than originally assumed. Plant compounds may also be undergoing simultaneous or subsequent metabolism by different enzymes, or metabolism prior to excretion or sequestration. Finally,

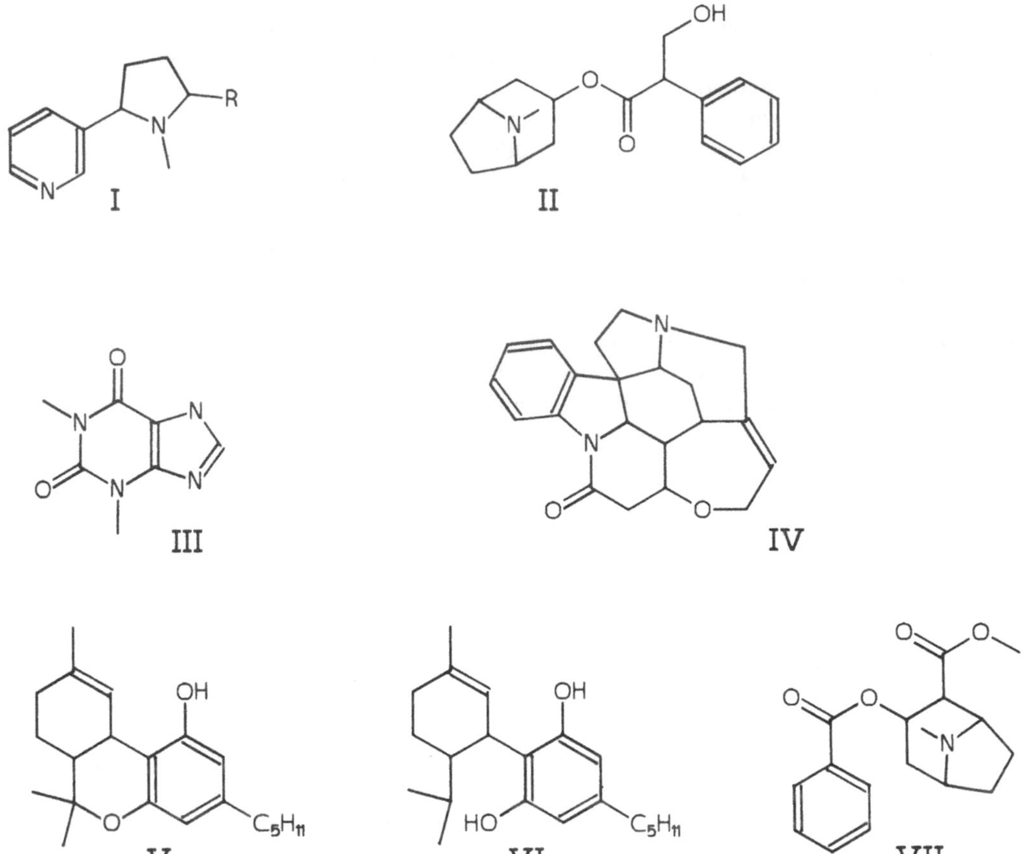

FIGURE 1.   (I) R=H: nicotine; R=O: cotenine. (II) Atropine. (III) Caffeine. (IV) Strychnine. (V) Δ-Tetra-hydro-canabinol. (VI) Canabidiol. (VII) Cocaine.

in some cases a metabolite which is produced is actually more toxic than the initial compound; these cases will also be discussed.

DETOXIFICATION OF HIGHER PLANT SUBSTANCES

ALKALOIDS

The two groups of the alkaloids which have been most extensively investigated are the nicotinoids (Figure 1) and pyrrolizidine alkaloids (Figure 2). The detoxification of nicotine by insects can involve all three of the major types of detoxification: relative impermeability, resistant target site, and metabolism. Nicotine is excreted in an an unaltered state by the tobacco hornworm, *Manduca sexta*.[13,14] Both the cabbage looper, *Trichoplusia ni*, and the tobacco budworm, *Heliothis virescens*, also excrete nicotine is an unaltered state.[15]

The nervous system of *M. sexta* is more resistant to nicotine than that of *Bombyx mori*, apparently due its impermeability to ionic compounds.[16] The central nervous system of *M. sexta* is also relatively (100 ×) resistant to nicotine compared to the nervous system of the American cockroach, *Periplaneta americana*.[17] The efflux of nicotine from the nervous system of *M. sexta* differs from that of *P. americana* in that it is biphasic as opposed to triphasic, and the slow phase is much slower than the two slow phases of *P. americana*.[18] The efflux of nicotine that has penetrated the nervous system of *M. sexta* is slower than other organic compounds, apparently due to specific physiological processes that are able

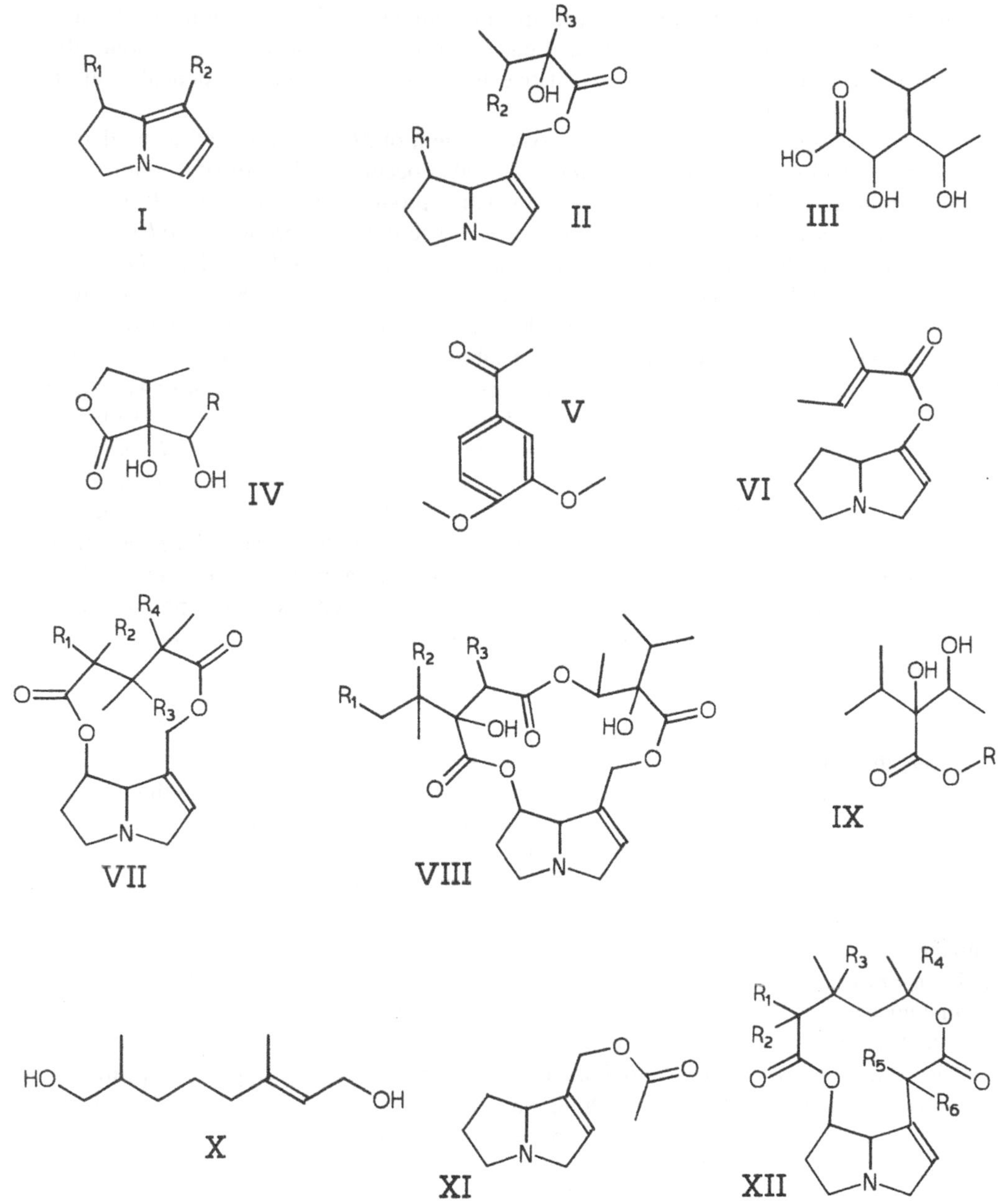

FIGURE 2. (I) R1=O, R2=CH3: danaidone; R1=H, R2=O: danaidal; R1=OH, R2=O: hydroxydanaidal; R1, R2=acetate, no = at position 1: rectonecine diacetate; R1=OH, R2=CH2OH #17 (from 39). (II) R1,R2=OH, R3=isopropyl: heliotrine; R1=H, R2=OH, R3=isopropyl: helurine; R1=H, R2=OH, R3=isopropyl: supenine, #3 from 39; R1,R2=OH, R3=isopropyl: indicine, and lycopsamine, #1 from 39 (isomers); R1=acetate, R2=OH, R3=isopropyl: #2 (from 39); R1,R2=OH, R3=ethyl: #6 (from 39); R1,R2=OH, R3=sec-butyl: #7 (from 39). (III) Trache-lanthic, viridifloric acid (isomers). IV, lactone pheromone of Ithomiinae, *Telleruo* sp. (V) Pheromone from *Danaus* or *Amauris*. (VI) #8 (from 39). (VII) R1=CH3, R2=H, R3,R4=OH: monocrotaline and #4 (from 39); R1=isopropyl, R2=H, R3,R4=OH: trichodesmine; R1=CH3, R2,R4=H, R3=OH: crispatine. (VIII) R1,R2,R3=H: #9; R1=CH3,R2,R3=H: #10 (from 39); R1,R2=H, R3=OH: #11 (from 39); R1=CH3, R2=H, R3=OH: #12 (from 39); R1=CH3, R2,R3=OH: #13 (from 39); all alkaloids from *Parsonia spiralis* found in *Danaus* spp. (IX) #18 (from 39) R=CH3, R=H, from hair pencils of Danainae and Ithomiinae; R=H, R=OH. (X) #19, hair pencil extract (from 39). (XI) Possible danaidal precursor. (XII) R1=CHCH3, R,R3,R6=H, R4=OH, R5=methyl: senecionine; R1=CHCH3, R,R3,R6,R5=H, R4=OH: intergermine; R1=-O-CHCH3-, R2,R3,R4,R6=H, R4=OH: jacobine, jacozine (isomers); R1=OH, R2=CHOHCH3, R3,R4,R5,R6=H, R4=OH: jacoline, jaconine (isomers); R1=CHCH3, R3=methyl, R4=OH, R5,R6=H: seneciphylline.

to immobilize the nicotine relative to other organic compounds.[19] Since arginine, which is also charged and of similar size, is flushed out five times more slowly than nicotine, the removal of nicotine may involves a facultative transport system that is coupled with an alkaloid uptake system.[19]

Some metabolism of nicotine by the nervous system of *M. sexta*, as well as by that of the American cockroach, *Periplaneta americana*, also occurs.[20] The primary product is a nicotine conjugate, although an intermediate oxidative product may also be involved; other minor metabolites are also produced by *M. sexta*.[20] The cigarette beetle, *Lasioderma serricorne*, the differential grasshopper, *Melanoplus differentialis*, the house fly, *Musca domestica*, and the tobacco wireworm, *Conoderus vespertinus*, all metabolize nicotine to a variety of metabolites, primarily cotenine, depending on whether it is topically or orally applied.[15] The excretion of nicotine and its metabolites by *M. domestica* is more limited than in the other insects which feed on nicotine-containing plants and are also known to metabolize it.[15] Cotenine is produced by isolated microsomes from *M. domestica*.[21] Cotenine (primarily) and other metabolites are also produced by the American cockroach, *Periplaneta americana*, and possibly the German cockroach, *Blatella germanica*, topically treated with nicotine.[22] However, orally treated southern armyworms, *Spodoptera eridania*, produce at least nine other metabolites which are found in the feces.[22] NADPH oxidation by microsomal midgut preparations from *Spodoptera frugiperda* as an indicator of oxidative metabolism occurs in the presence of nicotine, atropine, caffeine, and strychnine.[23] The production of oxidized metabolites from nicotine may be due to the unspecific monooxygenases.[24]

A comparison of the plant and insect extracts from *Arctia caja* fed on two strains of *Cannabis sativa* indicated that Δ-tetrahydrocannabinol (THC) is stored, possibly decarboxylated, or converted to the acid form by larvae fed on a strain high in THC (Figure 2).[25] However, those larvae fed on the strain high in cannabidiol may store it unchanged as the glucoside conjugate, or as the acid, and relatively more is excreted relative to those larvae fed on the high THC strain.[25] The grasshopper *Zonocerus elegans* also stores more THC when fed on the THC plant strain, compared to those fed on the cannabidiol strain, which sequesters less.[25] Overall, however, more of both compounds is excreted than is stored in both insects regardless of the plant strain that is fed upon, although much of the two cannabinoids is probably metabolized prior to excretion.[25] The lepidoptera *Eloria noyesi* feeds on *Erythroxylum coca*, which contains cocaine.[26] These insects excrete most of the cocaine unchanged, although some is stored by the larvae and carried through to the adult (with higher levels in the females).[26]

The work on pyrrolizidine alkaloids has recently been discussed by Boppré.[27] Apparently, many butterflies in the Danainae or Ithomiinae actively seek out dead or dying plants, especially those in genus *Heliotropium*, for their pyrrolizidine alkaloids.[27-29] These pyrrolizidine alkaloids are metabolized and/or sequestered by the different butterflies and may be used by the males as pheromones after being stored in their hair pencils (or corremata). However, any hydrolysis that does occur should detoxify the pyrrolizidine alkaloids.[30] Some of the plant alkaloids may also serve as attractants for these butterflies.[31,32]

For example, the hair pencils of *Danaus hamatus hamatus* and *Euploea tulliolus tulliolus* contain danaidone, danaidal, and hydroxydanaidal (pheromones)[33] as well as lycopsamine.[34] The plant precursors indicine or lycopsamine may be converted to the active pheromone by male Ithomiinae through initial hydrolysis followed by a hydroxylation and lactonization of the resulting acid moiety.[35] This possibility was confirmed when males of *D. glippus berenice* were fed on, or injected with a solution containing a pyrrolizidine alkaloid precursor (retonicine diacetate), and converted it to danaidone (the pheromone).[36] Several species of *Euploea* contain danaidone, while several species of *Danaus* contain danaidone as well as danaidal.[37] These compounds are probably derived from *Heliotropium amplexicaule* pyrrolizidine alkaloids and thus are precursors for the ketones (in *D. chrysippus*).[37] *Heliotropium* sp. may contain the precursors for the pyrrolizine compounds found in several other species of *Danaus* and *Amauris* (such as a ketone and a related 3,4-dimethoxyacetophenone).[38] In addition, *D. philene* and *D. affinis* also contain danaidone, danaidal, and hydroxydanaidal, while *D. chrysippus* contains danaidone and hydroxydanaidal.[39] Contact between the hair pencils and wing glands of *D. chrysippus* is necessary for production of danaidone in the hair pencils, which is apparently accomplished by enzyme transfer from the wing glands to the hair pencils.[40]

However, other butterflies or moths still obtain the precursors for their pheromones from the host as larvae. The hair pencils of *Utetheisa pulchelloides* collected from *Heliotropium europaeum* or *Echium lycopsis* contain 1-formyl-7-hydroxy-6,7-dihydro-5H-pyrrolizine, while the body of the moths contain the alkaloids heliotrin (major), heliurine, and supinine, which are presumably plant-derived.[41] *Utetheisa lotrix*, which presumably feed on *Crotalaria mucronata* contain the same pyrrolizine alkaloid as the other species, as well as the des-7-hydroxy derivative, although the ratio varies.[41] The larvae of *U. ornatrix* also sequester pyrrolizidine alkaloids (which are apparently converted to the active pheromones by the males) from their host *Crotalaria* sp.[42] Larvae of *Euploea treitschkei aenea* are responsible for obtaining the pheromone precursors (which may be the hydrolysis products) from *Parsonsia spiralis*.[39] The females of *E. treitschkei aenea* contain both alkaloid precursors and hydrolysis products.[39]

Other caterpillars simply store pyrrolizidine alkaloids, which are not converted to pheromones in the adults. Larvae of the cinnabar moth *Tyria jacobaeae* metabolize and sequester alkaloids from *Senecio vulgaris* and *S. jacobaea* to a variety of metabolites.[43,44] However, there are some differences in the compounds stored in pupae vs. adults.[43,44] Pupae of those fed on *S. jacobae* contained primarily jacobine, jacoline, intergerimine, senecionine, seneciphyline, and an unidentified metabolite, while adults contained primarily intergerrimine, senecionine, seneciphylline, and the same unidentified metabolite, although they occur in proportions different from those in the pupae.[43] Pupae of those larvae fed on *S. vulgaris* contain the same compounds as adults fed on *S. jacobae*, although in different proportions.[43] The relative concentrations in the insects are similar to those found in the host plants.[43] Seneciophylline is stored as the derivative of an N-oxide compound found in *S. jacobaeae*.[45] However, the white ermine moth *Spilosoma lubricipeda* found on *Senecio vulgaris*, sequesters seneciphylline, senecionine, and integerrimine (some as N-oxides), while the buff ermine, *S. lutea*, sequesters integerrimine, seneciphylline, senecionine, jacobine, jacoline, jaconine, and jacozine when reared in *S. jacobeae*, and senecionine and seneciphylline when reared on *S. vulgaris*.[46] *Diaphora mendica* stores senecionine, integerrimine, and seneciphylline when reared on *S. vulgaris*.[46] The Arctiid *Amphicallia bellatrix* sequesters primarily crispatine and trichodesmine from *Crotalaria semperflorans*[47] and trichodesmine from *Crotalaria lantana*.[46] *Argina cridaria* sequesters monocrotaline from *Crotalarus* sp. Even *D. plexippus*, which is more commonly known to sequester cardenolides from *Asclepias* spp. (see below), may sequester pyrrolizidine alkaloids if fed on plants coated with these compounds.[48]

Other insects also store pyrrolizidine alkaloids in some form. The grasshopper *Zonocerus variegatus* sequesters alkaloids from *Crotalaria* sp., including monocrotaline.[49] However, *Zonocerus variegatus* converts monocrotaline *to* the N-oxide before sequestering it.[45] The Lygaeidid *Neacoryphus bicrucis* stores pyrrolizidine alkaloids from *Senecio* spp. which apparently confer unpalatibility, although some metabolism prior to sequestration also apparently occurs.[50]

As indicated from the preceding discussion, the alkaloids which are sequestered in the highest relative concentration by insects may not be the ones that occur in the highest amounts in the host plant. The relative polarity of cardenolides appears to influence which ones are stored by insects (see discussion on cardenolides), so the polarity of alkaloids may

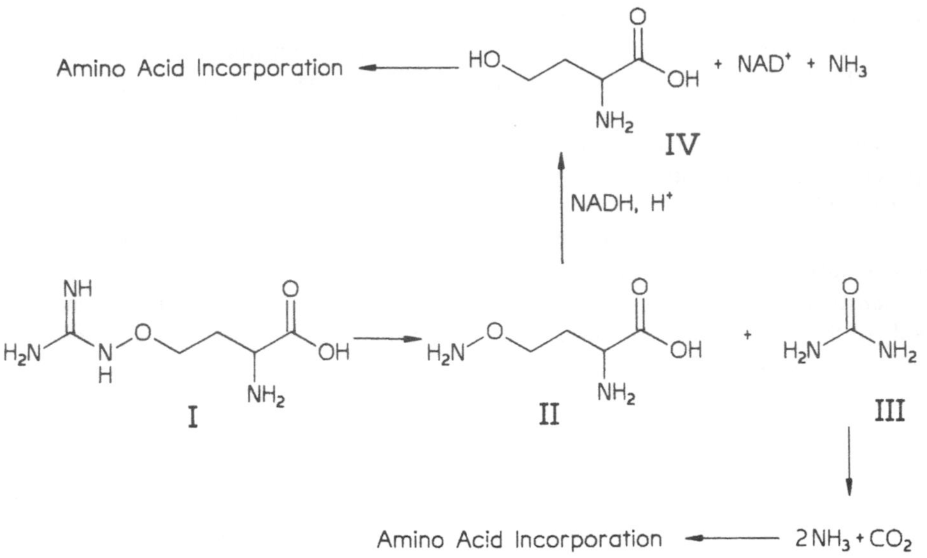

FIGURE 3.   (I) Canavanine. (II) Canaline. (III) Urea. (IV) Homoserine.

also play a role in determining which ones are stored by insects. In addition, Malpighian tubules may play a general excretory role in alkaloid detoxification.[51]

ANALOGS OF ESSENTIAL NUTRIENTS, GROWTH REGULATORS, AND THEIR INHIBITORS

The detoxification of canavanine, an analog of the amino acid arginine, has been investigated in some detail. The bruchid *Caryedes brasiliensis*, which encounters canavanine in feeding on *Dioclea megacarpa*, has several means of detoxifying this compound (Figure 3). The beetle contains an arginyl-transfer-RNA synthetase (arginine-tRNA ligase, E.C. 6.1.1.19[5]) that is capable of discriminating between canavanine and the analogous arginine,[52] thereby limiting its incorporation into proteins.[53] A weevil, *Sternechus tuberculatus*, which feeds on the canavanine containing legume *Canavalia brasiliensis*, also possesses an arginine-tRNA ligase that is highly discriminatory.[54] Canavanine is hydrolyzed to canaline and urea[55] by an arginase (E.C. 3.5.3.1)[5] which is much more effective in hydrolyzing canavanine that the arginases of other insects tested.[56] The tobacco budworm, *Heliothis virescens*, is also apparently able to metabolize canavanine.[57]

Caryedes brasiliensis also converts canaline to homoserine and ammonia by reductive deamination.[58] The urea formed from canaline hydrolysis is then hydrolyzed by a highly active urease (rare in insects) (E.C. 3.5.1.5) to ammonia and CO_2.[58] The ammonia released by urea hydrolysis is excreted[59] or subsequently incorporated into other amino acids (also a part of detoxification),[60] where the initial step is apparently formation of glutamine from glutamic acid or asparagine from aspartic acid.[61] Both glutamic acid dehydrogenase (glutamate dehydrogenase, E.C. 1.4.1.1), which converts 2-oxoglutarate to glutamate, and glutamine synthetase (glutamate-ammonia ligase, E.C. 6.3.1.2), which converts glutamate to glutamine, are involved in this process, and proline is used to spare the tricarboxylic acid cycle intermediate (2-oxoglutarate) used in the conjugation.[62] The homoserine is also incorporated into amino acids.[60,61] The ammonia incorporation may be performed by associated microorganisms,[60,61] some of which can apparently use the toxic amino acids as nitrogen sources.[61] See Rosenthal[54] for a comprehensive review of the subject of canavanine detoxification.

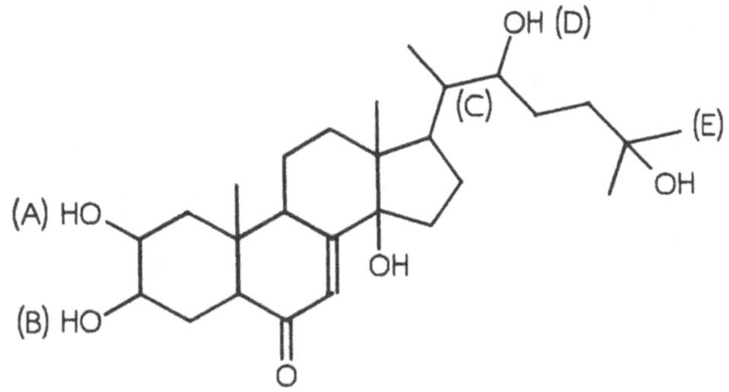

FIGURE 4. Sites of metabolic attack of ecdysone (adapted from 63): (A) conjugation; (B) conjugation, oxidation, and reduction (epidermization); (C) hydroxylation, side chain cleavage; (D) conjugation; (E) hydroxylation.

Insect hormones and analogs are also found in plants. Due to the extensive amount of research which has involved these chemicals, only a few pertinent examples will be considered herein. The reader is directed to the Chapter entitled Insect Growth Regulators, Part A for more information. Both the molting hormone 20-hydroxy ecdysone as well as analogs are found in plants, and they may be metabolized in the same way as the naturally occurring one in insects is metabolized.[63] (Figure 4). Injected ecdysone is dehydroxylated by *Schistocerca gregaria*.[64] Both orally applied and injected, ecdysone is metabolized in the same manner by *Bombyx mori*, although the midgut limits penetration relative to *beta*-sitosterol.[65] The plant analog ponasterone can be initially metabolized to 20-hydroxyecdysone (an activation reaction), or converted to inokosterone, which is less effective[66] (Figure 5). However, the conversion of the two intermediates to posterone constitutes the major detoxification step.[66] As might be expected, ingested ecdysone may be metabolized in other ways when it is injected (which would simulate its natural presence in insects).[67] Injected ecdysone is converted to 20-hydroxyecdysone by *Locusta migratoria*, while orally administered ecdysone is converted to 3-dehydroecdysone and conjugates.[67] Juvenoids would presumably be metabolized in a manner identical to the routes found for them when they are naturally occurring in insects[68] (Figure 6).

The antijuvenile hormones, the precocenes, have been extensively studied in many insects, and to detail the modes of detoxification in all of the insects that have been tested in is beyond the scope of this discussion. Differential metabolism does appear to determine whether or not they are effective in a particular insect species.[70] A generalized scheme of metabolism is provided in Figure 7.[70] A thorough review has been provided by Pratt.[70] The reader is directed to the Chapter entitled Insect Growth Regulators, Part A for more recent developments.

CARDENOLIDES

Due to the extensive and complex studies which have involved detoxification of cardenolides (Figure 8) by insects, they will generally be considered on a case by case basis. Digitoxin, although not encountered naturally, may be detoxified by the milkweed bug, *Oncopeltus fasciatus* through a fairly complicated process. A hemolymph protein carrier may move it (or other cardenolides) to the site of storage, where it is hydroxylated or conjugated to two unidentified metabolites.[71] Further investigation indicated that digitoxin is metabolized by the midgut and Malpighian tubules, but not by the hemolymph or gland fluid; the saliva or midgut bacteria may also be involved in metabolism.[72] However, ouabain

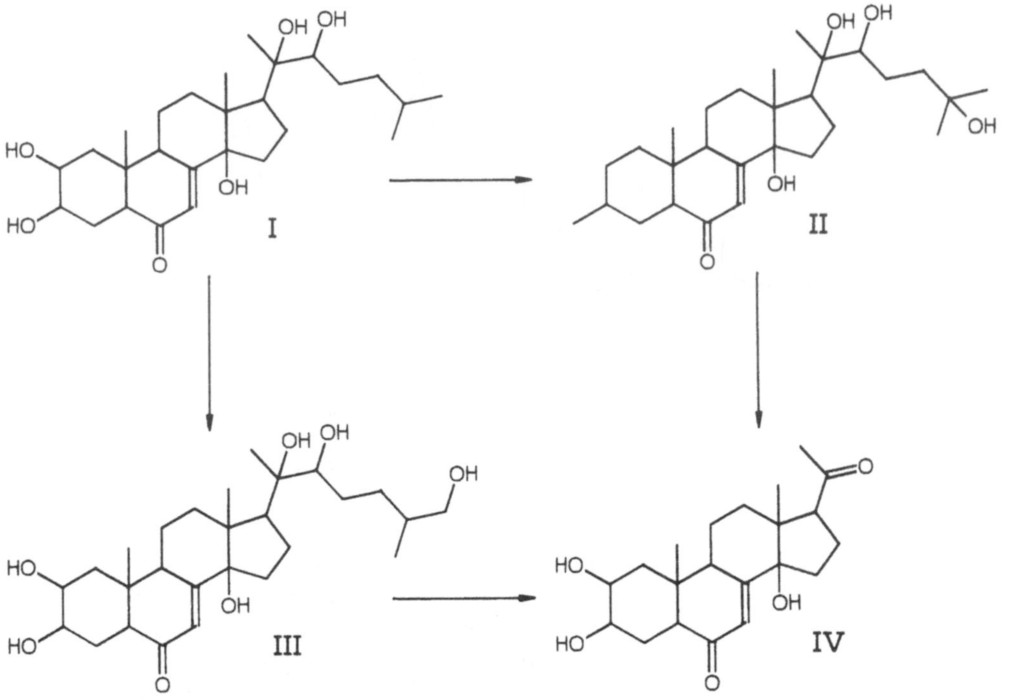

FIGURE 5.   (I) Ponasterone. (II) 20-Hydroxyecdysone. (III) Inokosterone. (IV) Poststerone.

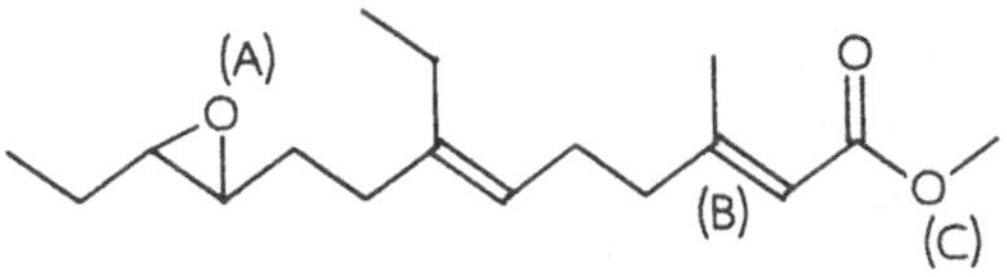

FIGURE 6.   Sites of initial metabolic attack of juvenile hormone I (adapted from References 1 and 69): (A) diol formation; (B) epoxidation; (C) hydrolysis.

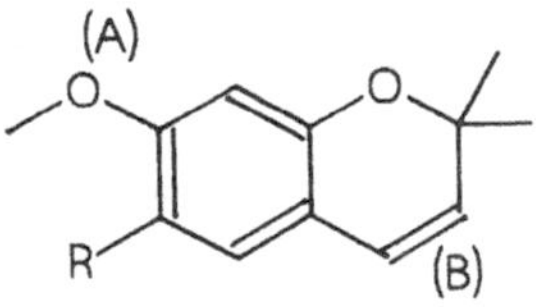

FIGURE 7.   R=H: precocene I; R=OCH3: precocene II; Sites of metabolic attack of precocenes (adapted from 66): (A) *O*-demethylation; (B) hydroxylation, diol formation, conjugation; or epoxidation [possibly followed by degradation to the diol or destructive alkylation of cellular components (target site being the corpora allatum)].

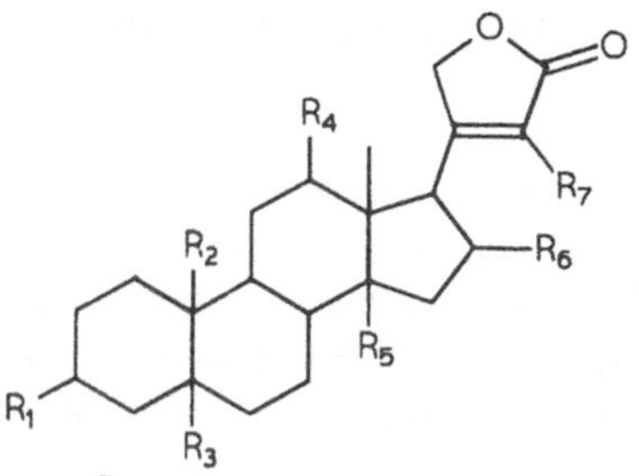

FIGURE 8. (I) R1-4,R6,R7=H; R5=OH: digitoxin; R1,R5=OH; R2-4, R6,R7=H: digitoxigenin and uzarigenin (isomers); R1,R4,R5=PH, R2=O=, R3,R6,R7=H: calotropogenin; R1=glucose-acetyldigitoxose-digitoxose-digitox-ose-O-, R5=OH, R2-4,R6,R7=H: lanatoside-A; R1=glucose-acetyldigitoxose-digitoxose-digitoxose-O-, R2=methyl, R5,R6=OH, R3,R4,R7=H: lanatoside-B; R1=cymarose, R3,R5=OH, R2=C=O, R4,R6,R7=H: strophanthidin-K, R1=glucose-acetyldigitoxose-digitoxose-digitoxose-O-, R2=methyl, R4,R5=OH, R3,R6,R7=H: lanatoside-C; R1=6-deoxy-4-methyl-glucose-O, R2=methyl, R5=OH, R3,R4,R6,R7=H: neriifolin, R1=O-oleandrose, R5=OH, R6=acetate, R2-4,R7=H: oleandrin; R1=O-diginose, R5=OH, R7=isoamylate, R2-4,R6=H: digoxin; R1=O-digitalose, R5=OH, R6=acetate, R2-4, R7=H: neritaloside; R1=O-diginose, R5=OH, R2-4,R6,R7=H: odoroside-A; R1=O-diginose, R5=OH, R6=acetate, R2-4,R7=H: nerigoside; R1=O-digitalose, R5=OH, R2-4,R6,R7=H: odoroside-H; R1=O-digitalose, R5,R6=OH, R2-R4,R7=H: strospeside. (II) Ouabain, Glu=glucose. (III) R1=H, R2=CH3, R3==O: uscharidin; R1=H, R2=CH3, R3=H: calactin/calotropin (isomers); R1,R3=OH, R2=methyl: calotoxin. (IV) Digi-togenin. (V) Adynerine, Dig=diginose. (VI) Labriformin. (VII) R=O: labriformidin; R=OH: desglucosyrioside; =O, is OH, syriobioside. (VIII) Aspecioside.

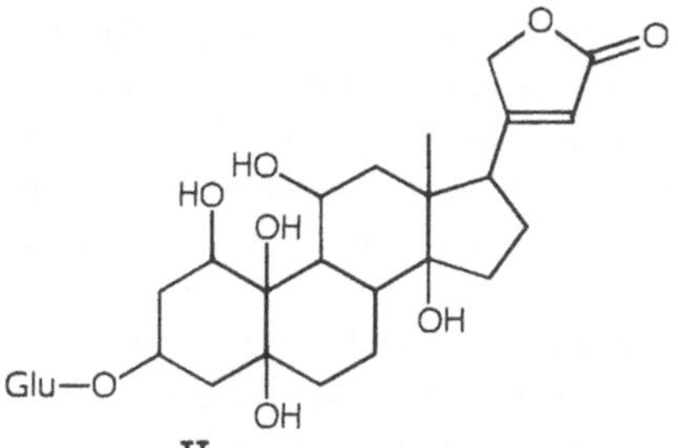

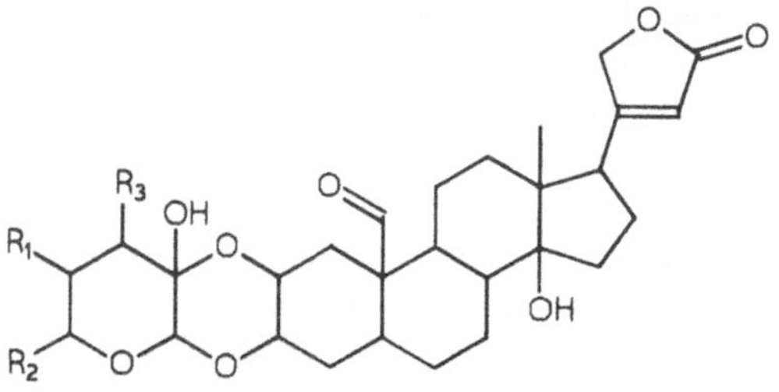

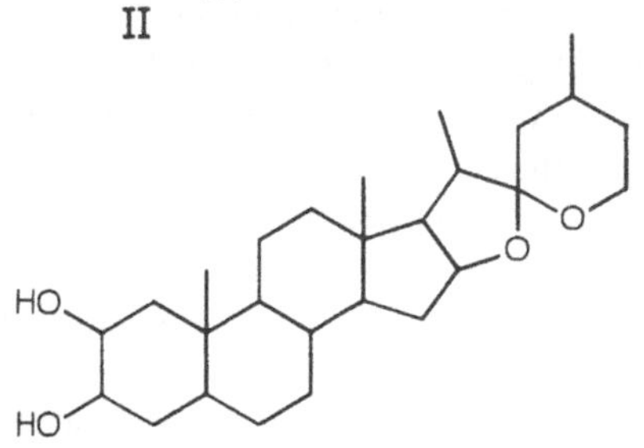

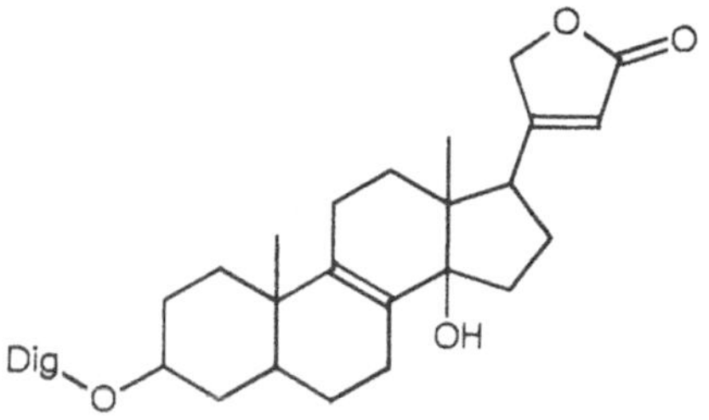

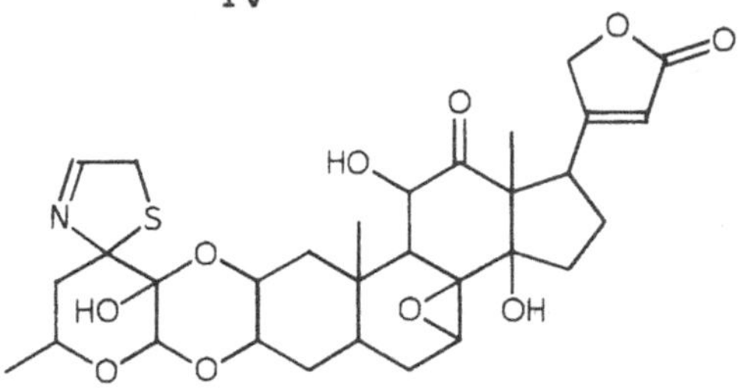

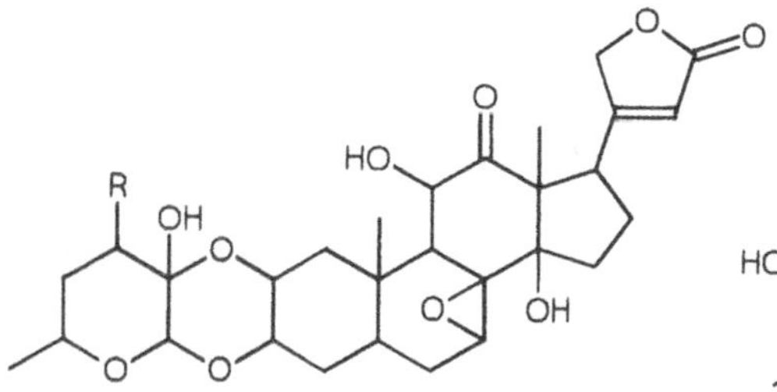

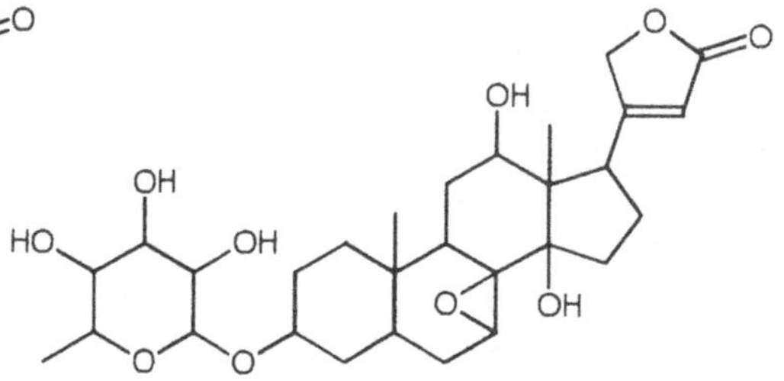

is only metabolized by the Malpighian tubules.[73] The Malpighian tubules also reabsorb ouabain in a specially differentiated area.[73]

When this insect is fed on *Asclepias syriaca*, the less polar metabolites are eliminated in the hemolymph or feces, while the more polar forms are stored in the middorsal abdominal gland in nymphs and adults, and in ventral metathoracic glands in adults.[71] The epidermis of *O. fasciatus* has two layers, the inner layer of which possess cardenolide resistant cells that are highly vacuolated, and probably contain the cardenolides.[74] Transport of cardenolides into this area may be a passive process, since structural features characteristic of energy-driven transport are not present.[74]

All stages (including eggs) of *O. fasciatus* contain cardenolides and females have higher levels than males, which may reflect the presence of cardenolides in eggs,[71] although this sexual differentiation may vary.[75] When *O. fasciatus* are fed on different species of *Asclepias*, the relative quantities of cardenolides in the different hosts are reflected in the insects,[76-79] although cardenolides are concentrated more in insects that feed on plants containing relatively low levels.[79] Little differences in weight of larvae are noted, regardless of the cardenolide content of the host.[76,79-81] Thus, sequestering may be an energy-independent process, since no adverse effects are found regardless of the cardenolide content of the host.[79] Feeding *O. fasciatus* such cardenolides as ouabain, digitoxin, digoxin, lanatosides A and B, and strophanthidin K indicated that they were stored in an emulsion phase of the glands, although some digitoxin (which is relatively nonpolar) may also concentrate in lipophilic areas such as the fat body; hemolymph concentration may be limited to $3.5 \times 10^{-7} M$ (the maximum found for ouabain).[78] Associated proteins may assist in retaining the cardenolides in the emulsion phase.[74] Lanatoside C and neriifolin are also stored by *O. fasciatus*.[81] Although the gut of *O. fasciatus* is permeable to both ouabain and digitoxin, it is more permeable (14 ×) to digitoxin, so the lower permeability to ouabain may be the limiting factor for the rate of sequestration.[72] This permeability can be decreased by the presence of Na^+ ions.[82] However, the guts of *Shistocerca gregaria* and *Periplaneta americana* are impermeable to ouabain.[72] Those cardenolides that are stored by the milkweed bug are primarily polar compounds[72] (but see Reference 81), although the range in polarity can be wide.[83] Finally, as a further means of detoxification, *O. fasciatus* is tolerant to high levels of cardenolides in the hemolymph, and the nervous system Na^+,K^+-ATPase (Na^+,K^+-transporting ATPase, E.C. 3.6.1.37) is relatively resistant to ouabain compared to that of *S. gregaria*.[84]

Storage of cardiac glycosides is very widespread in other species of Lygaeinae as well.[85] The lygaeids *Caenocoris nerii* and *Spilostethus pandurus*, which feed on oleander, *Nerium oleander*, do not store oleandrin, but do store other cardenolides.[86] The *C. nerii* store primarily adigoside and neritaloside, with smaller amounts of nerigoside, odoroside-A, and strospeside, while *S. pandurus* stores primarily odoroside-A, along with nerigoside.[86] For a comprehensive discussion of the detoxification of cardenolides by lygaeids, see Scudder et al.[83]

In studies of other insects, of the six cardenolides identified from *Asclepias curassavica*, only calactin and calotropin, with a trace of uzarigenin, are found in the defensive secretions of the grasshopper *Poekilocerus bufonis*, so metabolism of uscharidin (one of the plant precursors) is probably occuring[87] (see below). Other species of *Poekilocerus*, such as *P. hieroglyphicus*, and *P. pictus*, as well as *Phymaeteus viridipes*, also store calactin and calotropin.[88] The scale insect *Aspidiotus nerii*, which feeds on oleander, *Nerium oleander*, contains adynerine, odoroside-H, and strospeside.[89] The oleander aphid, *Aphis neri*, contains the same cardenolides, but oleandrin, which occurs in the host *Nerium oleander*, is not found.[90] When this aphid is fed on the milkweed *Asclepias curassavica*, it contains primarily proceroside, along with the calotropin, but not calactin.[91] Thus, this insect is capable of selectively storing cardenolides, regardless of the host involved. However, oleandrin is sequestered by the Ctneuchid *Syntomeida epilais*, along with adynerin, nerigoside, and strospeside.[92] Cardiac glycosides are also sequestered by *Arctia caja,* although some metabolism may occur during gut passage.[47]

Unlike *Hyalophora cercopia,* the high levels of potassium ions in the hemolymph of *M. sexta* and *D. plexippus* antagonizes the binding of ouabain to the Na^+/K^+-transporting ATPase in their neural tissue, which is sensitive to ouabain in all three species.[93] In addition, in *D. plexippus* this enzyme is 300 × less sensitive to ouabain binding than the one in *M. sexta* and *H. cecropia.*[93] This sensitivity reduction may protect the neural tissues while cardenolides are undergoing excretion, storage, or metabolism.[93] No binding to the ATPase may occur in the gut since the ATPase is not present.[94]

Since cardenolide sequestration and metabolism has been extensively studied in butterflies, they will be considered as a group. Many of the methods of detoxification are similar to those found for *O. fasciatus.* Although not naturally encountered, digitoxin and ouabain are metabolized by the monarch butterfly, *Danaus plexippus.*[95] Larvae of *D. plexippus* fed specifically on cardenolides convert uscharidin to calactin and calotropin, which are also present in the frass, along with another unidentified metabolite.[96] Digitoxigenin is not stored, but is converted to unidentified more polar compounds which are stored; both are found in the feces.[96] Calactin and calotropin are stored intact, along with a small amount of unidentified metabolite; these same compounds, plus other unidentified metabolites, are also found in the frass.[96] Uzarigenin is stored as is, along with an unidentified metabolite, and found in the frass.[96] Since higher levels of uscharidin are found in adults than in the larvae, the unidentified metabolite may be reconjugated.[96] As in *O. fasciatus,* there is no apparent cost for the sequestration process in *D. plexippus,* although more polar glycosides are stored preferentially.[96] Gut preparations metabolize uscharidin to calotropin and calactin, while fat body preparations only produce calotropin.[97] Apparently hydrolysis of cardiac glycosides occurs first in these insects, followed by a ketone reduction which requires NADPH.[97] This reductase is found in the soluble fraction of a 100,000 *g* centrifugal spin, and is insensitive to typical inhibitors of unspecific monooxygenases.[97] Thus, unspecific monooxygenases are apparently not involved in this metabolism, but a ketone reductase may be.[97] NADPH oxidation by microsomal preparations of *Spodoptera frugiperda* midguts, as an indicator of oxidative metabolism, occurs in the presence of digitoxin.[23]

The detoxification of naturally encountered cardenolides by *D. plexippus* has also been extensively studied. This insect is known to store host-derived cardenolides such as calotropagenin, calotoxin, calotropin, calactin, and uzarigenin.[98] Females store higher levels of cardenolides than males, which may reflect the levels found in eggs. Although the highest levels are found in the wings, the highest emetic potential (and hence most effective predator defense) is in the body.[99] The scales contain high levels of cardenolides, while the hemolymph also contains high levels.[100] The fat body and gut have moderate levels of cardenolides, while the wing muscles have only traces.[100] The adults of *D. plexippus* reared on *Asclepias currassavica* stored different proportions of cardenolides than those adults developing from larvae fed on *Gomphocarpus* sp.[95] Those fed on *A. currassavica* are more efficient in retaining cardenolides through into the adults.[95] However, in both cases the more polar cardenolides are stored, although some are excreted by the larvae, and the cardenolides which are stored reflect those present in the host plants.[95] Cardenolides are stored in varying levels by different subspecies of *D. plexippus.*[101] The storage of cardenolides may again be a passive process which requires little energy.[102]

Larvae fed on *Asclepias eriocarpa* metabolize labriformin and labriformidin to desglucosyrioside (possibly by preliminary hydrolysis followed by reduction), which is stored.[103] Many other plant cardenolides are also stored, and apparently the more polar ones are stored as is, while less polar forms (as determined by the nature of the group attached to the C-3′ of the sugar moiety) are metabolized.[103] The levels of the stored cardenolides are the median of the highs and lows found in the host plants, so the maintenance of cardenolide concen-

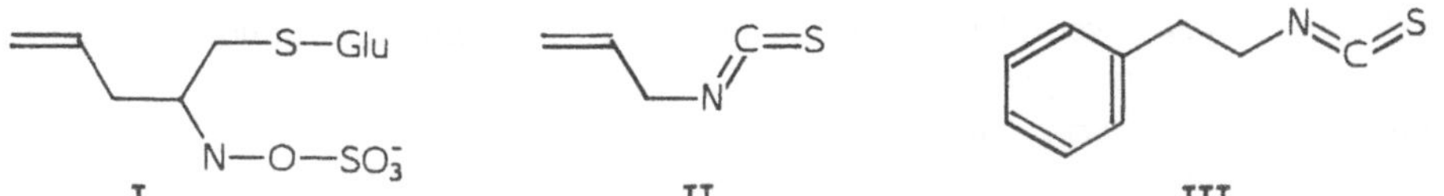

FIGURE 9.   (I) Sinigrin. (II) Allylisothiocyanate. (III) Phenylethylisothiocyanate.

trations appears to occur in a precise manner that balances rates of absorption, distribution, excretion, and storage.[103] Females tend to store higher levels than males, which probably reflects their presence in eggs carried by the females.[103] For those larvae fed on *A. speciosa*, more polar forms are again stored, and some metabolism of more nonpolar forms occurs.[104] Although desglucosyrioside (which may have been formed from labriformin and labriformidin) and uzarigenein are detected, in general less metabolism occurs and concentrations are higher (ca. 2 ×) than in the host plant.[104] Labriformin is converted to labriformidin and then to desglucosyrioside by the larvae.[105] Syrioside and aspecioside are also detected at higher levels than in the plants.[106] Finally, for those larvae fed on *A. californica*, uscharidin is apparently metabolized to calactin and calotropin, since no uscharidin is found in the butterflies, and many polar forms are stored intact at levels circa 3.5 × those in the host plant.[107] Once again, the cardenolides are more concentrated than in the plants.[107] From these three studies, it appears that only the forms with a certain polarity are stored, while less polar forms are metabolized or excreted.[103,104,107] In addition, the levels in the butterflies stay relatively constant regardless of the plant species fed upon, so that those fed on *A. eriocarpa* have levels median to the plant ranges (which is relatively high in cardenolides), while the levels are high compared to the host for those larvae fed on the other plant species which have lower levels.[103,104,107] Apparently, the process of sequestration in this insect is a carefully regulated phenomenon.[103,104,107]

CYANOGENIC COMPOUNDS

The investigation of the detoxification of actual cyanogenic compounds (Figure 9) has been limited, although an enzyme which may be involved (thiosulfate sulfurtransferase, or rhodanese) has been studied in some detail. Actually, an activation reaction may occur in *Papillio polyxenes* and *S. eridania*, where allyl glucosinolates may be hydrolyzed to the more toxic allyl isothiocyanates by the insects.[108] Allyl isothiocyanate is presumably sequestered by larvae of *Pieris brassicae* and retained in pupae, adults, and eggs.[109] Allyl isothiocyanate is also found in pupae of *Pieris rapae*.[109] In addition, the precursor sinigrin and an unidentified mustard glycoside are also stored in pupae of *P. brassicae*.[109] NADPH oxidation by microsomal midgut preparations of *S. frugiperda* as an indicator of oxidative metabolism occurs in the presence of sinigrin and 2-phenylethyl isothiocyanate.[23]

Hydrogen cyanide may also be liberated when plant tissues are damaged by insects.[110] The grain insect, *Sitophilus granarius*, complexes most of the HCN with amino acids which are found as part of a polypeptide; the radiolabelled carbon is primarily contained in aspartic acid.[111] However, other underivitized amino acids also contains radiolabeled carbon atoms, as does cyanocobalamin (vitamin B-12), all of which are excreted in the feces.[111] Some of the cyanide is also expired as CO_2.[111] Similarly, larvae of *Heliothis zea* and *Spodoptera eridania* and adult *Oncopeltus fasciatus* can metabolize HCN to *beta*-cyanoalanine, asparagine, aspartic acid, and SCN.[112]

HCN is apparently detoxified by thiosulfate sulfurtransferase in the butterfly *Polyommatus icarus* and the weevil *Hypera plantaginis*.[113] The activity of thiosulfate sulfurtransferase has also been demonstrated in the horse botfly, *Gasterophilus intestinalis*,[114] the blow fly, *Calliphora vomitoria*,[115] the southern armyworm, *Spodoptera eridania*,[8,116] and many other insect species.[8,9] *Apanteles zygaenarum zenillia* and *Zenilla* sp., parasites of the cyan-

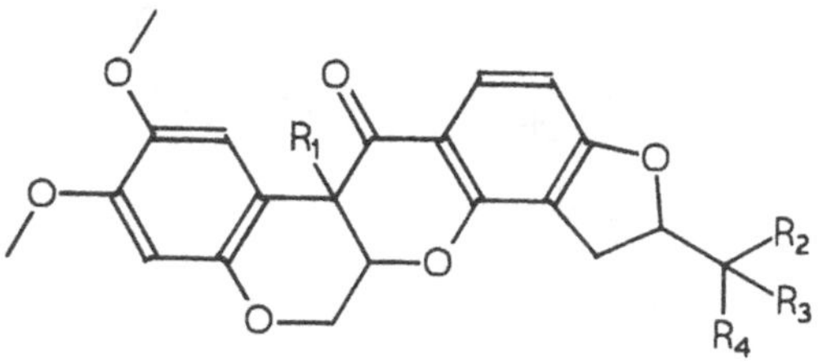

FIGURE 10. R1=H, R2=CH3, R3=CH2: rotenone; R1=H, R2=CH2OH, R3=CH2: 8'-hydroxyrotenone; R1=H, R2=CH2OH, R3=CH2, R4=OH: 6',7'-dihydro-6',7'-dihydroxyrotenone; R1=OH, R2=CH3, R3=CH2: rotenolone I and II (isomers).

ogenic plant feeding burnet moth *Zygaenia lonicerae* (which contains hydrocyanic acid that is apparently not plant derived), also contain thiosulfate sulfurtransferase, although other parasites do not.[110,117] However, some initial studies indicated that some insects that feed on cyanogenic plants contain very low levels of rhodanese, suggesting that other methods of detoxification may be involved.[113] Thiosulfate sulfurtransferase activity is not inducible in *S. eridania* (which can feed on cyanogenic plants),[8] and the cytochrome oxidase is not resistant to cyanogenic compounds.[118] In addition, further studies have demonstrated that thiosulfate sulfurtransferase is widely distributed in insects that do not feed on cyanogenic plants, as well as in those that do.[8,9] Thus, enzymes other than thiosulfate sulfurtransferase may be responsible for detoxifying cyanogenic compounds in insects.[9,118] One possibility is *beta*-cyanoalanine synthetase, which occurs in plants[9,118] (L-3-cyanoalanine synthetase, E.C. 4.4.1.9). This enzyme complexes HCN with cysteine to form hydrogen sulfide and L-3-cyanoalanine,[5] and the cyanoalanine may have water added to form asparagine, as catalyzed by 3-cyanoalanine hydratase (E.C. 4.2.1.65[5]). The previous discussion on HCN metabolism by *S. granarius*,[111] *O. fasciatus*, *S. eridania*, and *H. zea*,[112] where the radiolabelled carbon of HCN is ultimately incorporated into asparagine, suggests that this is a good possibility. Another possibility is glutathione transferase, which is inducible and present in insects at high levels and has the cyanide moiety as an acceptable substrate.

FLAVONOIDS AND RELATED COMPOUNDS

The detoxification of the commercially available insecticide rotenone has been investigated in some detail. A variety of oxidative metabolites are produced in *M. domestica (in vivo* and *in vitro*),[119,120] the German cockroach, *Blatella germanica*, (whole body, *in vitro*), a long horned beetle, *Xylotrupes dichotomus* (midgut, *in vitro*), and *P. americana*[120] (Figure 10). The metabolites consist primarily of rotenone I and 6',7'-dihydro-6',7'-dihydroxyrotenone, although rotenone II and 8-hydroxyrotenone are also produced.[119,120] There is some variability in the proportion of metabolites produced according to the tissues investigated, and the fat body of *P. americana* is more metabolically active than the gut.[120] The chemical structure of the initial metabolites, and the effectiveness of oxidative enzyme inhibitors such as piperonyl butoxide, MGK 264, syneprin 500, and SKF 525A in inhibiting metabolism by *P. americana* fat body preparations suggests that the unspecific monooxygenases are responsible for the initial metabolism.[121] However, subsequent conjugation may be more responsible for detoxification, since many of the initial metabolites are still toxic to mice (8-hydroxyrotenone slightly more so that rotenone itself), or significantly inhibit the activity of NADH oxidase.[120] The hydroxylation products may be formed by initial attack on the isoprenoid side chain and ring junction.[121]

Early work suggested that flavonoid pigments may be conjugated by insects.[122] However,

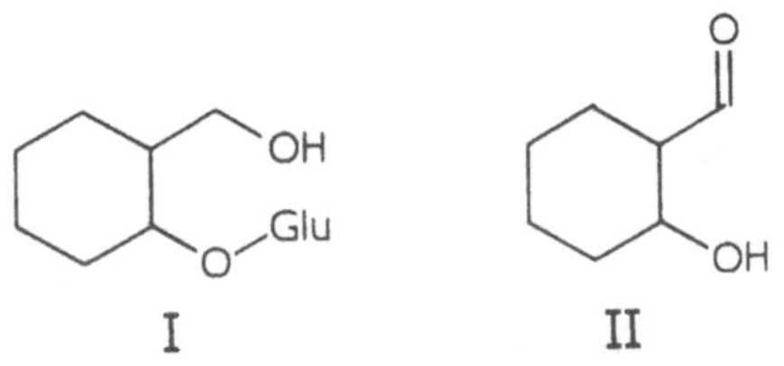

FIGURE 11.   (I) Salicin. (II) Salicyl aldehyde.

salicin is oxidized in one position, and hydrolyzed in another one by *Melasoma populi*[123] and the brassy willow beetle, *Phyllodecta (Phratora) vitellinae*[124] (Figure 11). The resulting salicylaldehyde is only produced when *P. vitellinae* (or *Chrysomela tremulae*) is fed on salicin containing plants or on diets containing salicin.[125,126] The glucose that is hydrolyzed from the salicin is apparently used as a nutrient source.[126,127] The activity of β-glucosidase is higher in the defensive glands of *Chrysomela tremulae* than in the gut, so the glands may be the location of most of the metabolism.[126,128] The reader is directed to the review by Rowell-Rahier and Pasteels[128] for more information on salicin detoxification. The polyphenoloxidase (phenolase, catechol oxidase, E.C. 1.10.3.1) from saliva of many insects (especially the Hemiptera) may generally detoxify (oxidize) all the phenolic compounds in cases where it is able to counteract the phenolase-quinone content of the host.[129]

A wing pigment of the marbled white, *Melanargia galatea*, was initially found to be the same as one in the host plant, *Dactylis glomerata*.[130] Investigation of the small heath butterfly, *Coenonympha pamphilus*, indicated that it sequesters the host plant flavonoids tricin, orientin, and their glycosides in the wings[131] as does the marbled white, *Melanargia galatea*.[132] Further work on *M. galatea* demonstrated that it sequesters most of the flavonoids from the apparent host *Festuca rubra*, including apigenin, apidgen 7-glucoside, isoorientin, isoorientin 7-glucoside, isovitexin, isovitexin 7-glucoside, luteolin 7-glucoside, luteolin 7-di- and triglucoside, tricin (also found in the feces), tricin 4-glucoside, -7-glucoside, and -7-diglucoside, vitexin, and vitexin 7-glucoside[133,134] (Figure 12). These flavonoids may all be used as antimicrobials by the insect.[135] The 4′-conjugate of tricin (possibly a phosphate or sulfate) is produced by the insect, and is among the most common of the sequestered flavonoids along with tricin, tricin 7-glucoside, orientin 7-glucoside, and vitexin 7-glucoside.[133,134] The presence of 4′- and 7-glucosides, and the absence of 5′-glucosides suggests that the 5′-glucosides may be hydrolyzed, while the sterically hindered 4′ and 7-glucosides are not.[133,134] This information apparently explains why the 4′- and 7-glucosides are stored and why the 5′-glucosides are not.[133,134] Many of the same flavonoids are stored by other species of *Melanargia*.[133] Flavonoids sequestered by *M. galathea* from other potential host plants include tricin, tricin conjugate, and orientin from *Agrostis tenuis*; tricin, the tricin conjugate, luteolin, and isovitexin from *Bromus erectus*; tricin, tricin conjugate, and quercitin from *Festuca arundinacea*; tricin, tricin conjugate, and kaempferol from *Festuca pratensis*; tricin and kaempferol from *Lolium perenne*; and tricin, tricin conjugate, and orientin from *Phleum pratense*.[133] The zebra swallowtail, *Eurytides marcellus*, only sequesters quercetin-3-glucoside from its host *Asimina triloba*, which also contains other quercetin derivatives.[136] Other Papilionidae sequester quercetin and kaempferol glucosides.[136] The level of flavonoids also influences the intensity of yellow color in the wings of *M. galatea*.[135]

The lepidopterous insects, *Agonopterix* and *Depressaria* spp., which feed on umbelliferous plants that may contain linear furanocoumarins (Figure 13) (such as xanthotoxin) have an interesting method of detoxification.[137] Although these compounds are activated by ultraviolet light,[138] these caterpillars feed in rolled leaves which block the ultraviolet light, and so apparently prevent conversion of linear furanocoumarins to the active form.[137] Other

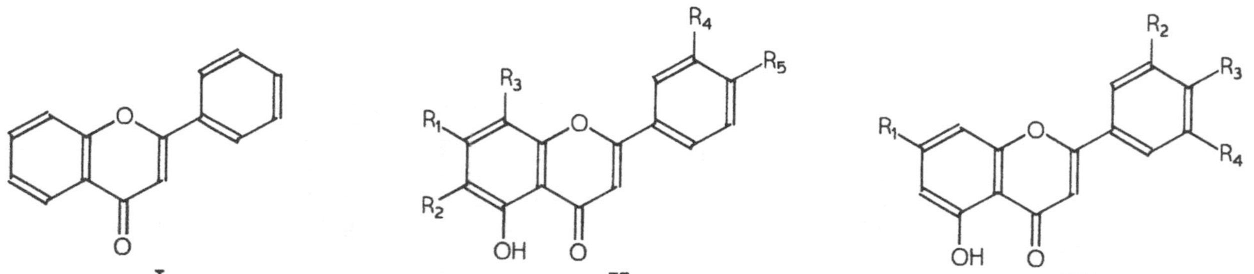

FIGURE 12. (I) Flavone. (II) R1,R5=OH, R2-4=H: apigenin; R1=glucose, R5=OH, R2-4=H: apigenin-7-glucoside; R2=glucose, R1,R4,R5=OH, R3=H: iso-orientin: R1,R2=glucose, R4,R5=OH, R3=H: iso-orientin-7-glucoside; R2=glucose, R1,R5=OH, R3, R4=H: isovitexin; R1,R2=glucose, R5=OH, R3,R4=H: isovitexin-7-glucoside; R1,R5=OH, R2-4=H: kaempherol; R1=glucose, R5=OH, R2-4=H: luteolin-7-glucoside; R1=glucose-glucose, R5=OH, R2-4=H: luteolin-7-diglucoside; R1=glucose-glucose-glucose, R5=OH, R2-4=H: luteolin-7-triglucoside; R3=glucose, R3-5=OH, R2=H: orientin; R1,R3=glucose, R4,R5=OH, R2=H: orientin-7-glucoside; R1,R4,R5=OH, R2,R3=H: quercitin; R3=glucose, R1,R5=OH, R2,R4=H: vitexin; R1,R3=glucose, R5=OH, R2,R4=H: vitexin-7-glucoside. (III) R1,R3=OH, R2,R4=OCH3: tricin; R1=glucose, R3=OH,R2,R4=OCH3: tricin-7-glucoside; R3=glucose, R1=OH, R2,R4=OCH3: tricin-7-diglucoside; R3=glucose, R1=OH, R2,R4=OCH3: tricin-4'-glucoside.

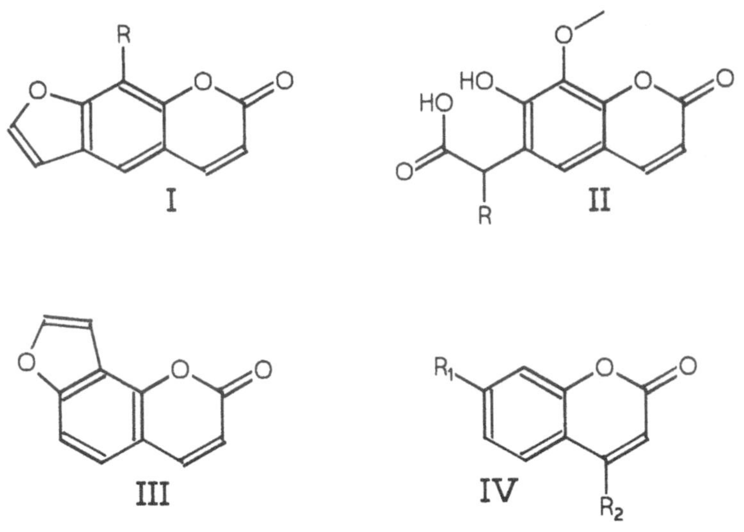

FIGURE 13. (I) R=H: psoralen; R=OHC3: xanthotoxin. (II) R=H, or R=OH, xanthotoxin metabolites. (III) Isopsoralen. (IV) R1,R2=H: coumarin; R1=OH, R2=H: umbelliferone; R1=OH, R2=methyl; methylumbelliferone.

methods of detoxification also exist for furanocoumarins (or other light-activated substances) including dark pigmentation and quenching of excited states of the chemicals or reactive forms of oxygen with such compounds as carotenoids, amines, or furans (depending on the chemical involved).[139] The metabolism of xanthotoxin is one of the more complete stories on the detoxification of a plant compound by insects. Xanthotoxin is metabolized to four major metabolites by *Papilio polyxenes* and *Spodoptera frugiperda*.[140-142] The two primary ones have been identified (Figure 13), and are probably produced by oxidative enzymes.[140-142] The metabolism occurs at a much higher rate in *P. polyxenes* and elimination occurs much more rapidly, with the result that the levels of the xanthotoxin and its metabolites in this insect species are much lower than in *S. frugiperda*.[140-142] Both the midgut (most active) and other tissues appear to be involved in the metabolism of xanthotoxin in *P. polyxenes*, while the midgut of *S. frugiperda* does not appear to be involved in the metabolism.[140,142] However, polar metabolites are formed in greater amounts by *S. frugiperda* than in *P. polyxenes*.[140,142] The enhancement of activity in the presence of NADPH, and inhibition of activity in the presence of piperonyl butoxide, suggests that unspecific monooxygenases are responsible for the metabolism in both insect species.[141] The oxidative enzyme heptachlor epoxidase, as well as the conjugating enzyme glutatione transferase, are induced in *S. frugiperda* by xanthotoxin,[143] and NADPH oxidation (an indicator of oxidative metabolism) occurs in the presence of xanthotoxin as well,[23] suggesting that the metabolism of xanthotoxin is part of a carefully regulated metabolic scheme. In addition, the toxicity of xanthotoxin to *P. polyxenes* is increased by coapplication of the naturally co-occurring methylene-dioxyphenyl compound myristicin, which inhibits oxidative enzymes, suggesting that this compound would inhibit metabolism in insects feeding on plants that contain both xanthotoxin and myristicin.[144] Psoralen and isopsoralen are also metabolized by the gut and carcass of *P. polyxenes*, although isopsoralen is metabolized more slowly than the other two compounds, and less of the metabolism is performed by the gut.[140,145] The metabolites are analogous to those produced from xanthotoxin.[145] Although relatively more isopsoralen is retained in the body for a longer period of time than psoralen (2 to 3 ×)[145] or xanthotoxin, this factor does still not appear to explain the relatively higher toxicity of isopsoralen,[140] although differences in metabolism are certainly a contributing factor.[145] The related com-

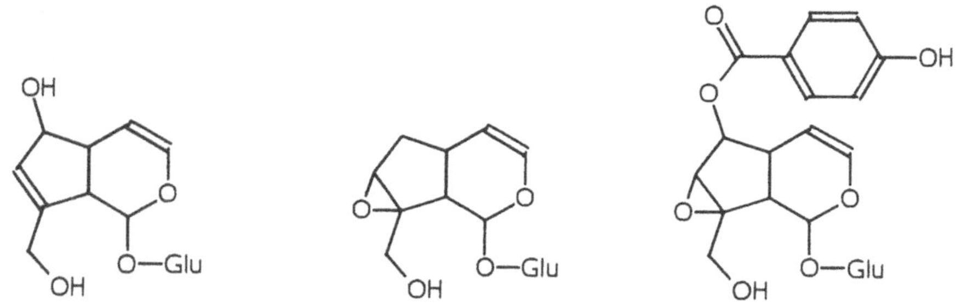

FIGURE 14.   (I) Acubin. (II) Catalpol. (III) Catalposide; Glu=glucose.

pound, 4-methylumbelliferone, is conjugated with glucose at higher rate in the fat body than the gut of *Locusta migratoria*.[146] NADPH oxidation by microsomal preparations from the midgut of *S. frugiperda* as an indicator of oxidative metabolism also occurs in the presence of flavone, rotenone, coumarin, and umbelliferone.[23]

The iridoid glycosides, acubin, catalpol, and catalposide (Figure 14) are sequestered by *Euphydryas phaeton*.[147] When this insect is fed on *Plantago lanceolata*, more acubin than catalpol is sequestered, while the opposite is true when the insects are reared on *Chelone glabra*.[147] Adults may selectively sequester caltalpol, or convert acubin to catalpol.[147] On the other hand, neither *Junonia coenia* nor *Ceratomia catalpae* retain iridoid glycosides in the adults.[147] Instead, they are eliminated in the meconium of the pupae, and so only catalpol is sequestered in the larvae or pupae.[147] None of the iridoid glycosides are sequestered by *Lymantria dispar;* rather, they are eliminated in the feces.[147]

LIGNINS, TANNINS, AND RELATED PHENOLICS

Although lignins and tannins can reduce digestibility of proteins by binding to them, lepidopterous insects may have relatively high gut pHs to limit binding[4] (but see discussion on tannin toxicity to locusts[148]). Lignin apparently is hydrolyzed to guaiacol (*o*-methoxy-phenol) (Figure 15) by *Locusta migratoria*, and subsequently alkylated to locustol (2-meth-oxy-5-ethylphenol), a gregarization pheromone.[149] Other metabolites include the *m*-, and *p*-methoxy analogs of guaiacol.[149] However, bacteria may actually be responsible for the metabolism, since the antibiotic thipyrimeth prevents the formation of locustol.[150] Lignin is apparently demethylated by *Kalotermes flavicollis*[151] Lignin and related compounds (such as ferulic acid), are apparently demethylated, broken down to phenolics, and even metabolized all of the way to CO_2 by the termite *Nasutitermes exitiosus*, with the nongaseous compounds being present in the feces to some degree.[152] However, once the lignins are broken down into their component parts, microorganisms may be responsible for the rest of the metabolism.[152]

The condensed tannin quebracho is bound to the peritrophic membrane in *Shistocerca gregaria, Locusta migratoria, Chartoicetes terminifera,* and *Zonocerus varieagatus* when it is excreted in the feces.[153] Resistance to condensed tannin is somewhat higher in locusts which naturally encounter high levels.[154] Protein binding by tannins or other toxic effects apparently does not occur to anywhere near the same degree in these grasshoppers as it can in the Lepidoptera.[148] Midgut caecal pockets may contribute to the ability of the tannin to be excreted in conjunction with the peritrophic membrane by temporarily storing the tannin.[155] Tannic acid is hydrolyzed to gallic acid both *in vitro* and *in vivo* in *S. gregaria* (Figure 16).[156] After hydrolysis gallic acid (or a derivative) is incorporated into the cuticle of *Anacridium melanorhoden*.[157-159] The gallic acid may also be excreted, held in the gut, or

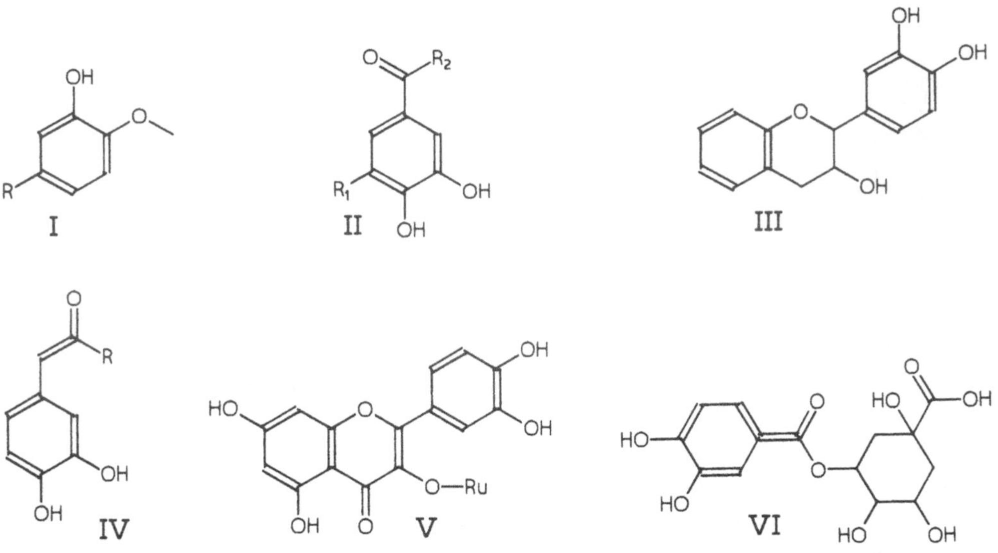

FIGURE 15. (I) R=H: guaiacol; R=C2H5: locustol. (II) R1,R2=OH: gallic acid; R1=H, R2=OH: protocatechuic acid; R1,R2=H: protocatechualdehyde. (III) Quebracho. (IV) R=OH: caffeic acid; R=OCH3: ferulic acid. (V) Rutin, Ru=rutinose. (VI) Chlorogenic acid.

expired as CO_2.[158] During the incorporation process, the gallic acid may be transferred through the hemolymph as a glucoside conjugate.[158] The gallic acid,[158] as well as proto-catechuic and caffeic acid, may apparently substitute for phenylalanine in protein tanning.[160] The condensed tannin quebracho is excreted in an unmetabolized form in both *Shistocerca gregaria* and *Locusta migratoria*.[148] However, both the impermeability of the peritrophic membrane and relatively higher rates of hydrolysis appear to detoxify tannic acid in *S. gregaria*, as opposed to *L. migratoria*, which is susceptible to the effects of tannic acid.[148] Beetles and sawflies which fed on *Eucalyptus* sp. also bind tannin to the peritrophic membrane, while caterpillars sequester it in midgut cells.[161]

The uptake into the hemolymph (2 to 5% through the gut) of rutin and chlorogenic acid is limited to *H. zea*, since most is excreted biphasically in an unaltered form in both acutely or chronically dosed insects, although some metabolism does occur.[162] Metabolites, which are the products of hydrolysis, include quinic acid from chlorogenic acid, and rhamnose, glucose, and rutinose from rutin.[162]

TERPENOIDS

Many coniferous terpenoids (Figures 16 and 17) are metabolized by the various bark beetles that feed on these hosts. For an explanation of the roles of the pheromonal products, the reader is directed to other discussions.[163,164]

Feeding bark beetles exposed to oleoresins, such as *Dendroctonus brevicomis*, *D. frontalis*, *D. valens*, and *D. ponderosae*, sometimes produce more metabolites than those which are not feeding.[165] Camphene is oxidized to 6-hydroxycamphene and camphenol by *D. brevicomis* and *D. frontalis* males and females.[166] However, 3-carene may be converted to 1-methyl-5-(-hydroxyisopropyl)-cyclohexa-1,3-diene by *Ips pini* and *I. paraconfusus* males.[166] Myrcene is oxidized to myrcenol by both males and females of *Dendroctonus brevicomis*,[166] *D. frontalis*,[165] and *D. ponderosae*.[167] Myrcene is also converted to the sex pheromones ipsdienol (and ipsenol) by *D. brevicomis* males,[166-168] as well as male *D. ponderosae*,[165,167] and *D. pseudotsugae*.[165] Myrcene is converted to ipsdienol and then is reduced to ipsenol in male *Ips grandicollis* and *I. calligraphus*, while only ipsdienol is produced by *I. avulsus*

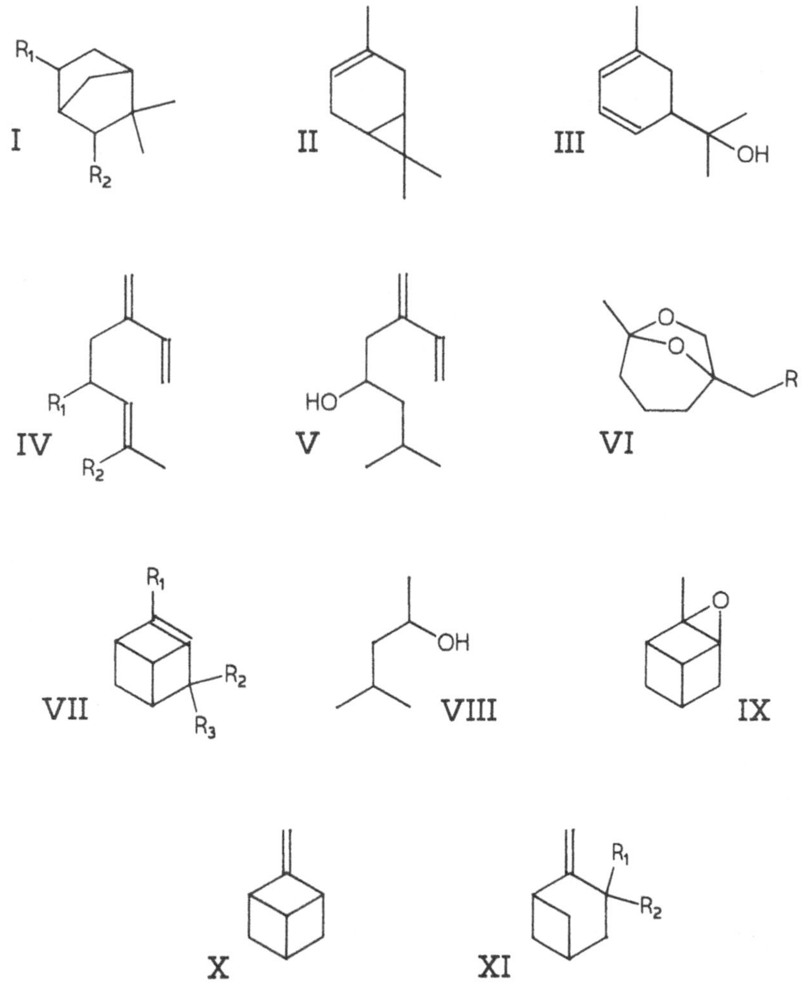

FIGURE 16. (I) R1=H, R2=CH2: camphene; R1=OH, R2=CH3: camphenol; R1=OH, R2=C=C: 6-hydroxycamphene. (II) 3-carene. (III) carene metabolite. (IV) R1=OH, R2=CH3: ipsdienol; R1=H, R2=CH3: myrcene; R1=H, R2=OH: myrcenol. (V) ipsenol. (VI) R=CH3: brevicomin; R=H: frontalin. (VII) R1-3=H: α-pinene; R1,R2=H, R3=OH: cis-verbenol; R1,R3=H, R2=OH: trans-verbenol; R1=H, R2=O: verbeneone; R1=C=C, R2,R3=H: myrtenal; R1=CH2OH, R2,R3=H: myrtenol. (VIII) 4-methyl-2-pentanol. (IX) Pinene oxide. (X) β-pinene. (XI) R1=H, R2=OH: cis-pinocarveol; R1=OH, R2=H: trans-pinocarveol; R1=O: pinocarvone.

and *I. paraconfusus*.[169-172] Finally, α-pinene is oxidized to *cis*- and *trans*-verbenol and myrtenal by males, and 4-methyl-2-pentenol and myrtenol by both male and female *Dendroctonus frontalis*, or pinocarveol and pinocarvone.[173] α-Pinene presence also increases the production of verbenol in *D. ponderosae*,[174] and the production of verbenols, myrtenol, *cis*-pinen-2-ol, and 4-methyl-2-pentanol in *D. frontalis*.[165] *Ips paraconfusus* convert (−)-α-pinene to (+)-*cis*-verbenol, while (+)-α-pinene is converted to (+)-*trans*-verbenone.[175] (−)-α-Pinene is also converted to *trans*-verbenol and myrtenol by both sexes,[176] a process which is limited in immature adults.[177] The production of *trans*-verbenol increases in larvae of *D. frontalis* and *D. terebrans* when they are exposed to α-pinene, and although the pupae do not produce detectable products (except a possible conjugate intermediate), the emerging males do produce more verbenol.[178] Both sexes of *D. ponderosae* also convert α-pinene to

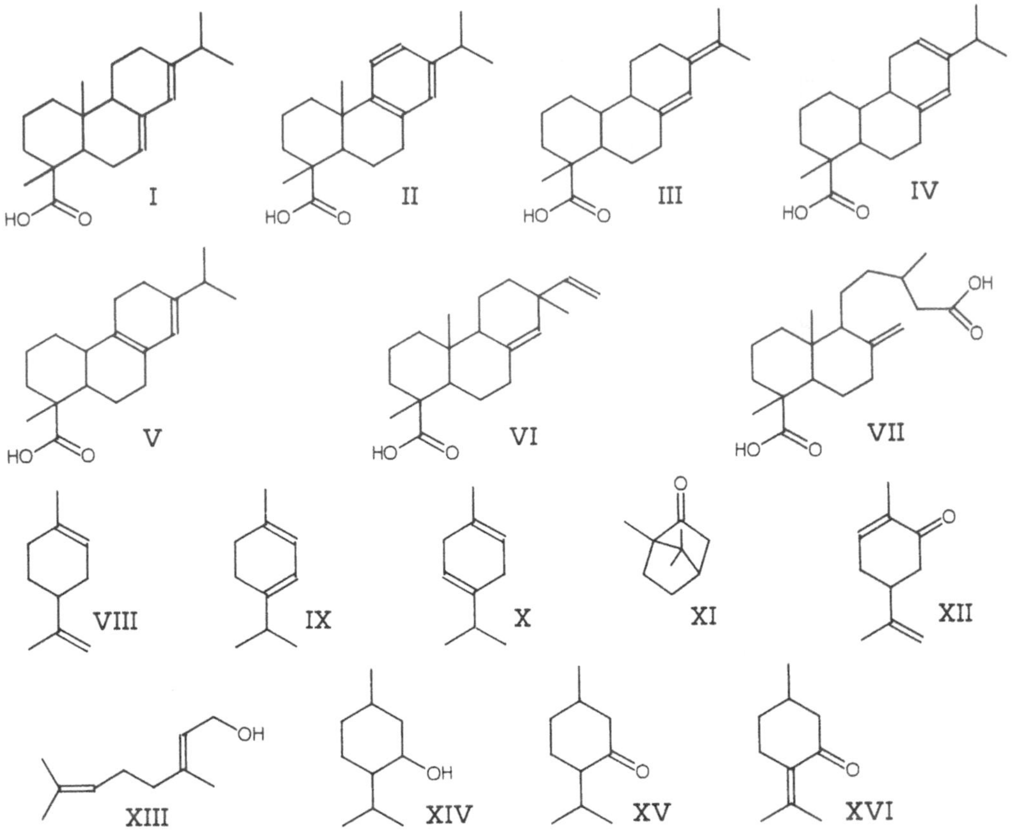

FIGURE 17. (I) Abietic acid. (II) Dehydroabietic acid. (III) Neoabietic acid. (IV) Levopimaric acid. (V) Palustric acid. (VI) Pimaric acid. (VII) Pinifolic acid. (VIII) Limonine. (IX) α-terpinene. (X), γ-terpinene. (XI) Camphor. (XII) *d*-carvone. (XIII) Geraniol. (XIV) Menthol. (XV) Menthone. (XVI) Pulegone.

trans-verbenol, a process that is stereospecific in *D. brevicoris*.[179] Many of the terpene metabolites appear to be produced internally and finally result in frontalin (*D. frontalis*) or brevicomin (*D. brevicomis*).[165] The bacterium *Bacillus cereus*, which can be pathogenic to insects, and was isolated from the guts of *Ips paraconfusus*, can convert α-pinene to *cis*- and *trans*-verbenol.[180] Insects fed streptomycin no longer produce the metabolites,[181] so *B. cereus* or other bacteria may be responsible for some of the metabolism in the gut.[180,181] However, cultures of mycangial fungi from *D. frontalis* are capable of converting *trans*-verbenol to verbenone, so they may be also responsible for some of the metabolism by these beetles.[182] β-Pinene is converted to *trans*-pinocarveol and pinocarvone by males of *Dendroctonus frontalis*.[173] Microsomal preparations from *Dendroctonus terebrans* convert α-pinene to α-pinene oxide (an epoxide) and other unidentified metabolites.[183] The α-pinene induces its own metabolism to unidentified metabolites, although neither the level of cytochrome P-450 nor α-pinene oxide production increases.[183] The α-pinene oxide may be a precursor to the aforementioned beetle pheromones.[183] NADPH oxidation, as an indicator of oxidative metabolism, occurs in the microsomal midgut preparations from *S. eridania* in the presence of α- and β-pinene (autoinducible), (+)- and (−)-limonene, and α- and γ-terpinene and pulegone,[184,185] and also in microsomal midgut preparations from *S. frugiperda* in the presence of (+)- and (−)-camphene, (+)-camphor, d-carvone, geraniol, (+)-limonene, (−)-methanol, 1-menthone, α-myrcene, (+)-α-, (−)-α-, (+)-β-, (−)-β-pinene and pulegone.[23] However, the sawfly *Neodiprion sertifer* stores α- and β-pinene, as well

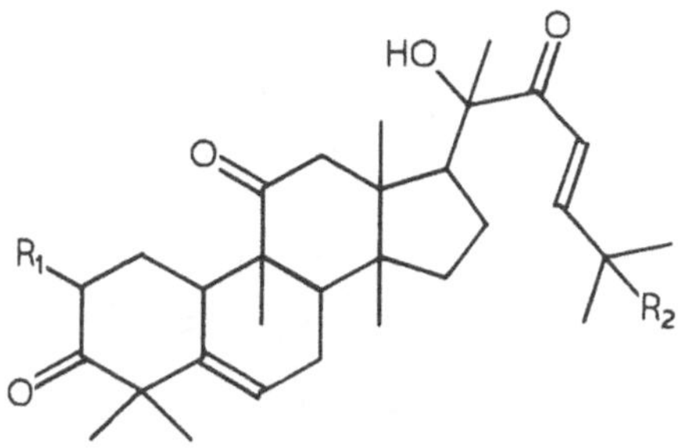

FIGURE 18. R1,R2=OH: cucurbitacin-D; R1=OH, R2=OAc; cucurbitacin-B; R1=glucose, R2=OH: cucurbitacin-D conjugate.

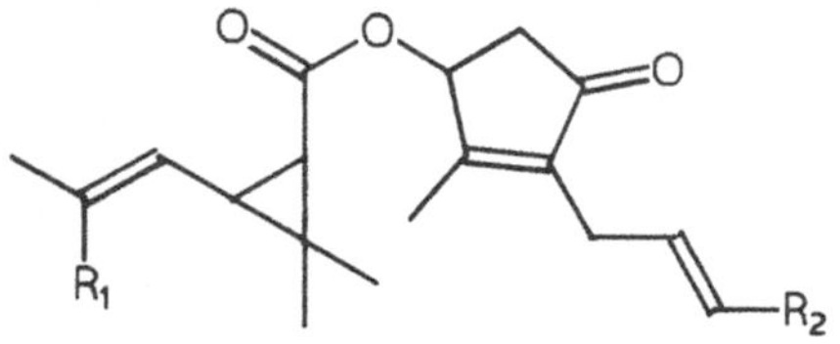

FIGURE 19. R1,R2=H: cinerin I: R1=H, R2=2-propene; pyrethrin I; R1=COOH, R2=H: cinerin I metabolite; R1=COOH, R2=propene: pyrethrin I metabolite.

as other resinous substance (including pimaric acid, levopimaric acid, abietic acid, dehydroabietic acid, neoabietic acid, palustric acid, and pinifolic acid), in diverticular pouches of the foregut, which it uses as a defensive secretion.[186]

Cucurbitacin D (Figure 18) is conjugated by *Diabrotica balteata, D. undecimpunctata howardi, D. virgifera virgifera*, and *Acalymma vittatum*, and the conjugates are sequestered in the hemolymph.[187] The same insect species, as well as *D. cristata*, produce the same array of metabolites of cucurbitacin B, but proportions vary.[188] The *A. vittatum* and *D. v. virgifera* excrete most of it unchanged, but *D. balteata, D. cristata*, and *D. u. howardi* form mostly polar metabolites prior to excretion.[188] Various metabolites are also found in the gut and hemolymph, which may be formed by hydrolysis (to curcurbitacin D) followed by conjugation.[188] The *D. v. howardi* fed with cucurbitacin D also form primarily polar metabolites, which are mostly excreted.[188] Cucurbitacin D is conjugated with glucose in *D. v. virgifera*.[189] For a discussion on the chemical ecology of the cucurbitacins, see Metcalf.[190]

The detoxification of the commercially available pyrethrins (Figure 19) has been investigated in some insect species. Since early research was somewhat hampered by the purity of the radiolabeled substrates,[191] only the most recent research will be discussed, although this is primarily limited to *Musca domestica*. Topically treated *M. domestica* convert pyrethrin I and cinerin I to compounds with intact ester linkages and a small amount of acid metabolites.[192] Further investigation of the *M. domestica* detoxification, both *in vivo* and *in vitro*, demonstrated that many metabolites are produced, but they primarily consist of O-demethylated pyrethrin I, depending on the starting material (Figure 18).[191,193] The production of oxidative metabolites is limited by use of sesamex[192] and piperonyl butoxide.[191] However, the O-demethylated pyrethrin I is still toxic, and further reactions involving conjugation of the alcohols and oxidation of aldehydes to acids may be more important in detoxification.[191] Pyrethrum-associated NADPH oxidation, as an indicator of oxidative metabolism, occurs in microsomal midgut preparations from *S. eridania*.[185]

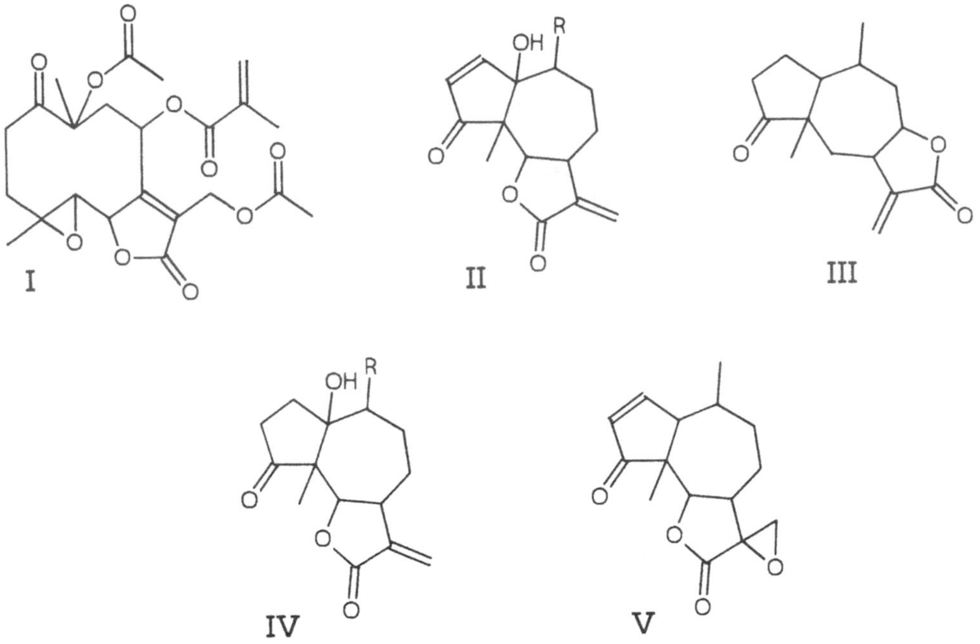

FIGURE 20.   (I) Glaucolide-A. (II) R=CH3: parthenin; R=CH20Ac: conchosin-B. (III) Confertin. (IV) R=CH3: Cononopilin; R=CH2OAc: tetraneurin-A. (V) stromonin=B.

The detoxification of some sesquiterpene lactones (Figure 20) by insects has also been investigated. The eucalyptus oil sesqueterpenoids are stored in the gut and used defensively by sawfly larvae,[194] but many insects excrete them unchanged.[195] Glaucolide-A is converted to oxidative metabolites by *S. frugiperda*.[196] The gut appears to be an effective barrier to sesquiterpene lactone (conchosin-B, confertin, coronopillin, parthenin, stramonin-B, and tetraneurin-A) penetration into the hemolymph of *Melanoplus sanguinipes*, since injected compounds are toxic, but ingested ones are not.[197]

MISCELLANEOUS COMPOUNDS
A heteropolysaccharide from *Phaseolus vulgaris* is hydrolyzed by the bruchid *Acanthoscelides obtectus* to a central core (which contains arabinose, xylose, and rhamnose), and side chains which contain arabinose, glucose, and galactose.[198] This insect is also resistant to the core polysaccharide.[198] This compound prevents feeding by *Callosobruchus chinensis*, which can only slightly hydrolyze it, and which is sensitive to the hydrolyzed core.[198] However, *C. chinensis* (and presumably other insects) is able to hydrolyze the oligosaccharide chain of toxic saponins.[199] Glycosidases may be responsible for the degradation of legume saponins by locusts.[199]

Aristolochic acids (Figure 21) are found in the host plants of some Papilionidae. *Pachlioptera aristolochiae* sequesters aristolochic acid, which is retained in the adult, from its hosts, *Aristolochia clematis* and *A. rotunda*.[200] *Zerynthia polyxena* sequesters aristolochic acids Ia and C from its host, *Aristolochia clematitis*,[201] and *Battus archidamus* sequesters aristolochic acids I and IVa from *Aristolochia chiliensis*.[202] Quinones, phenols, and terpenes may be sequestered by *Romalea microptera*.[203]

Methylazoxymethanol, the aglycone of cycasin, is converted to cycasin by glycosylation and then sequestered by *Seirarctia echo*.[204] In addition, hypericin, which occurs in *Hypericum hirsutum*, is sequestered by *Chrysolina brunsvicensis*.[205] The dihydromatricaria acid secreted by the soldier beetle, *Chauliognathus lecontei*, may also be derived from the Composite

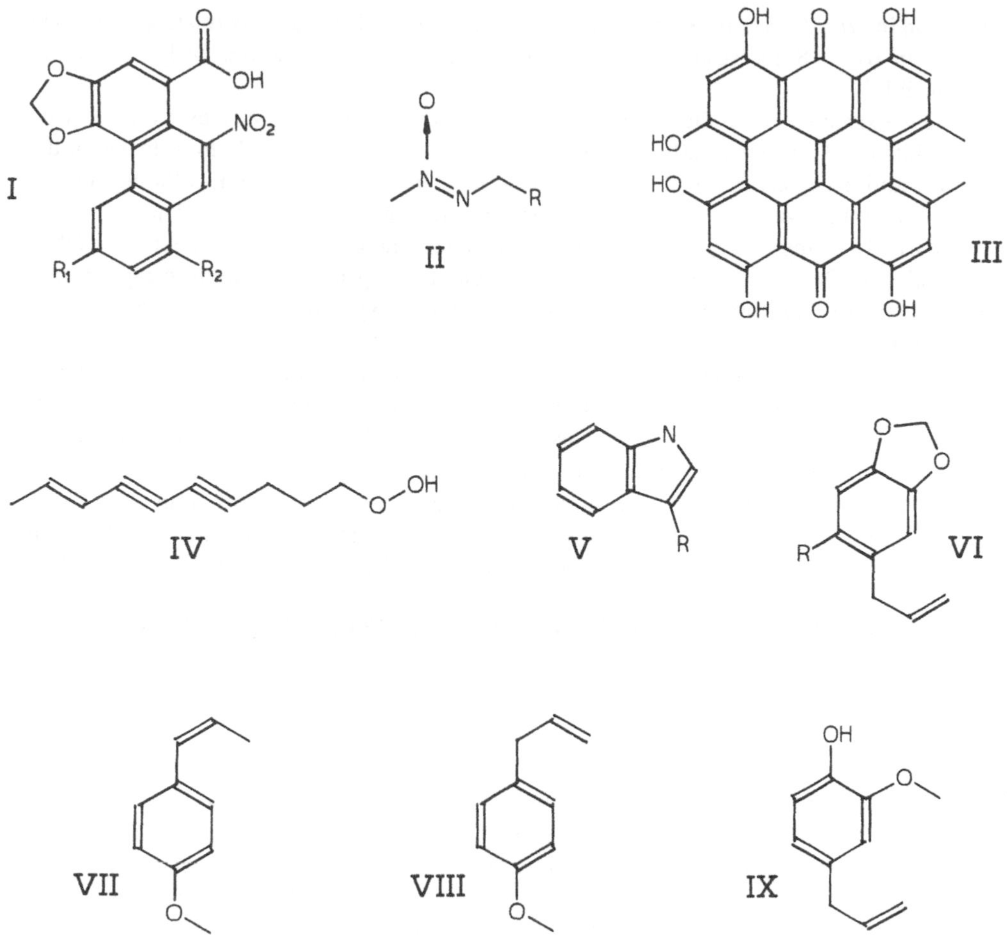

FIGURE 21.   (I) R1=H, R2=OCH3: aristolochic acid I; R1=H, R2=OH: aristolochic acid Ia; R1=OH, R2=H; aristolochic acid C; R1=OH, R2=OCH3; aristolochic acid IVa. (II) R=H: methylazoxymethanol; R=glucose: cycasin. (III) Hypericin. (IV) Dihydromatricaria acid. (V) R=CH2CN: indole-3-acetonitrile; R=CH2OH: indole-3-carbinol. (VI) R=H: safrole; R=OCH3: myristin. (VII) Anethole. (VIII) Estragole. (IX) Eugenol.

hosts.[206] NADPH oxidation by microsomal midgut preparations from *S. frugiperda*, as an indicator of microsomal metabolism, occurs in the presence of indole-3-acetonitrile, indole-3-carbinol (autoinducible), myristicin, safrole, *trans*-anethole, estragole, and eugenol.[23]

DETOXIFICATION OF LOWER PLANT SUBSTANCES

The investigation of the detoxification of lower plant compounds by insects has been very limited. Interactions between insects and chemicals from mosses, liverworts, or algae have apparently not been tested. However, many species of insects use fungi as a food source, and many fungi contain toxic chemicals, so some detoxification of fungal compounds must be taking place. Many species of Phoridae, Mycetophilidae, Anthomyiidae, and Muscidae feed on many species of Boletales, which can contain a variety of toxic compounds.[207] Several species of *Drosophila* are able to feed on amanitin-containing fungi, which may help them escape from nematode parasites that infest other species of mycetophilous *Drosophila*.[208] Some stored grain insects are able to feed successfully on species of molds that are known to produce mycotoxins.[209] In fact, the trichothecene, T-2 toxin, enhances egg production in *Tribolium* confusum.[210] *Periplaneta americana* is apparently relatively resistant to aflatoxin B1.[211] Some strains of *Drosophila melanogaster* are also relatively resistant to aflatoxin B1.[212]

A few examples of studies of detoxification of lower plant substances by insects do exist. Aflatoxin B1 is apparently transported to different tissues (initially high levels in the fat body) and the excretory system of *M. domestica* by the hemolymph, although its accumulation in the ovaries apparently is responsible for causing sterility.[213] The dichloride of aflatoxin B1 is apparently activated to a form which attacks the guanyl moieties of DNA in the fruit fly, *Drosophila melanogaster*.[214] Fungus-feeding insects can more rapidly metabolize (primarily hydrolyze) the trichothecene monoacetoxyscirpenol than can nonfungus-feeding insects.[215]

With the increasing interest in the avermectins and their analogs as insecticides, some studies on the detoxification of these actinomycete-derived compounds has been undertaken. Avermectin B1$_a$ orally applied to the tobacco budworm, *Heliothis virescens*, is metabolized more slowly than that applied to the corn earworm, *H. zea*, and the fall armyworm, *Spodoptera frugiperda*.[216] However, the similarity in metabolism and distribution as well, in the *H. zea* and *S. frugiperda*, does not explain the much greater susceptibility of *H. zea* to avemectin B1$_a$ than *S. frugiperda*, so differences in target site sensitivity may be involved.[216]

FACTORS WHICH INFLUENCE DETOXIFICATION

DISTRIBUTION OF DIFFERENT MODES OF DETOXIFICATION IN INSECTS OF DIFFERENT FEEDING STRATEGIES

Evolutionary theory suggests that a species should maximize its efficiency in dealing with its environment. Of course, a major part of a herbivorous insect's environment consists of its host (or hosts). Thus, in the interaction of an insect with its host (or of specific interest here, the interaction of the insect with its host toxins), the insect species in question should seek to be as efficient as possible. It is well known that different species of insects can have a wide variation in the number of hosts utilized, some being confined to one species of plant, while others will naturally feed on a wide variety of unrelated species of plants. Extending along the lines of this argument then, a monophagous insect may have different means of detoxifying a host's toxins, since it only encounters a few, than a polyphagous insect, which may require a detoxification scheme which is capable of dealing with the wider range of toxins encountered in its wide range of hosts. The insect species involved in each respective feeding strategy (i.e. monophagous or specialist vs. polyphagous or generalist) would attempt to maximize its efficiency of dealing with the host plant toxins that it encounters, which should differ according to the feeding strategy or the insect. While this rationale has generally been investigated in regard to the distribution, activity, or types of enzymes responsible for detoxification (see following discussion), other forms of detoxification will also be considered in the same context in the following discussion.

However, it is necessary to temper the preceding theory, since there are factors other than relative detoxification capability which influence the plant an insect species may feed on in nature (and hence its natural host range). First of all, a polyphagous insect may have a wide host range over its natural distribution, but have localized feeding specialization[217] or preference.[218] Such a specialization may result in physiological differences which are represented by the biotypes which can develop on insect-resistant strains of crop plants.[219] This physiological difference can be manifested in the activity of enzymes associated with detoxification within a population.[220] The differences appear to be fixed in different populations as well and may be related to the types of toxins which are encountered in the different hosts.[221] Thus, as will be discussed later, the host plants that an insect actually feeds on and is adapted to in a particular area, regardless of its apparent host range, should be considered in drawing conclusions on relative strategies of detoxification, since these factors may affect the true host range which is relevant to the comparison. In addition, the plants themselves may vary widely in their toxin content, which may influence the levels of toxins that are sequestered.[222] This host toxin variation may also influence the levels of detoxification enzymes which are inducible.

Second, and somewhat the opposite of the first parameter, because an insect species is found in nature to feed only on a few plant species, it does not mean that it is incapable of successfully coping with the toxins of, and thus developing on, a wider range of hosts. The natural host range may be affected by the ovipositional selectivity of the parent, which may be influenced by such factors as host density and distribution, the insect species population itself, and the presence of competing herbivores.[217] Species of Heliconidae collected from only a few hosts can successfully develop and reproduce on co-occurring nonhosts,[223-225] as can some flea beetles.[224]

Third, and somewhat unrelated to the first two factors, is the cost of feeding on a certain host, in terms of length of development period, size of the resulting adult, reproductive capacity of adults, and amount of host material consumed.[226] The cost may be apparently nonexistent for insects which appear well adapted to their hosts (i.e. *Oncopeltus fasciatus*[76] or *Danaus plexippus*),[96] as well as insects which do not feed on a particular host in nature,[225] or costly to insects that appear well adapted to a host, as well as those apparently not adapted to a host.[226] Some of this variability may be a function of the type of detoxification method used by the insect in question. For example, methods of detoxification which are passive would have a constant or fixed cost. These methods may include impermeability of the gut or sequestration processes which would function the same regardless of the toxin, provided the toxin is a suitable candidate for the detoxification process. On the other hand, active forms of detoxification, such as enzymatic detoxification involving regulated induction, would have a variable additive cost depending on the qualitative and quantitative amounts of enzymes involved.

Another consideration which involves the general classes of enzymes involved in detoxification is the energy efficiency of the different enzymes involved. While presumably able to attack a wide range of different plant toxins, unspecific monooxygenases require a two-enzyme complex, a cofactor (O_2), and a high energy substrate (NADPH). Thus, the cost of supporting this system is relatively high. The glutathione transferases require a cofactor (glutathione), but are still fairly general. The hydrolytic enzymes, although more specific, essentially require nothing other than a suitable substrate, since the cofactor water is ubiquitous. In addition, the relative activity of these enzyme systems, as identified by general indicator substrate metabolism, is different. For example, in comparing the rates of aldrin epoxidation (oxidation), 1,4-chloro, 2-nitrobenzene-glutathione conjugation (conjugation), and 1-naphthyl acetate hydrolysis (hydrolysis), the relative rates of metabolism in the fall armyworm, *Spodoptera eridania*, are reported in picomolar,[227] nanomolar,[143] and micromolar[228] rates, respectively, although the methods of enzyme preparation vary. Obviously, when an ester of any form needs to be detoxified, hydrolysis is the most energy-efficient method. In fact, hydrolysis appears to be a fairly widespread method of detoxification for plant toxins which are appropriate substrates (Table 1). Even though some of these appear to be activation reactions, subsequent metabolism may detoxify them. Thus, with the exception of the pyrethrins (where a secondary alcohol participates in the ester bond, which could sterically hinder hydrolysis), all of the compounds that have bonds susceptible to hydrolysis are hydrolyzed initially, even though they also may have sites suitable for the activity of oxidative or conjugating enzymes. Of course, oxidative enzymes are invaluable in detoxifying compounds that have no sites for attack by other detoxifying enzymes, such as those containing unsubstituted aromatic rings. Comparative studies of the metabolism of plant toxins susceptible to more than one form of metabolism would be useful in obtaining more information on the relative importance of different detoxifying enzymes.

TABLE 1
Examples of Different Classes of Enzymes Involved in Plant Substance Detoxification by Insects (Including Activation)

Substance	Enzyme mechanism	Ref.
Alkaloids		
Cannabidiol	Conjugation	25
	Oxidation	25
Nicotine	Conjugation	20
Nicotine	Oxidation	15, 22
Pyrrolizindine alkaloids	Hydrolysis	30
Δ-tetrahydrocannabinol	Oxidation	25
	Reduction	25
Essential compounds and analogs		
Canaline	Hydrolysis	58
Canaline	Reduction	58
Canavanine	Hydrolysis	55
Ecdysone	Conjugation	63
Ecdysone	Oxidation	63
Ecdysone	Oxidation (A)	63
Ecdysone	Reduction	64
Juvenile hormone	Hydrolysis	69
Juvenile hormone	Oxidation	69
Ponasterone	Oxidation (A)	66
Ponasterone	Oxidation	66
Precocenes	Oxidation (A)	70
Precocenes	Oxidation	70
Cardenolides		
Digitoxin	Oxidation	71
Labriformin	Hydrolysis	103, 105
Labriformidin	Reduction	103, 105
Uscharin	Hydrolysis	97
Uscharidin	Reduction	97, 103
Uscharidin	Conjugation	96
Cyanogenic compounds		
Allylglucosinolates	Hydrolysis (A)	108
HCN (CN-)	Conjugation	111, 112
CN-	Conjugation	113
Flavonoids		
Flavonoids (in general)	Oxidation	129
5'-Glucoside flavonoids	Hydrolysis	133, 134
Isopsoralen	Oxidation	145
4-Methylumbelliferone	Conjugation	146
Psoralen	Oxidation	145
Rotenone	Oxidation	119—121
Salicin	Hydrolysis	124—128
Salicin	Oxidation	124—128
Xanthotoxin	Oxidation	140—142
Tannins, lignins, and related compounds		
Chlorogenic acid	Hydrolysis	162

TABLE 1 (continued)
Examples of Different Classes of Enzymes Involved in Plant
Substance Detoxification by Insects (Including Activation)

Substance	Enzyme mechanism	Ref.
Lignin	Hydrolysis	149, 162
Lignin	Oxidation	152
Rutin	Hydrolysis	162
Tannic acid	Hydrolysis	156
Terpenoids		
Camphene	Oxidation	166
3-Carene	Oxidation	166
Cucurbitacin B	Conjugation	188
Cucurbitacin D	Conjugation	189
Glaucolide-A	Oxidation	196
Myrcene	Oxidation	165—172
α-Pinene	Oxidation	165, 173—179
β-Pinene	Oxidation	173
Pyrethrin I	Oxidation	191, 193
Miscellaneous compounds		
Heteropolysaccharide	Hydrolysis	198
Saponins	Hydrolysis	199
Methylaxozymethanol	Conjugation	204

Note: Only those examples of metabolism where the product has been identified have been included. Cases of metabolism where activation occurs are followed by an (A).

Thus, it should be kept in mind that the apparent host range may or may not reflect a particular insect's ability to deal with host plant (or nonhost plant) toxins. In addition, the total detoxification scheme may involve a number of detoxification mechanisms (some of which may not have been discovered or investigated for a particular insect species) which can vary in cost. The detoxification complex can also vary according to the host an insect has evolved on (whether this involves sibling species considerations or not), and the total scheme may not necessarily be reflected by an investigation that involves only one or two mechanisms of detoxification.

The activity and distribution of different detoxifying mechanisms in insects of different feeding strategies will now be considered, with primary emphasis on specific vs. generalist insects. Previous investigations of relative detoxification strategies associated with insects of different feeding strategies have involved a consideration of relative enzymatic activity. However, other mechanisms of detoxification, which are useful for comparison, also exist in insects with different feeding strategies. Most insects have common forms of detoxification, such as excreting a compound before it can be absorbed, fat bodies which act as lipophilic sinks, and Malpighian tubules which serve as excretory organs.[222] Toxins may also be eliminated as part of the insect exuvia when the insect molts.[222] The gut clearing associated with insect molts may prevent toxins from entering the insect at times when the ability of the insect to detoxify chemicals is temporarily disrupted.

Other more specialized forms of detoxification can occur in both monophagous and polyphagous insects. For example, both the monophagous Lygaeid *O. fasciatus* and the polyphagous Lygaeid *Spilostethus pandurus* sequester cardenolides.[25] In addition, the monophagous grasshopper *Poekilocerus bufonis* and the polyphagous grasshoppers *P. pictus* and *Phymaeteus viridipes* all sequester cardenolides.[25] The general feeder *Arctia caja* sequesters both pyrrolizidine alkaloids and cardenolides.[47] Similarly, the specific feeding insect *Manduca sexta*, and more general feeders such as *Trichoplusia ni* and *Heliothis virescens*, all excrete nicotine unchanged.[15] However, in spite of the occurrence of sequestering ability and relative gut impermeability in insects which are both monophagous and polyphagous, there are examples of detoxification mechanisms that appear specifically designed around a particular toxin which is encountered by a monophagous insect species. These mechanisms include the target site insensitivity of *M. sexta* to nicotine,[20] and of the Na^+/K^+-transporting ATPase of *D. plexippus*[93] and *O. fasciatus*[84] to cardenolides, as well as the selectivity of the arginine tRNA ligase belonging to *Carydes brasiliensis*.[52]

The original information which stimulated the consideration of the detoxification mechanisms of insects with different feeding strategies was concerned with the levels of aldrin epoxidase activity in different species of caterpillars. Results of the study indicated that, in general, the activity of aldrin epoxidase is higher in caterpillars which are polyphagous vs. oligophagous vs. monophagous.[229] However, a more extensive study demonstrated no differences in aldrin epoxidase activity of caterpillars with differences in host range.[230] In both of these studies, within group variability was greater than that between the groups. One factor that may be involved in causing intragroup variability is the effect of enzyme induction, which may vary according to the host fed upon, and which could influence the relative activity of the oxidative enzymes involved (see following section). Alternatively, other enzyme classes may be involved in detoxification, or as stated above, other modes of detoxification may be relatively more important than enzymatic metabolism. For example, as indicated previously, the general feeding insect *Heliothis virescens* does not metabolize nicotine, but rather excretes it unchanged.[15] In addition, as discussed previously, thiosulfate sulfurtransferase, which could be responsible for cyanide detoxification, is widely distributed in insects which do not feed on plants which contain cyanogenic compounds, as well as in those insects that do feed on these plants.[9]

Another possibility is that the distribution of the classes of enzyme activity in the insects of different feeding strategies is related to the sites on the allelochemicals present in the host which are accessible to attack by the different detoxifying enzyme classes. For example, the relatively monophagous *M. histeronica* which feeds on plants in the *Brassicae*, have five times higher levels of glutathione transferase than *Oebolus pugnax*, which feeds on the Poaceae.[231] The Brassicae are known to contain relatively high levels of cyanogenic compounds,[11] which are appropriate substrates for glutathione *S*-transferase, while cyanogenic compounds are relatively infrequent in the Poaceae.[232]

However, one example of enzyme activity which does appear to be related to the feeding strategy of herbivorous insects is the activity of *trans*-epoxide hydrolase, and the ratio of *trans/cis* epoxide hydrolase activity, in the herbivorous insects vs. that of predaceous, parasitic, or saprophagous insects.[12,233-235] The *trans*-epoxide hydrolase activity in herbivorous vs. other insects is up to 11-fold higher, while the ratio of *trans/cis*-epoxide hydrolase activity is up to 6.4-fold higher.[233] Polyphagous chrysomelids also have a higher ratio of activity of these enzymes than do monophagous ones,[233] while herbivorous coccinellids have higher *trans*-epoxide hydrolase activity and a higher ratio of the isomeric enzyme activity than predaceous ones.[235] The distribution of the *trans*-epoxide hydrolase in herbivorous insects may be related to the presence of *trans*-olefins in plants, which are rarer in prey animals.[235] This difference in enzyme distribution may be a mechanism which can be exploited in designing insecticides which are more specific to herbaceous insects.[235] In addition, phloem-sucking insects also have a relatively lower *trans/cis* epoxide hydrolase activity ratio, as well as a lower aldrin epoxidase activity, then do chewing herbivorous insects, which would be consistent with their limited exposure to the toxins encountered by the chewing insects.[12] The logic behind the lower enzyme activity in predaceous vs. herbivorous insects is that the predators have less of a need for the toxin-metabolizing enzymes than the herbivorous insects, since they have less contact with plant toxins.[12] However, predators which feed by liquifying the contents of the prey and then imbibing them, such as the Pentatomidae, would encounter the same dose of chemicals which occurs in the gut of the prey. This may explain the relatively high levels of esterase and glutathione transferase found in predaceous Pentatomidae, which are comparable to those levels found in general and specific herbivorous Pentatomidae.[231]

Another area of detoxification which has seen little investigation and which should be considered in the context of the present discussion, is that of the presence of, and interactions between, multiple chemicals that occur in host plants. An excellent discussion of this aspect is provided by Berenbaum.[144] This area has briefly been mentioned in the discussion of xanthotoxin metabolism, where the co-occurring methylenedioxyphenyl compound myristicin synergizes xanthotoxin toxicity.[144] Methylenedioxyphenyl compounds, which may inhibit oxidative enzymes, are widely distributed in different plant families.[144] Other plant substances that may act as synergists include saponins, thiocyanates, compounds with acetylenic triple bonds, o-anisaldehyde, theobromine, and capillin.[144] Analogs of compounds may also act in a synergistic manner.[144] Nontoxic (at naturally occurring levels) substances from fungi can also synergize the toxicity of co-occurring fungal mycotoxins or other xenobiotics.[236] Considering the complexity of chemical interactions which may occur whether an insect feeds on one or several hosts in cases where detoxification relies primarily on metabolism, even monophagous insects may still need a variety of enzymes to deal with the variety of toxic plant chemicals they may encounter. This is an additional factor that may explain why some monophagous insects may still have levels of oxidative, hydrolytic, or conjugating enzyme activity similar to some polyphagous insects.

Thus, to understand better the activity of detoxification mechanisms of insects with different feeding strategies, it is necessary to consider all detoxification mechanisms involved for each insect, and in the context of the plants (and their allelochemicals) involved as hosts. One possibility is to perturb individually the different systems involved, to test the effect relative to other systems. Synergists are useful in investigating the enzymes involved in detoxification, provided one realizes that the sensitivity of the same enzyme in different insects or their tissues may vary, or that an inhibitor may affect more than one enzyme system. Some possibilities useful for investigating other systems of detoxification might be to use chemicals that alter gut or nervous system permeability, or chemicals that protect the target site (if known). While this is indeed a formidable task, concentrating on a single mode of detoxification may lead to results which do not reflect all of the interactions which are occurring.

THE INDUCTION PHENOMENON

Probably no other area concerning the subject of detoxification of plant compounds by insects is as intriguing as the enzyme induction phenomenon which occurs in insects (as well as other organisms), since it provides a mechanism for conserving resources. This phenomenon involves the apparent increased synthesis of detoxifying enzymes after exposure to a particular substrate. Although induction generally refers to increases in enzyme levels due to *de novo* protein synthesis, for the purpose of the following discussion induction will refer to increased enzyme *activity* stimulated by an inducing chemical, even though allosteric enhancement or enzyme protection may also be a causal factor. Studies on enzyme induction in insects generally do not establish a link between enhanced enzyme activity and increased synthesis of the enzyme responsible for metabolism. In addition, there are few examples where the *plant-derived* chemical that induces a particular enzyme is subsequently shown to be the substrate for the induced enzyme (e.g., α-pinene[183,185] and indole-3-carbinol[23]). This does not mean that such events only rarely occur, but that research on induction has a tremendous head start over that on subsequent metabolism of inducers. This phenomenon is also important due to the ramifications involved in the potential reduction of effectiveness of insecticides.[23] The enzymes induced by plant feeding can increase the rate of metabolism of aldrin (aldrin epoxidase), (see below for references), phorate,[237] and several pyrethroid insecticides.[238] Temperature can also influence the level of induction in insects.[239] However, the adaptation of the same insect species to different species of host plants appears to affect their levels of detoxifying enzymes (including those that metabolize insecticides) in a more permanent fashion.[222] Although work with mammals has often served as a guideline for studies on enzyme induction in insects, the receptors (Ah) involved in aryl hydroxylase induction in mammals appear to be different from those in insects.[240] Since plant chemical associated induction has recently been reviewed by Yu,[23] who has done most of the work in this area, the following discussion will be somewhat limited, and the reader is directed to this review[23] for more information.

The enzymes induced by specific plant allelochemicals include unspecific monooxygenases, such as aldrin epoxidase,[143,185,241-245] heptachlor epoxidase,[143] aniline hydroxylase,[246] biphenyl hydroxylase,[247] parathion desulfurase,[23] phorate sulfoxidase,[238] and 4-chloro-N-methylaniline N-demethylase.[23,143,185,246] NADPH oxidation, which is probably associated with unspecific monooxygenase activity, is also induced in insects.[23,118,185] However, the recent work by Marty and Krieger,[97] which demonstrates NADPH-dependent reduction of uscharidin by an enzyme which is apparently not an unspecific monooxygenase, suggests that NADPH oxidation can reflect activity of other enzymes as well, depending on the enzyme source (see prior discussion on cardenolide metabolism). Other enzymes which are induced by plant extracts or chemicals include glutathione transferase,[143,185,228,242,248] aryl sulfotransferase,[185] 4-nitrophenyl acetate hydrolase,[249] 1-naphthyl acetate hydrolase,[229] cis-permethrin and fenvalerate hydrolase,[249] acephate amidase,[250] and styrene epoxide hydrolase.[23] The range of plant allelochemical inducers includes many compounds which can be metabolically detoxified by insects, such as cyanogenic compounds,[228,246] flavonoids,[23,143,229,238,247] and terpenoids.[23,118,185,240,241,243-245,247]

Although a wide variety of substrates have been tested for the induction of different enzymes on one system, these substrates are not always substrates for the enzymes induced (especially in regard to the hydrolases). Nontoxic components from plants may also be responsible for the enzyme induction in some cases.[250] In considering the most effective inducers, two factors are evident. Although terpenoids are effective inducers of unspecific monooxygenases and ineffective inducers of glutathione transferases,[23,242] some compounds that are effective inducers of one enzyme system are the more effective inducers of most or all of the other enzyme systems (i.e., indole-3-carbinol, indole-3-acetonitrile, and flavone). Second, many of the most effective inducers are nonpolar, and appear to contain a carbon-carbon double bond conjugated with a carbon oxygen bond (e.g., indole-3-carbinol, flavone, naphthylflavone, and xanthotoxin).

The induction of several different enzymes by the same chemicals suggests a common phenomenon, which is similar in concept to the single locus which appears to control the enzymatic response in insecticide-resistant insects.[251] The fact that induction by nonsubstrates can occur, and the fact that not all potential substrates will induce the appropriate enzymes, indicate another possibility which may play a role. As long as the inducer (whether toxic or not) co-occurs with the toxins needing to be metabolized, the induction still occurs at the proper time. This development may be a result of previous evolutionary difficulties with the coordination of the receptor site and the induced enzyme, due to the tremendous variety of potentially toxic structures which are involved. A few compounds now appear to be the most effective as general inducers (at least in *S. frugiperda*), although this function may now be something of an evolutionary artifact. The positive physiological effects of plant hormones on grasshoppers, which can influence reproduction,[252] and the previously reported inductive effects of plant hormone (indole-3-acetic acid) analogs,[23] suggest that these (or the metabolite indole-3-carbaldehyde,[253] which apparently has not been tested) could be one class of compounds involved in generalized induction.

SITES OF DETOXIFICATION

Different sites of detoxification of plant chemicals by insects will be examined in this section, although the reader is directed to the previous section on plant compound detoxification for more details. The gut is the most obvious site of detoxification, since it is the first area that the plant material comes in intimate contact with the insect. However, some preliminary detoxification may be performed by the saliva of the insect, which in sucking insects may be injected into plant material prior to consumption.[129] If the permeability of the gut is low, or the rate of passage through the gut is high, high rates of excretion of unmetabolized toxin can result. For example, the gut of *Shistocerca gregaria* is relatively impermeable to ouabain as compared to that of *Oncopeltus fasciatus*.[72] The gut (especially the midgut) is also a source of detoxifying enzymes, and can be the major source of metabolism of such compounds as furanocoumarins in *P. polyxenes*.[140,142] The hindgut of many bark beetles appears to be associated with terpene metabolism.[183] Special modifications of the gut allow for the sequestration of defensive compounds in the sawfly *Neodiprion sertifer*.[186] The peritrophic membrane, which lines the gut, is associated with the impermeability of locust guts to tannins,[154] as well as binding the tannins prior to excretion.[154] Special caecal areas can assist in holding the bound tannin prior to excretion.[155] Even the pH of the gut can apparently prevent protein binding of such toxins as tannins in the Lepidoptera.[4]

However, if the chemicals penetrate the gut, there are other lines of defense. The toxin would next enter the hemolymph, which appears to play a limited role in detoxification. The K^+ ion level in the hemolymph does antagonize the binding of Na^+/K^+-transporting ATPase in *M. sexta* and *D. plexippus*.[93] The hemolymph of *O. fasciatus* contains a carrier protein for cardenolides.[71] Cucurbitacins are sequestered in the hemolymph of cucurbit-feeding Chrysomelids.[187] Of course, the circulation of the hemolymph can deliver the toxins to other areas of detoxification as well as to the target site.

Another site of detoxification is the fat body, which can serve as a general lipophilic sink,[222] or a source of metabolic detoxification.[254] The fat body of *P. americana* is the most active tissue tested in rotenone metabolism.[120] The fat body of *Locusta migratoria* is more active than the gut in conjugating 4-methylumbelliferone.[146] The Malpighian tubules also can serve as general excretory organs, and appear to be involved in general alkaloid detoxification.[51] They can also be a site of high detoxifying enzyme activity.[255] Interestingly enough, special modifications of the Malpighian tubules also assist *O. fasciatus* in cardenolide sequestration.[73]

The target site itself may have its own line of defense. The Na^+/K^+-transporting ATPases of *O. fasciatus* are resistant to the effects of cardenolides.[84] The nervous system of *M. sexta* is resistant to the effects of nicotine due to its metabolic ability[20] and efficient excretion.[17-19] Once the target sites of more plant toxins in insects are identified, more information in this area should be available.

However, detoxification of some plant toxins by certain insects may involve some highly specialized structures or unusual areas of the insect, including those that have already been mentioned in relation to the gut. Many of these are involved in sequestration. For example, *O. fasciatus* has a bilayer epidermis which possesses cardenolide-resistance cells that contain the cardenolides that are sequestered.[74] Specialized glands are also involved in sequestration of cardenolides in *O. fasciatus*, which promote storage in an emulsion phase.[78] Another novel structure is the hair pencil of Danainae and Ithomiinae males, which stores pyrrozolidine alkaloid-derived pheromones, and in some cases must make contact with the wing glands to promote metabolism.[40] The wings of *D. plexippus* are one area of storage of cardenolides,[99] while the wings of *M. galatea* serve as the storage site for flavonoids.[132-135] The meconium of *Junonia coenia* and *Ceratomia catalpae* is a source of elimination of iridoid glucosides.[147]

However, not all of the detoxification may be performed by the tissues of the insect. Microorganisms associated with the gut may be responsible for the metabolism of digitoxin in *O. fasciatus*[72] and the metabolism of lignin in *Nasutitermes exitosus*.[152] Microorganisms may also be involved in canavanine metabolism by *Carydes brasiliensis*.[61] Microorganisms associated with bark beetles appear to do some of the conversion of terpenoids to pheromones,[180-182] one of which is the potential insect pathogen *Bacillus cereus*.[180] In fact, microorganisms are widely associated with herbivorous insects, and they may play a wider role in detoxification of plant substances than formerly suspected.[256] For example, intracellular symbiotic microorganisms from mycetomes of several insects produce potentially detoxifying hydrolytic, oxidative, and conjugating enzymes *in situ*, which are often more active than those associated with the gut, as indicated by histochemical analysis.[257] In addition, some insects use host(fungus)-derived enzymes for digestion.[258] Whether fungus-feeding or herbivorous insects can also use host-derived enzymes for detoxification purposes apparently remains to be tested. The known presence of cyanide metabolizing enzymes in plants, and the information that indicates insect-derived metabolites may be produced by these enzymes (see previous discussion), suggests that plant-derived enzyme detoxification by insects is a possibility.

ADDITIONAL BENEFITS ASSOCIATED WITH DETOXIFICATION

Detoxification of plant chemicals by insects is in itself a benefit to the insect, since this process enables the insect species to exploit the host (see Chapter entitled Insect Feeding Deterrents). However, as previously discussed, behavioral factors may also play a role in limiting the species of plants insects feed on, even though they may be physiologically acceptable hosts.[223-225,259] Some insects have taken the detoxification process one step further, and found some additional benefit from the detoxified chemicals.

Probably the most well known additional benefit occurs in insects that sequester the toxins or their metabolites, which renders them less palatable or toxic to predators. This ability is found in a diverse number of insects and is associated with many different plant chemicals. For example, the unpalatability of insects that feed on milkweed and store the cardenolides has been demonstrated in several cases,[25] including invertebrates.[260] The presence of cucurbitacins in the cucumber beetles has been shown to make them unpalatable to mantids.[187] Cardenolides in *D. chrysippus* can also protect against parasitism.[261] Stored flavonoids,[135] including salicylaldehyde,[123] may provide protection against pathogens. The protective phenomenon of detoxified plant chemicals is probably present in many of the herbivorous insects that have aposematic colorations.[262] For more information on this facet of detoxification, the reader is directed to the review by Rowell-Rahier and Pasteels.[128]

Another use for the byproducts of detoxification is nutritional supplementation. Certainly insects that hydrolyze glucoside or other sugar conjugates should be able to absorb the sugars for nutritional purposes. For example, the beetles that hydrolyze salicin apparently use the liberated glucose as a nutrient source.[126] The bruchid *Carydes brasiliensis* can use the nonprotein amino acid canavanine as a nitrogen source once it is metabolized.[60,61] In addition, the grasshopper *Anacridium melanorhodon* is able to use tannin-derived phenols and related compounds as dietary components which are eventually incorporated into the cuticle.[157-160] Stored flavonoids may assist in cellular repair after harmful irradiation.[263]

One of the more highly evolved uses for plant toxins is for host or mate location. The cucurbitacins serve as attractants for various species of the cucumber beetles,[187,190] and pyrrolizidine alkaloid degradation products appear to attract the male butterflies that use them as a source of pheromones.[31,32] For more information on this subject, the reader is directed to the Chapter entitled Plant Stimulants and Attractants. As previously discussed, other insects, such as bark beetles, Danaid butterflies, and locusts, convert toxins to pheromones. The reader is directed to Volumes of this series for more information on pheromones.

HINDSIGHT AND FUTURE PROSPECTS FOR RESEARCH ON THE DETOXIFICATION OF PLANT SUBSTANCES BY INSECTS

Since the author was last involved in a discussion of the metabolic aspects of detoxification[1] some new additions have served to fill previous gaps. The number of plant chemicals actually tested in metabolic studies has increased. A fairly comprehensive picture is known concerning the detoxification of nicotine by *Manduca sexta* and cardenolides by *Oncopeltus fasciatus*. More information is known about nonoxidative enzymes that are responsible for plant chemical detoxification, including information that these enzymes may also be induced by differential plant feeding or plant chemicals. Most intriguing is the discovery of Mullin and associates that the ratio of isomer specific epoxide hydrolases in plant feeders is different from insects that have other feeding strategies. Some of the confusion concerning the induction of oxidative enzymes has been abated since it has been determined that many forms may exist that respond in different manners, and which may have different substrate specificity. There are a few more examples of induction of an enzyme which actually is responsible for the metabolism of the inducer, although the appropriate metabolites have not been identified. Similarly, although as of yet there is no direct demonstration, there is evidence that xanthotoxin can induce oxidative enzymes in *Spodoptera eridania*,[76,143] which may be responsible for the production of metabolites by this insect.[140-142] In addition, a co-occurring oxidative enzyme inhibitor (myristicin) can increase toxicity of xanthotoxin.[144] More recent research has indicated that the enzyme specificity and distribution in insects with different feeding strategies is not as clear-cut as once believed, and that factors other than plant toxins may be more important in determining host range of a particular insect species.

However, compared to insecticides, examination of the detoxification of plant substances by insects is still very limited, especially in the area of identification and characterization of the enzymes involved (other than tests involving inhibitors or differential centrifugal sedimentation). No purification studies have been performed which identify molecular weights or substrate specificity of enzymes involved in detoxification of plant substances. The induction story is still unclear, although significant progress has been made. The few examples which show that microorganisms are apparently involved in plant toxin detoxification are particularly intriguing and warrant further investigation. The number of studies on the metabolism of chemicals from nonvascular plants by insects is very limited.

Recent refinements in analytical techniques should soon be applied to investigation of plant chemical detoxification by insects. The 2-D NMR spectroscopy may allow for a continuous monitoring of *in vivo* metabolism. Tandem GC-MS/MS, including computer-accessed libraries of spectra of compounds, may allow for easier identification of sequestered products or metabolites, and thus a corresponding increase in studies of this nature. Techniques used in reconstitution studies of purified unspecific monooxygenases performed in mammals can be applied to insects to determine more about the different forms and their substrate specificity. Continued work in the area of detoxification of plant substances by insects should increase the information leading to the understanding of the relative importance of different methods in insects exposed to various chemicals, both alone and in combination. This information is useful in the identification and use of plant-derived substances or synthetic insecticides for insect control. Through understanding how chemicals toxic to insects are detoxified, it may be possible to counteract or use these strategies for more effective insect control, or even for factors yet unknown which will be of benefit to mankind.

REFERENCES

1. **Dowd, P. F., Smith, C. M., and Sparks, T. C.,** Detoxification of plant toxins by insects, *Insect Biochem.,* 13, 453, 1983.
2. **Berenbaum, M. R.,** Target site insensitivity in insect/plant interactions, in *Molecular Aspects of Insect-Plant Associations,* Brattsten, L. B. and Ahmad, S., Eds., Plenum Press, New York, 1986, 257.
3. **Duffey, S. S.,** Sequestration of plant natural products by insects, *Annu. Rev. Entomol.,* 25, 447, 1980.
4. **Berenbaum, M.,** Adaptive significance of midgut pH in larval Lepidoptera, *Am. Nat.,* 115, 138, 1980.
5. Nomenclature Committee of the International Union of Biochemistry, *Enzyme Nomenclature,* Academic Press, Orlando, FL, 1984, 1.
6. **Ahmad, S., Brattsten, L. B., Mullin, C. A., and Yu, S. J.,** Enzymes involved in metabolism of plant allelochemicals, in *Molecular Aspects of Insect-Plant Interactions,* Plenum Press, New York, 1986, 73.
7. **Gould, F.,** Mixed function oxidases and herbivore polyphagy: the devil's advocate position, *Ecol. Entomol.,* 9, 29, 1984.
8. **Long, K. Y., and Brattsten, L. B.,** Is rhodanese important in the detoxification of dietary cyanide in southern armyworm (*Spodoptera eridania* Cramer) larvae?, *Insect Biochem.,* 12, 367, 1982.
9. **Beesley, S. G., Compton, S. G., and Jones, D. A.,** Rhodanese in insects, *J. Chem. Ecol.,* 11, 45, 1985.
10. **Wilkinson, C. F.,** Xenobiotic conjugation in insects, in *Xenobiotic Conjugation Chemistry,* Paulson, G. D., Caldwell, J., Hutson, D. H., and Menn, J. J., Eds., American Chemical Society, Washington, D.C., 1986, 48.
11. **Hegnauer, R.,** *Chemotaxonomie der Pflanzen, Band 3, Acanthaceae-Cyrillaceae,* Birkhauser Verlag, Basel, 1964, 1.
12. **Mullin, C. A.,** Detoxification enzyme relationships in arthropods of differing feeding strategies, in *Bioregulators for Pest Control,* Hedin, P. E., Ed., American Chemical Society, Washington, D.C., 1985, 267.
13. **Self, L. S., Guthrie, F. E., and Hodgson, E.,** Adaptation of tobacco hornworms to the ingestion of nicotine, *J. Insect Physiol.,* 10, 907, 1964.
14. **Schmeltz, I.,** Nicotine and other tobacco alkaloids, in *Naturally Occurring Insecticides,* Jacobson, M., and Crosby, D. G., Eds., Marcel Dekker, New York, 1971, 99.
15. **Self, L. S., Guthrie, F. E., and Hodgson, E.,** Metabolism of nicotine by tobacco feeding insects, *Nature,* 204, 300, 1964.
16. **Yang, R. S. H. and Guthrie, F. E.,** Physiological responses of insects to nicotine, *Ann. Entomol. Soc. Am.,* 62, 141, 1969.
17. **Morris, C. E.,** Electrophysiological effects of cholinergic agents on the CNS of a nicotine-resistant insect, the tobacco hornworm *(Manduca sexta), J. Exp. Zool.,* 229, 361, 1984.
18. **Morris, C. E.,** Efflux of nicotine and its CNS metabolites from the nerve cord of the tobacco hornworm, *Manduca sexta, J. Insect Physiol.,* 29, 953, 1983.
19. **Morris, C. E.,** Efflux patterns for organic molecules from the CNS of the tobacco hornworm, *Manduca sexta, J. Insect Physiol.,* 29, 961, 1983.
20. **Morris, C. E.,** Uptake and metabolism of nicotine by the CNS of a nicotine-resistant insect, the tobacco hornworm *(Manduca sexta), J. Insect Physiol.,* 29, 807, 1983.
21. **Self, L. S.,** Adaptation of insects to tobacco, *Diss. Abstr.,* 26, 1133, 1965.
22. **Guthrie, F. E., Ringler, R. L., and Bowery, T. C.,** Chromatographic separation and identification of some alkaloid metabolites of nicotine in certain insects, *J. Econ. Entomol.,* 50, 821, 1957.
23. **Yu, S. J.,** Consequences of induction of foreign compound-metabolizing enzymes in insects, in *Molecular Aspects of Insect-Plant Interactions,* Brattsten, L. B. and Ahmad, S., Eds., Plenum Press, New York, 1986, 153.
24. **Robinson, T.,** The evolutionary ecology of alkaloids, in *Herbivores: Their Interaction with Secondary Plant Metabolites,* Rosenthal, G. A., and Janzen, D. H., Eds., Academic Press, Orlando, FL, 1979, 413.
25. **Rothschild, M., Rowan, M. G., and Fairbairn, J. W.,** Storage of cannabinoids by *Arctia caja* and *Zonocerus elegans* fed on chemically distinct strains of *Cannabis sativa, Nature,* 266, 650, 1977.
26. **Blum, M. S., Rivier, L., and Plowman, T.,** Fate of cocaine in the Lymantriid *Eloria noyesi,* a predator of *Erythroxylum coca, Phytochemistry,* 20, 2499, 1981.
27. **Boppré, M.,** Insects pharmacophagously utilizing defensive plant chemicals (pyrrolizidine alkaloids), *Naturwissenschaften,* 73, 17, 1986.
28. **Boppré, M. and Schneider, D.,** Insects and pyrrolizidine alkaloids, in *Proc. 5th Int. Symp. Insect-Plant Relat., Wageningen, 1982,* Pudoc, Wageningen, The Netherlands, 1982, 373.
29. **Edgar, J. A.,** Pyrrolizidine alkaloids in insect-plant co-evolution, *Toxicon,* Suppl. 3, 97, 1983.
30. **Mattocks, A. R.,** Toxicity and metabolism of *Senecio* alkaloids, in *Phytochemical Ecology,* Harborne, J. B., Ed., Academic Press, Orlando, FL, 1972, 179.
31. **Pliske, T. E.,** Attraction to Lepidoptera of plants containing pyrrolizidine alkaloids, *Environ. Entomol.,* 4, 455, 1975.

32. **Pliske, T. E., Edgar, J. A., and Culvenor, C. C. J.,** The chemical basis of attraction of Ithomiine butterflies to plants containing pyrrolizidine alkaloids, *J. Chem. Ecol.,* 2, 255, 1976.

33. **Edgar, J. A., Culvenor, C. C. J., and Smith, L. W.,** Dihydropyrrolizine derivatives in the "hair pencil" secretions of danaid butterflies, *Entomol. Exp. Appl.,* 27, 761, 1971.

34. **Edgar, J. A. and Culvenor, C. C. J.,** Pyrrolizidine ester alkaloid in danaid butterflies, *Nature,* 248, 614, 1974.

35. **Edgar, J. A., Culvenor, C. C. J., and Pliske, T. E.,** Isolation of a lactone, structurally related to the esterifying acids of pyrrolizidine alkaloids from the costal fringes of male Ithomiinae, *J. Chem. Ecol.,* 3, 263, 1976.

36. **Schneider, D., Boppré, M., Schneider, H., Thompson, W. R., Boriak, C. J., Petty, R. L., and Meinwald, J.,** A pheromone precursor and its uptake in male *Danaus* butterflies, *J. Comp. Physiol.,* 97, 245, 1975.

37. **Edgar, J. A., Culvenor, C. C. J., and Robinson, G. S.,** Hairpencil dihydropyrrolizines of Danainae from the New Hebrides, *J. Aust. Entomol. Soc.,* 12, 144, 1973.

38. **Meinwald, J., Boriack, C. J., Schneider, D., Boppré, M., Wood, M. F., and Eisner, T.,** Volatile ketones in the hair-pencil secretion of danaid butterflies, *Amauris* and *Danaus, Experientia,* 30, 721, 1974.

39. **Edgar, J. A.,** Pyrrolizidine alkaloids sequestered by Solomon Island Danaine butterflies. The feeding preferences of Danainae and Ithomiinae, *J. Zool., London,* 196, 385, 1982.

40. **Boppré, M., Petty, R. L., Schneider, D., and Meinwald, J.,** Behaviorally mediated contacts between scent organs: another prerequisite for pheromone production in *Danaus chrysippus* males (Lepidoptera), *J. Comp. Physiol.,* 126, 97, 1978.

41. **Culvenor, C. C. J. and Edgar, J. A.,** Dihydropyrrolizine secretions associated with coremata of *Utethesia* moths (Family Arctiidae), *Experientia,* 38, 627, 1972.

42. **Connor, W. E., Eisner, T. E., Vander Meer, R. K., Guerrero, A., and Meinwald, J.,** Precopulatory sexual interaction in an arctiid moth (*Utetheisa ornatrix*): role of a pheromone derived from dietary alkaloids, *Behav. Ecol. Sociobiol.,* 9, 228, 1981.

43. **Alpin, R. T., Benn, M. H., and Rothschild, M.,** Poisonous alkaloids in the body tissues of the cinnabar moth *Callimorpha jacobaeae* (L.), *Nature,* 219, 747, 1968.

44. **Alpin, R. T. and Rothschild, M.,** Poisonous alkaloids in the body tissues of the garden tiger moth (*Arctia caja* L.) and the cinnabar moth [*Tyria* (= *Callimorpha jacobaeae* L.)], (Lepidoptera), in *Toxins of Animal and Plant Origin, Vol. 2,* deVries, A. and Kochva, K., Eds., Gordon & Breach, New York, 1972, 579.

45. **Blum, M. S.,** Detoxification, deactivation, and utilization of plant compounds by insects, in *Plant Resistance to Insects,* Hedin, P. E., Ed., Academic Press, Orlando, FL, 1983, 265.

46. **Rothschild, M., Alpin, R. T., Cockrum, P. A., Edgar, J. A., Fairweather, P., and Lees, R.,** Pyrrolizidine alkaloids in arctiid moths (Lep.) with a discussion of host plant relationships and the role of these secondary plant substances in the Arctiidae, *Biol. J. Linn. Soc.,* 12, 305, 1979.

47. **Rothschild, M. and Alpin, R. T.,** Toxins in tiger moths (Arctiidae: Lepidoptera), in *Chemical Releasers in Insects,* Tahori, A. S., Ed., Gordon & Breach, New York, 1971, 177.

48. **Rothschild, M. and Edgar, J. A.,** Pyrrolizidine alkaloids from *Senecio vulgaris* sequestered and stored by *Danaus plexippus, J. Zool., London,* 186, 347, 1978.

49. **Bernays, E. A. and Chapman, R. F.,** Plant chemistry and acridoid feeding behavior, in *Biochemical Aspects of Plant and Animal Coevolution,* Harborne, J. B., Ed., Academic Press, Orlando, FL, 1978, 99.

50. **McLain, D. K. and Shure, D. J.,** Host plant toxins and unpalatability of *Neacoryphus bicrucis* (Hemiptera: Lygaeidae), *Ecol. Entomol.,* 10, 291, 1985.

51. **Maddrell, S. H. P. and Gardiner, B. Ö. C.,** Excretion of alkaloids by Malpighian tubules of insects, *J. Exp. Biol.,* 64, 267, 1976.

52. **Rosenthal, G. A., Dahlman, D. L., and Janzen, D. H.,** A novel means for dealing with L-canavanine, a toxic metabolite, *Science,* 192, 256, 1976.

53. **Rosenthal, G. A. and Janzen, D. H.,** Avoidance of nonprotein amino acid incorporation into protein by the seed predator, *Caryedes brasiliensis* (Bruchidae), *J. Chem. Ecol.,* 9, 1353, 1983.

54. **Rosenthal, G. A.,** Biochemical insight into insecticidal properties of L-canavanine, a higher plant protective allelochemical, *J. Chem. Ecol.,* 12, 1145, 1986.

55. **Rosenthal, G. A., Janzen, D. H., and Dahlman, D. L.,** Degradation and detoxification of canavanine by a specialized predator, *Science,* 196, 658, 1977.

56. **Rosenthal, G. A. and Janzen, D. H.,** Arginase and L-canavanine metabolism by the bruchid beetle, *Caryedes brasiliensis, Entomol. Exp. Appl.,* 34, 336, 1983.

57. **Berge, M. A., Rosenthal, G. A., and Dahlman, D. L.,** Tobacco budworm, *Heliothis virescens* (Noctuidae) resistance to L-canavanine, a protective allelochemical, *Pestic. Biochem. Physiol.,* 25, 319, 1986.

58. **Rosenthal, G. A., Dahlman, D. L., and Janzen, D. H.,** L-canaline detoxification, a seed predator's biochemical mechanism, *Science,* 202, 528, 1978.

59. **Rosenthal, G. A. and Janzen, D. H.,** Nitrogenous excretion by the terrestrial seed predator, *Caryedes brasiliensis, Biochem. Syst. Ecol.,* 9, 219, 1981.

60. **Rosenthal, G. A., Hughes, C. G., and Janzen, D. H.,** L-canavanine, a dietary nitrogen source for the seed predator *Caryedes brasiliensis* (Bruchidae), *Science,* 218, 353, 1982.

61. **Rosenthal, G. A.,** Biochemical adaptations of the bruchid beetle, *Caryedes brasiliensis* to L-canavanine, a higher plant allelochemical, *J. Chem. Ecol.,* 9, 803, 1983.

62. **Rosenthal, G. A. and Janzen, D. H.,** Ammonia utilization by the bruchid beetle, *Caryedes brasiliensis* (Bruchidae), *J. Chem. Ecol.,* 11, 539, 1985.

63. **Koolman, J.,** Ecdysone metabolism, *Insect Biochem.,* 12, 225, 1982.

64. **Carlisle, D. B., and Ellis, P. S.,** Bracken and locust ecdysones: their effects on molting in the desert locust, *Science,* 159, 1472, 1968.

65. **Hikino, H., Ohizumi, Y., and Takemoto, T.,** Detoxification mechanism of *Bombyx mori* against exogenous phytoecdysone, *J. Insect Physiol.,* 21, 1953, 1975.

66. **Hikino, H., Ohizumi, Y., and Takemoto, T. O.,** Steroid metabolism in *Bombyx mori* I, catabolism of ponasterone A and ecdysterone in *Bombyx mori, Hoppe-Seyler's Z. Physiol. Chem.,* 356, 309, 1975.

67. **Feyereisen, R., Lagueux, M., and Hoffman, J. A.,** Dynamics of ecdysone metabolism after ingestion and injection in *Locusta migratoria, Gen. Comp. Endocrinol.,* 29, 319, 1976.

68. **Highnam, C. K. and Hill, L.,** *The Comparative Endocrinology of the Invertebrates,* University Park Press, Baltimore, Maryland, 1977, 1.

69. **Hammock, B. D. and Quistad, G. B.,** The degradative metabolism of juvenoids by insects, in *Progress in Pesticide Biochemistry,* Hutson, D. H. and Roberts, T. R., Eds., John Wiley & Sons, New York, 1981, 1.

70. **Pratt, G. E.,** The mode of action of pro-allatocidins, in *Natural Products for Innovative Pest Management,* Whitehead, D. L. and Bowers, W. S., Eds., Pergamon Press, New York, 1983, 323.

71. **Duffey, S. S. and Scudder, G. G. E.,** Cardiac glycosides in *Oncopeltus fasciatus* (Dallas) (Hemiptera: Lygaeidae). I. The uptake and distribution of natural cardenolides in the body, *Can. J. Zool.,* 52, 283, 1974.

72. **Scudder, G. G. E. and Meredith, J.,** The permeability of the midgut of three insects to cardiac glycosides, *J. Insect Physiol.,* 28, 689, 1982.

73. **Meredith, J., Moore, L., and Scudder, D. D. E.,** The excretion of ouabain by Malpighian tubules of *O. fasciatus, Am. J. Physiol.,* 246, R705, 1984.

74. **Scudder, G. G. E. and Meredith, J.,** Morphological basis of cardiac glycoside sequestration by *Oncopeltus fasciatus* (Dallas) (Hemiptera: Lygaeidae), *Zoomorphology,* 99, 87, 1982.

75. **Moore, L. V. and Scudder, G. G. E.,** Selective sequestration of milkweed (*Asclepias* sp.) cardenolides in *Oncopeltus fasciatus* (Dallas) (Hemiptera: Lygaeidae), *J. Chem. Ecol.,* 11, 667, 1985.

76. **Isman, M. B.,** Dietary influence of cardenolides on larval growth and development of the milkweed bug, *Oncopeltus fasciatus, J. Insect Physiol.,* 23, 1183, 1977.

77. **Isman, M. B., Duffey, S. S., and Scudder, G. G. E.,** Variation in cardenolide content of the lygaeid bugs, *Oncopeltus fasciatus* and *Lygaeus kalmi kalmi* and of their milkweed hosts (*Asclepias* spp.) in central California, *J. Chem. Ecol.,* 3, 613, 1977.

78. **Duffey, S. S., Blum, M. S., Isman, M. B., and Scudder, G. G. E.,** Cardiac glycosides: a physical system for their sequestration by the milkweed bug, *J. Insect Physiol.,* 24, 639, 1978.

79. **Vaughan, F. A.,** Effect of gross cardiac glycoside content of seeds of common milkweed, *Asclepias syriaca,* on cardiac glycoside uptake by the milkweed bug *Oncopeltus fasciatus, J. Chem. Ecol.,* 5, 89, 1979.

80. **Chaplin, S. J. and Chaplin, S. B.,** Growth dynamics of a specialized milkweed seed feeder (*Oncopeltus fasciatus*) on seeds of familiar and unfamiliar milkweeds (*Asclepias* sp.), *Entomol. Exp. Appl.,* 29, 345, 1981.

81. **Jones, C. G., Hoggard, M. P., and Blum, M. S.,** Is sequestration structure-specific in the milkweed bug, *Oncopeltus fasciatus?, Comp. Biochem. Physiol. C.,* 76, 283, 1983.

82. **Yoder, C. A., Leonard, D. E., and Lerner, J.,** Intestinal uptake of ouabain and digitoxin in the milkweed bug, *Oncopeltus fasciatus, Experientia,* 32, 1549, 1976.

83. **Scudder, G. G. E., Moore, L. V., and Isman, M. B.,** Sequestration of cardenolides in *Oncopeltus fasciatus:* morphological and physiological adaptations, *J. Chem. Ecol.,* 12, 1172, 1986.

84. **Moore, L. V. and Scudder, G. G. E.,** Ouabain-resistant Na,K-ATPases and cardenolide tolerance in the large milkweed bug, *Oncopeltus fasciatus, J. Insect Physiol.,* 31, 72, 1986.

85. **Scudder, G. G. E. and Duffey, S. S.,** Cardiac glycosides in the Lygaeinae (Hemiptera: Lygaeidae), *Can. J. Zool.,* 50, 35, 1972.

86. **von Euw, J., Reichstein, T., and Rothschild, M.,** Heart poisons (cardiac glycosides) in the Lygeid bugs *Caenocoris nerii* and *Spilostethus pandurus, Insect Biochem.,* 1, 373, 1971.

87. **von Euw, J., Fishelson, L., Parsons, Reichstein, T., and Rothschild, M.,** Cardenolides (heart poisons) in a grasshopper feeding on milkweeds, *Nature,* 214, 35, 1967.

88. **Reichstein, T.,** Cardenolide (herzwirksame glykoside) als abwehrstoffe bei insekten, *Naturwiss. Rundsch.,* 20, 499, 1967.

89. **Rothschild, M., von Euw, J., and Reichstein, T.,** Cardiac glycosides in a scale insect (*Aspidiotus*), a ladybird (*Coccinella*) and a lacewing (*Chrysopa*), *J. Entomol. (R. Soc. London)*, A48, 89, 1973.

90. **Rothschild, M., von Euw, T., and Reichstein, T.,** Cardiac glycosides in the oleander apid, *Aphis nerii*, *J. Insect Physiol.*, 16, 1141, 1970.

91. **Rothschild, M. and Reichstein, T.,** Some problems associated with the storage of cardiac glycosides by insects *Nova Acta Leopold.*, 7 (Suppl.), 507, 1976.

92. **Rothschild, M., von Euw, J., and Reichstein, T.,** Cardiac glycosides (heart poisons) in the polka-dot moth *Syntomeida epilais* Walk. (Ctenuchidae:Lep.) with some observations on the toxic qualities of *Amata* (= *Syntomis*) *phegea* (L.), *Proc. R. Soc. London, Ser. B.*, 183, 227, 1973.

93. **Vaughan, G. L. and Jungreis, A. M.,** Insensitivity of lepidopteran tissues to ouabain, physiological mechanisms for protection from cardiac glycosides, *J. Insect Physiol.*, 23, 585, 1977.

94. **Jungreis, A. M. and Vaughan, G. L.,** Insensitivity of Lepidoptera tissues to ouabain: absence of ouabain binding and Na^+-K^+ ATPases in larval and adult midgut, *J. Insect Physiol.*, 23, 503, 1977.

95. **Roeske, C. N., Seiber, J. N., Brower, L. P., and Moffit, C. M.,** Milkweed cardenolides and their comparative processing by monarch butterflies, Danaus plexippus L., *Rec. Adv. Phytochem.*, 10, 93, 1976.

96. **Seiber, J. N., Tuskes, P. M., Brower, L. P., and Nelson, C. J.,** Pharmacodynamics of some individual milkweed cardenolides fed to larvae of the monarch butterfly (*Danaus plexippus* L.), *J. Chem. Ecol.*, 6, 321, 1980.

97. **Marty, M. A. and Krieger, R. I.,** Metabolism of uscharidin, a milkweed cardenolide, by tissue homogenates of monarch butterfly larvae, *Danaus plexippus* L., *J. Chem. Ecol.*, 10, 945, 1984.

98. **Reichstein, T., von Euw, J., Parsons, J. A., and Rothschild, M.,** Heart poisons in the monarch butterfly, *Science*, 161, 861, 1968.

99. **Brower, L. P. and Glazier, S. C.,** Localization of heart poisons in the monarch butterfly, *Science*, 189, 19, 1975.

100. **Nishio, S. and Blum, M. S.,** unpublished data cited in Blum, M. S., *Chemical Defenses of Arthropods*, Academic Press, Orlando, FL, 1981, 213.

101. **Rothschild, M. and Marsh, N.,** Some peculiar aspects of danaid/plant relationships, *Entomol. Exp. Appl.*, 24, 437, 1978.

102. **Dixon, C. A., Erickson, J. M., Kellett, D. N., and Rothschild, M.,** Some adaptations between *Danaus plexippus* and its food plant, with notes on *Danaus chrysippus* and *Euploea core* (Insecta: Lepidoptera), *J. Zool. London*, 185, 437, 1978.

103. **Brower, L. P., Seiber, J. N., Nelson, C. J., Lynch, S. P., and Tuskes, P. M.,** Plant-determined variation in the cardenolide content, thin-layer chromatography profiles, and emetic potency of monarch butterflies, *Danaus plexippus* reared on the milkweed, *Asclepias eriocarpa* in California, *J. Chem. Ecol.*, 8, 579, 1982.

104. **Brower, L. P., Seiber, J. N., Nelson, C. J., Lynch, S. P., and Holland, M. M.,** Plant-determined variation in the profiles, and emetic potency of monarch butterflies, *Danaus plexippus* L. reared on milkweed plants in California: 2. *Asclepias speciosa*, *J. Chem. Ecol.*, 10, 601, 1984.

105. **Nelson, C. J., Seiber, J. N., and Brower, L. P.,** Seasonal and intraplant variation of cardenolide content in the California milkweed, *Asclepias eriocarpa*, and implications for plant defense, *J. Chem. Ecol.*, 7, 981, 1981.

106. **Seiber, J. N., Brower, L. P., Lee, S. M., McChesney, M. M., Cheung, H. T. A., Nelson, C. J., and Watson, T. R.,** Cardenolide connection between overwintering monarch butterflies from Mexico and their larval food plant, *Asclepias syriaca*, *J. Chem. Ecol.*, 12, 1158, 1986.

107. **Brower, L. P., Seiber, J. N., Nelson, C. J., Lynch, S. P., Hoggard, M. P., and Cohen, J. A.,** Plant-determined variation in cardenolide content and thin-layer chromatography profiles of monarch butterflies, *Danaus plexippus* reared on milkweed plants in California: 3. *Asclepias californica*, *J. Chem. Ecol.*, 10, 1823, 1984.

108. **Blau, P. A., Feeny, R., and Contardo, L.,** Allyl glucosinolate and herbivorous caterpillars: a contrast in toxicity and tolerance, *Science*, 200, 1296, 1978.

109. **Alpin, R. T., d'Arcy Ward, R., and Rothschild, M.,** Examination of the large white and small white butterflies (*Pieris* spp.) for the presence of mustard oil and mustard oil glycosides, *J. Entomol. Ser. A*, 50, 73, 1975.

110. **Jones, D. A.,** Cyanogenic glycosides and their function, in *Phytochemical Ecology*, Harborne, J. B., Ed., Academic Press, Orlando, FL, 1972, 103.

111. **Bond, E. J.,** The action of fumigants on insects. III. The fate of hydrogen cyanide in *Sitophilus granarius* (L.), *Can. J. Biochem. Physiol.*, 39, 1793, 1961.

112. **Duffey, S. S.,** Cyanide and arthropods, in *Cyanide in Biology*, Vennesland, B., Knowles, C. J., Conn, E. E., and Westley, J., Eds., Academic Press, Orlando, FL, 1981, 381.

113. **Parsons, J. and Rothschild, M.,** Rhodanese in the larva and pupa of the common blue butterfly [*Polyommatus icarus* (Rott.)] (Lepidoptera), *Entomol. Gaz.*, 15, 58, 1964.

114. **Bertran, E. C.,** Rhodanese in parasites, *An. Fac. Vet. Univ. Madrid,* 4, 334, 1952.

115. **Parsons, J. and Rothschild, M.,** Rhodanese in the blow-fly, *Calliphora vomitoria* L., *J. Insect Physiol.,* 8, 285, 1962.

116. **Brattsten, L. B.,** Ecological significance of mixed-function oxidations, *Drug Metab. Rev.,* 10, 35, 1979.

117. **Jones, D. A., Parsons, J., and Rothschild, M.,** Release of hydrocyanic acid from crushed tissues of all stages in the life-cycle of a species of Zygaeninae (Lepidoptera), *Nature,* 193, 52, 1962.

118. **Brattsten, L. B., Samuelian, J. H., Long, K. Y., Kincaid, S. A., and Evans, C. K.,** Cyanide as a feeding stimulant for the southern armyworm, *Spodoptera eridania, Ecol. Entomol.,* 8, 125, 1983.

119. **Fukami, Y., Yamamoto, I., and Casida, J. B.,** Metabolism of rotenone *in vitro* by tissue homogenates from mammals and insects, *Science,* 155, 713, 1967.

120. **Fukami, J., Shishido, T., Fukunaga, K., and Casida, J. E.,** Oxidative metabolism of rotenone in mammals, fish, and insects and its relation to selective toxicity, *J. Agric. Food Chem.,* 17, 1217, 1969.

121. **Fukami, J., Mitsui, T., Fukunaga, K., and Shishido, T.,** The selective toxicity of rotenone between mammal, fish, and insect, in *Biochemical Toxicology of Insecticides,* O'Brien, R. D., and Yamamoto, I., Eds., Academic Press, Orlando, FL, 1970, 159.

122. **Hollande, A. C.,** Colouration vitale due corps adipeux d'un insecte phytophage par un anthocyane absorbe avec la nourriture, *Arch. Zool. Exp. Gen.,* 51, 53, 1913.

123. **Pavan, M.,** Studi sugli antibiotici e insetticidi di origine animale I. -Sul principio attiro della larva di *Melasoma populi* L. (Col. Chrysomelidae), *Arch. Zool. Ital.,* 38, 157, 1953.

124. **Wain, R. L.,** The secretion of salicylaldehyde by the larvae of the brassy willow beetle (*Phyllodecta vitellinae* L.), *Ann. Rep. Hort. Res. Sta., Long Ashton,* 108, 1943.

125. **Rowell-Rahier, M. and Pasteels, J. M.,** The significance of salicin for a *Salix*-feeder, *Pharatora (Phyllodecta) vitellinae,* in *Proc. 5th Int. Symp. Insect-Plant Relat., Wageningen, 1982,* Pudoc, Wageningen, The Netherlands, 1982, 73.

126. **Pasteels, J. M., Rowell-Rahier, M., Braekman, J. C., and Dupont, A.,** Salicin from host plant as precursor of salicylaldehyde in defensive secretion of Chrysomeline larvae, *Physiol. Entomol.,* 8, 307, 1983.

127. **Rowell-Rahier, M. and Pasteels, J. M.,** Economics of chemical defense in Chrysomelinae, *J. Chem. Ecol.,* 12, 1189, 1986.

128. **Pasteels, J. M., Daloze, D., and Rowell-Rahier, M.,** Chemical defense in chrysomelid eggs and neonate larvae, *Physiol. Entomol.,* 11, 29, 1986.

129. **Miles, P. W.,** Insect secretion in plants, *Annu. Rev. Phytopathol.,* 6, 196, 1968.

130. **Thomson, D. L.,** Occurrence of the pigment of *Melanargia galatea* in *Dactylis glomerata, Biochem. J.,* 20, 1026, 1926.

131. **Morris, S. J. and Thomson, R. H.,** The flavonoid pigments of the small heath butterfly, *Coenonympha pamphilus* L., *J. Insect Physiol.,* 10, 377, 1964.

132. **Morris, S. J. and Thomson, R. H.,** The flavonoid pigments of the marbled white butterfly (*Melanargia galathea* Seltz), *J. Insect Physiol.,* 9, 391, 1963.

133. **Wilson, A.,** Flavonoid pigments in marbled white butterfly (*Melanargia galathea*) are dependent on flavonoid content of larval diet, *J. Chem. Ecol.,* 11, 1161, 1985.

134. **Wilson, A.,** Flavonoid pigments of butterflies in the genus *Melanargia, Phytochemistry,* 23, 1685, 1985.

135. **Wilson, A.,** Flavonoid pigments and wing color in *Melanargia galathea, J. Chem. Ecol.,* 12, 49, 1986.

136. **Wilson, A.,** Flavonoid pigments in swallowtail butterflies, *Phytochemistry,* 25, 1309, 1986.

137. **Berenbaum, M. R.,** Toxicity of a furanocoumarin to armyworms: a case of biosynthetic escape from insect herbivores, *Science,* 201, 532, 1978.

138. **Berenbaum, M. R.,** Introduction to interactions between insects and photoactive chemicals, *J. Chem. Ecol.,* 12, 809, 1986.

139. **Larson, R. A.,** Insect defenses against phototoxic plant chemicals, *J. Chem. Ecol.,* 12, 859, 1986.

140. **Bull, D. L., Ivie, G. W., Beier, R. C., Pryor, N. W., and Oertli, E. H.,** Fate of photosensitizing furanocoumarins in tolerant and sensitive insects, *J. Chem. Ecol.,* 10, 893, 1984.

141. **Ivy, G. W., Bull, D. L., Beier, R. C., Pryor, N. W., and Oertli, E. H.,** Metabolic detoxification: mechanism of insect resistance to plant psoralens, *Science,* 221, 374, 1983.

142. **Bull, D. L., Ivie, G. W., Beier, R. C., and Pryor, N. W.,** *In vitro* metabolism of a linear furanocoumarin (8-methoxypsoralen, xanthotoxin) by mixed-function oxidases of larvae of black swallowtail butterfly and fall armyworm, *J. Chem. Ecol.,* 12, 885, 1986.

143. **Yu, S. J.,** Interactions of allelochemicals with detoxification enzymes of insecticide-susceptible and resistant fall armyworms, *Pestic. Biochem. Physiol.,* 22, 60, 1984.

144. **Berenbaum, M. R.,** Brementown revisited: interactions among allelochemicals in plants, *Recent Adv. Phytochem.,* 19, 139, 1985.

145. **Ivie, G. W., Bull, D. L., Beier, R. C., and Pryor, N. W.,** Comparative metabolism of [³H] psoralen and [³H] isopsoralen by black swallowtail (*Papilio polyxenes* Fabr.) caterpillars, *J. Chem. Ecol.,* 12, 871, 1986.

146. **Smith, J. N. and Turbert, H. B.,** Enzymic glucoside synthesis in locust, *Nature,* 189, 600, 1961.

147. **Bowers, M. D. and Puttick, G. M.,** Fate of ingested iridoid glycosides in lepidopteran herbivores, *J. Chem. Ecol.,* 12, 169, 1986.

148. **Bernays, E. A.,** Tannins: an alternative viewpoint, *Entomol. Exp. Appl.,* 24, 44, 1978.

149. **Nolte, D. J., Eggers, S. H., and May, I. R.,** A locust pheromone: locustol, *J. Insect Physiol.,* 19, 1547, 1973.

150. **Nolte, D. J.,** The action of locustol, *J. Insect Physiol.,* 23, 899, 1977.

151. **Seifert, K.,** Die chemische Veranderung der Holzzelwand-komponenten unter dem Einfluss tierischer und pflanzlicher Schädlinge. 4. Mitteilung: Die Verdauung von Kie Fern-und Rotbuchenholz durch die Termite *Kalotermes flavicollis* Fabr., *Holzforschung,* 19, 105, 1962.

152. **Butler, J. H. A. and Buckerfield, J. C.,** Digestion of lignin by termites, *Soil Biol. Biochem.,* 11, 507, 1979.

153. **Bernays, E. A., Chamberlain, D. J., and McCarthy, P.,** The differential effects of ingested tannic acid on different species of Acridoidea, *Entomol. Exp. Appl.,* 28, 158, 1980.

154. **Bernays, E. A., Chamberlain, D. J., and Leather, E. M.,** Tolerance of acridids to ingested condensed tannin, *J. Chem. Ecol.,* 7, 247, 1981.

155. **Bernays, E. A.,** A specialized region of the gastric caecae in the locust, *Schistocerca gregaria, Physiol. Entomol.,* 6, 1, 1981.

156. **Bernays, E. A. and Chamberlain, D. J.,** A study of tolerance of ingested tannin in *Shistocerca gregaria, J. Insect Physiol.,* 26, 415, 1980.

157. **Bernays, E. A. and Woodhead, S.,** Plant phenols utilized as nutrients by a phytophagous insect, *Science,* 216, 201, 1982.

158. **Bernays, E. A. and Woodhead, S.,** Incorporation of dietary phenols into the cuticle in the tree locust *Anacridium melanorhodon, J. Insect Physiol.,* 28, 601, 1982.

159. **Bernays, E. A.,** The insect on the plant — a closer look, in *Proc. 5th Int. Symp. Insect-Plant Relat., Wageningen, 1982,* Purdoc, Wageningen, The Netherlands, 1982, 3.

160. **Bernays, E. A., Chamberlain, D. J., and Woodhead, S.,** Phenols as nutrients for a phytophagous insect *Anacridium melanorhodon, J. Insect Physiol.,* 29, 535, 1983.

161. **Bernays, E. A.,** Plant tannins and insect herbivores: an appraisal, *Ecol. Entomol.,* 6, 353, 1981.

162. **Isman, M. B. and Duffey, S. S.,** Pharmacokinetics of chlorogenic acid and rutin in larvae of *Heliothis zea, J. Insect Physiol.,* 29, 295, 1983.

163. **Birch, M. C.,** Chemical communication in pine bark beetles, *Am. Sci.,* 66, 409, 1978.

164. **Birch, M. C.,** Aggregation in bark beetles, in *Chemical Ecology of Insects,* Bell, W. J., and Cardé, R. T., Eds., Sinauer Associates, Sunderland, MA, 1984, 331.

165. **Hughes, P. R.,** *Dendroctonus:* production of pheromones and related compounds in response to host monoterpenes, *Z. Angew. Zool.,* 73, 294, 1973.

166. **Renwick, J. A. A., Hughes, P. R., Pitman, G. B., and Vite, J. P.,** Oxidation products of terpenes identified from *Dendroctonus* and *Ips* bark beetles, *J. Insect Physiol.,* 22, 725, 1976.

167. **Hunt, D. W. A., Borden, J. H., Pierce, H. D., Jr., Slessor, K. N., King, G. G. S., and Czyzewska, E. K.,** Sex-specific production of ipsdienol and myrcenol by *Dendroctonus ponderosae* (Coleoptera: Scolytidae) exposed to myrcene vapors, *J. Chem. Ecol.,* 12, 1579, 1986.

168. **Byers, J. A.,** Male specific conversion of the host plant compounds myrcene to the pheromone (+)-ipsdienol, in the bark beetle, *Dendroctonus brevicomis, J. Chem. Ecol.,* 8, 363, 1982.

169. **Hughes, P. R.,** Myrcene: a precursor of pheromones in *Ips* beetles, *J. Insect Physiol.,* 20, 1271, 1974.

170. **Hughes, P. R. and Renwick, J. A. A.,** Neural and hormonal control of pheromone biosynthesis in the bark beetle, *Ips paraconfusus, Physiol. Entomol.,* 2, 117, 1977.

171. **Byers, J. A., Wood, D. L., Browne, L. E., Fish, R. H., Piatek, B., and Hendry, L. B.,** Relationship between a plant compound, myrcene, and pheromone production in the bark beetle *Ips paraconfusus, J. Insect Physiol.,* 25, 477, 1979.

172. **Hendry, L. B., Piatek, B., Browne, L. E., Wood, D. L., Byers, J. A., Fish, R. H., and Hicks, R. A.,** *In vivo* conversion of a labelled host plant chemical to pheromones of the bark beetle *Ips paraconfusus, Nature,* 284, 485, 1980.

173. **Renwick, J. A. A., Hughes, P. R., and Ty, T. D.,** Oxidation products of pinene in the bark beetle *Dendroctonus frontalis, J. Insect Physiol.,* 19, 1735, 1973.

174. **Hughes, P. R.,** Effect of α-pinene exposure on *trans*-verbenol synthesis in *Dendroctonus ponderosae* Hopk., *Naturwissenschaften,* 60, 261, 1973.

175. **Renwick, J. A. A., Hughes, P. R., and Krull, I. S.,** Selective production of *cis-* and *trans*-verbenol from (−)- and (+)-α-pinene by a bark beetle, *Science,* 191, 199, 1976.

176. **Byers, J. A.,** Pheromone biosynthesis in the bark beetle, *Ips paraconfusus,* during feeding or exposure to vapors of host plant precursors, *Insect Biochem.,* 11, 563, 1981.

177. **Byers, J. A.,** Influence of sex, maturity, and host substances on pheromones in the guts of the bark beetles, *Ips paraconfusus* and *Dendroctonus brevicomis, J. Insect Physiol.,* 29, 5, 1983.

178. **Hughes, P. R.,** Pheromones of *Dendroctonus*: origin of *alpha*-pinene oxidation products present in emergent adults, *J. Insect Physiol.,* 21, 687, 1975.

179. **Byers, J. A.,** Bark beetle conversion of a plant compound to a sex-specific inhibitor of pheromone attraction, *Science,* 220, 624, 1983.

180. **Brand, J. M., Bracke, J. W., Markovetz, A. J., Wood, D. L., and Browne, L. E.,** Production of verbenol pheromone by a bacterium isolated from bark beetles, *Nature,* 254, 136, 1975.

181. **Byers, J. A. and Wood, D. L.,** Antibiotic-induced inhibition of pheromone synthesis in a bark beetle, *Science,* 213, 763, 1981.

182. **Brand, J. M., Bracke, J. W., Britton, L. N., Markovetz, A. J., and Barras, S. J.,** Bark beetle pheromones: production of verbenone by a mycangial fungus of *Dendroctonus frontalis, J. Chem. Ecol.,* 2, 195, 1976.

183. **White, R. A., Jr., Franklin, R. T., and Agosin, M.,** Conversion of α-pinene to α-pinene oxide by rat liver and the bark beetle *Dendroctonus terebrans* microsomal fractions, *Pestic. Biochem. Physiol.,* 10, 233, 1979.

184. **Brattsten, L. B.,** Cytochrome P-450 involvement in the interactions between plant terpenes and insect herbivores, in *Plant Resistance to Insects,* Hedin, P. E., Ed., American Chemical Society, Washington, D.C., 1983, 173.

185. **Brattstein, L. B., Evans, C. K., Bonetti, C. K., and Zalkow, L. H.,** Induction by carrot allelochemicals of insecticide-metabolizing enzymes in the southern armyworm *(Spodoptera eridania), Comp. Biochem. Physiol.,* 77C, 29, 1984.

186. **Eisner, T., Johnessee, J. S., Carrel, J., Hendry, L. B., and Meinwald, J.,** Defensive use by an insect of a plant resin, *Science,* 184, 996, 1974.

187. **Ferguson, J. E. and Metcalf, R. L.,** Plant-derived defense compounds for diabroticites (Coleoptera: Chrysomelidae), *J. Chem. Ecol.,* 11, 311, 1985.

188. **Ferguson, J. E., Metcalf, R. L., and Fischer, D. C.,** Disposition and fate of cucurbitacin B in five species of diabroticites, *J. Chem. Ecol.,* 11, 1307, 1985.

189. **Anderson, J. F., Plattner, R. D., and Weisleder, D.,** Metabolic transformations of cucurbitacins by *Diabrotica virgifera virgifera* LeConte and *D. undecimpunctata howardi* Barber, *Insect Biochem.* 18, 71, 1988.

190. **Metcalf, R. L.,** Coevolutionary adaptations of rootworm beetles (Coleoptera: Chrysomelidae) to cucurbitacins, *J. Chem. Ecol.,* 12, 1109, 1986.

191. **Yamamoto, I., Kimmel, E. C., and Casida, J. E.,** Oxidative metabolism of pyrethroids in house flies, *J. Agric. Food Chem.,* 17, 1227, 1969.

192. **Chang, S. C. and Kearns, C. W.,** Metabolism *in vivo* of C¹⁴-labelled pyrethrin I and cinerin I by house flies with special reference to the synergistic mechanism, *J. Econ. Entomol.,* 57, 397, 1964.

193. **Yamamoto, I. and Casida, J. E.,** O-demethyl pyrethrin II analogs from oxidation of pyrethrin I, allethrin, dimethrin, and phthalthrin by a house fly enzyme system, *J. Econ. Entomol.,* 59, 1542, 1966.

194. **Morrow, P. A., Bellas, T. E., and Eisner, T.,** Eucalyptus oils in the defensive oral discharge of Australian sawfly larvae (Hymenoptera: Pergidae), *Oecologia,* 19, 293, 1976.

195. **Morrow, P. A. and Fox, L. R.,** Effects of variation in *Eucalyptus* essential oil yield on insect growth and grazing damage, *Oecologia (Berlin),* 45, 209, 1980.

196. **Burnett, W. C., Jr., Jones, S. B., Jr., and Mabry, T. J.,** The role of sesquiterpene lactones in plant-animal coevolution, in *Biochemical Aspects of Plant and Animal Coevolution,* Harborne, J. B., Academic Press, Orlando, FL, 1978, 233.

197. **Isman, M. B.,** Toxicity and tolerance of sesquiterpene lactones in the migratory grasshopper, *Melanoplus sanguinipes* (Acrididae), *Pestic. Biochem. Physiol.,* 24, 348, 1985.

198. **Applebaum, S. W. and Guez, M.,** Comparative resistance of *Phaseolus vulgaris* beans to *Callosobruchus chinensis* and *Acanthoscelides obtectus* (Coleoptera: Bruchidae): the differential digestion of soluble heteropolysaccharides, *Entomol. Exp. Appl.,* 15, 203, 1972.

199. **Applebaum, S. W. and Birk, Y.,** Saponins, in *Herbivores: Their Interaction with Secondary Plant Metabolites,* Rosenthal, G. A., and Janzen, D. H., Eds., Academic Press, Orlando, FL, 1979, 539.

200. **von Euw, J., Reichstein, T., and Rothschild, M.,** Aristolochic acid-I in the swallowtail butterfly *Pachlioptera aristolochiae* (Fabr.) (Papilionidae), *Isr. J. Chem.,* 6, 659, 1968.

201. **Rothschild, M., von Euw, J., and Reichstein, T.,** Aristolochic acids stored by *Zerynthia polyxena* (Lepidoptera), *Insect Biochem.,* 2, 334, 1972.

202. **Urzu'a, A., Salgado, G., Cassels, B. K., and Eckhardt, G.,** Aristolochic acids in *Aristolochia chilensis* and the *Aristolochia*-feeder *Battus archidamus* (Lepidoptera), *Collect. Czech. Chem. Commun.,* 48, 1513, 1983.

203. **Eisner, T., Hendry, L. B., Peakall, D. B., and Meinwald, J.,** 2,5-dichlorophenol (from ingested herbicide?) in defensive secretion of grasshopper, *Science,* 172, 277, 1971.

204. **Teas, H. J.,** Cycasin synthesis in *Seirarctia echo* (Lepidoptera) larvae fed methylazoxymethanol, *Biochem. Biophys. Res. Commun.,* 26, 686, 1967.

205. **Rees, C. J. C.,** Chemoreceptor specificity associated with choice of feeding site by the beetle *Chrysolina brunsvicensis* on its foodplant *Hypericum hirsutum, Entomol. Exp. Appl.,* 12, 565, 1969.

206. **Meinwald, J., Meinwald, Y. C., Chalmers, A. M., and Eisner, T.,** Dihydromatricaria acid: acetylenic acid secreted by soldier beetle, *Science,* 160, 890, 1968.

207. **Bruns, T. D.,** Insect mycophagy in the Boletales: fungivore diversity and the mushroom habitat, in *Fungus-Insect Relationships: Perspectives in Ecology and Evolution,* Wheeler, Q., and Blackwell, M., Eds., Columbia University Press, New York, 1984, 91.

208. **Jaenike, J.,** Parasite pressure and the evolution of amanitin tolerance in *Drosophila, Evolution,* 39, 1295, 1985.

209. **Sinha, R. N.,** Fungus as food for some stored-product insects, *J. Econ. Entomol.,* 64, 3, 1971.

210. **Wright, V. F., De Las Casas, E., and Harein, P. K.,** The response of *Tribolium confusum* to the mycotoxins zearalenone (F-2) and T-2 toxin, *Environ. Entomol.,* 5, 371, 1976.

211. **Llewellyn, G. C., Sherertz, P. C., and Mills, R. R.,** The response of dietary stressed *Periplaneta americana* to chronic intake of pure aflatoxin B1, *Bull. Environ. Contam. Toxicol.,* 15, 391, 1976.

212. **Chinnici, J. P. and Llewellyn, G. C.,** Reduced aflatoxin toxicity in hybrid crosses of aflatoxin B1 sensitive and resistant strains of *Drosophila melanogaster* (Diptera), *J. Invertebr. Pathol.,* 33, 81, 1979.

213. **Al-Adil, K. M., Kilgore, W. W., and Painter, R. R.,** Toxicity and sterilization effectiveness of aflatoxin B1 and G1 and distribution of aflatoxin B1-^{14}C in house flies, *J. Econ. Entomol.,* 65, 375, 1972.

214. **Fahmy, M. J., Fahmy, O. G., and Swenson, D. H.,** Aflatoxin B1-2,3-dichloride as a mode of the active metabolite of aflatoxin B1 in mutagenesis and carcinogenesis, *Cancer Res.,* 38, 2608, 1978.

215. **Dowd, P. F. and Van Middlesworth, F. L.,** *In vitro* metabolism of the trichothecene 4-monoacetoxy-scirpenol by fungus- and non-fungus-feeding insects, *Experientia* 45, 393, 1989.

216. **Bull, D. L.,** Toxicity and pharmacodynamics of avermectin in the tobacco budworm, corn earworm, and fall armyworm (Noctuidae: Lepidoptera), *J. Agric. Food Chem.,* 34, 78, 1986.

217. **Fox, L. R. and Morrow, P. A.,** Specialization: species property or local phenomenon?, *Science,* 211, 887, 1981.

218. **Cates, R. G.,** Host plant predictability and feeding patterns of monophagous, oligophagous, and polyphagous insect herbivores, *Oecologia (Berlin),* 48, 319, 1981.

219. **Futuyma, D. J.,** Evolutionary interactions among herbivorous insects and plants, in *Coevolution,* Futuyma, D. J. and Slatkin, M., Eds., Sinauer Associates, Sunderland, MA, 1983, 207.

220. **Ahmad, S.,** Mixed-function oxidase activity in a generalist herbivore in relation to its biology, food plants, and feeding history, *Ecology,* 64, 235, 1983.

221. **Dowd, P. F., Pantoja, A., Smith, C. M., and Sparks, T. C.,** Insecticide susceptibility and enzyme activity in strains of the fall armyworm, *Spodoptera frugiperda* (J. E. Smith) adapted to corn and rice, submitted, 1989.

222. **Blum, M. S.,** *Chemical Defenses of Arthropods,* Academic Press, Orlando, FL, 1981, 411.

223. **Smiley, J.,** Plant chemistry and the evolution of host specificity: new evidence from *Heliconius* and *Passiflora, Science,* 201, 745, 1978.

224. **Smiley, J. T.,** The herbivores of *Passiflora:* comparison of monophyletic and polyphyletic feeding guilds, in *Proc. 5th Int. Symp. Insect-Plant Relat., Wageningen, 1982,* Pudoc, Wageningen, The Netherlands, 1982, 325.

225. **Smiley, J. T. and Wisdom, C. S.,** Determinants of growth rate on chemically heterogeneous host plants by specialist insects, *Biochem. Syst. Ecol.,* 13, 305, 1985.

226. **Schoonhoven, L. M. and Meerman, J.,** Metabolic cost of changes in diet and neutralization of allelo-chemicals, *Entomol. Exp. Appl.,* 24, 489, 1978.

227. **Yu, S. J.,** Induction of detoxifying enzymes by allelochemicals and host plants in the fall armyworm, *Pestic. Biochem. Physiol.,* 19, 330, 1983.

228. **Yu, S. J. and Hsu, E. L.,** Induction of hydrolases by allelochemicals and host plants in fall armyworm (Lepidoptera: Noctuidae) larvae, *Environ. Entomol.,* 14, 512, 1985.

229. **Krieger, R. I., Feeny, P. P., and Wilkinson, C. F.,** Detoxification enzymes in the guts of caterpillars: an evolutionary answer to plant defenses?, *Science,* 172, 579, 1971.

230. **Rose, H. A.,** The relationship between feeding specialization and host plants to aldrin epoxidase activities of midgut homogenates in larval Lepidoptera, *Ecol. Entomol.,* 10, 455, 1985.

231. **Dowd, P. F., Ryder, D. A., Orr, D. B., and Sparks, T. C.,** A comparison of esterase and glutathione *S*-transferase activity in Pentatomids with different feeding strategies, submitted, 1989.

232. **Hegnauer, R.,** *Chemotaxonomie der Pflanzen, Band 2, Monocotyledonoae,* Birkhauser Verlag, Basel, 1963, 1.

233. **Mullin, C. A. and Croft, B. A.,** *Trans*-epoxide hydrolase: a key indicator enzyme for herbivory in arthropods, *Experientia,* 40, 376, 1984.

234. **Croft, B. A. and Mullin, C. A.,** Comparison of detoxification enzyme systems in *Argyrotaenia citrana* (Lepidoptera: Tortricidae) and the ectoparasite, *Oncophanes americanus* (Hymenoptera: Braconidae), *Environ. Entomol.,* 13, 1330, 1984.

235. **Mullin, C. A. and Croft, B. A.,** An update on selective pesticides favoring arthropod natural enemies, in *Biological Control in Agricultural IPM Systems*, Hoy, M. A. and Herzog, D. C., Eds., Academic Press, Orlando, FL, 1985, 123.

236. **Dowd, P. F.,** Toxicological and biochemical interactions of the fungal metabolites fusaric acid and kojic acid with xenobiotics in *Heliothis zea* (F.) and *Spodoptera frugiperda* (J. E. Smith), *Pestic. Biochem. Physiol.* 32, 123, 1988.

237. **Yu, S. J.,** Microsomal sulfoxidation of phorate in the fall armyworm, *Spodoptera frugiperda* (J. E. Smith), *Pestic. Biochem. Physiol.,* 23, 273, 1985.

238. **Dowd, P. F.,** Permethrin and fenvalerate hydrolysis in *Pseudoplusia includens* (Walker) and *Heliothis virescens* (F.), Ph.D. Dissertation, Louisiana State University, Baton Rouge, LA, 363 pp., 1985.

239. **Brattsten, L. B., Gunderson, C. A., Fleming, J. T., and Nikbahkt, K. N.,** Temperature and diet modulate cytochrome P-450 activities in southern armyworm, *Spodoptera eridania* (Cramer), caterpillars, *Pestic. Biochem. Physiol.,* 25, 346, 1986.

240. **Denison, M. S., Hamilton, J. W., and Wilkinson, C. F.,** Comparative studies of aryl hydrocarbon hydroxylase and the *Ah* receptor in nonmammalian species, *Comp. Biochem. Physiol.,* 80C, 319, 1985.

241. **Yu, S. J., Berry, R. E., and Terriere, L. C.,** Host plant stimulation of detoxifying enzymes in a phytophagous insect, *Pestic. Biochem. Physiol.,* 12, 280, 1979.

242. **Yu, S. J.,** Induction of microsomal oxidases by host plants in the fall armyworm, *Spodoptera frugiperda* (J. E. Smith), *Pestic. Biochem. Physiol.,* 17, 59, 1982.

243. **Moldenke, A. F., Berry, R. E., and Terriere, L. C.,** Cytochrome P-450 in insects—V. Monoterpene induction of cytochrome P-450 and associated monoxygenase activities in the larva of the variegated cutworm *Peridroma saucia* (Hubner), *Comp. Biochem. Physiol.,* 74C, 365, 1983.

244. **Collins, P. J.,** Induction of the polysubstrate monooxygenase system of the native budworm *Heliothis punctiger* (Wallengren) (Lepidoptera: Noctuidae), *Insect Biochem.,* 15, 551, 1985.

245. **Christian, M. F. and Yu, S. J.,** Cytochrome P-450 dependent monooxygenase activity in the velvetbean caterpillar, *Anticarsia gemmatalis* Hubner, *Comp. Biochem. Physiol.,* 83C, 23, 1986.

246. **Brattsten, L. B., Wilkinson, C. F., and Eisner, T.,** Herbivore-plant interactions: mixed-function oxidases and secondary plant substances, *Science,* 196, 1349, 1977.

247. **Yu, S. J. and Ing, R. T.,** Microsomal biphenyl hydroxylase of fall armyworm larvae and its induction by allelochemicals and host plants, *Comp. Biochem. Physiol.,* 78C, 145, 1984.

248. **Riskallah, M. R., Dauterman, W. C., and Hodgson, E.,** Nutritional effects on the induction of cytochrome P-450 and glutathione transferase in larvae of the tobacco budworm, *Heliothis virescens, Insect Biochem.,* 16, 491, 1986.

249. **Dowd, P. F., Smith, C. M., and Sparks, T. C.,** Influence of soybean leaf extracts on ester cleavage in cabbage and soybean loopers (Lepidoptera: Noctuidae), *J. Econ. Entomol.,* 76, 700, 1983.

250. **Dowd, P. F., Rose, R. L., Smith, C. M., and Sparks, T. C.,** Influence of extracts from soybean *Glycine max* (L.) (Merr.) leaves on hydrolytic and glutathione *S*-transferase activity in the soybean looper, *Pseudoplusia includens, J. Agric. Food Chem.,* 34, 444, 1986.

251. **Plapp, F. W.,** Genetics and biochemistry of insecticide resistance in arthropods: prospects for the future, in *Pesticide Resistance: Strategies and Tactics for Management,* Committee on Strategies for the Management of Pesticide Resistant Pest Populations, National Academy Press, Washington, D.C., 1986, 74.

252. **Neumann, S. V.,** Plant growth hormones affect grasshopper growth and reproduction, in *Proc. 5th Symp. Insect-Plant Relationships, Wageningen, 1982,* Pudoc, Wageningen, The Netherlands, 1982, 57.

253. **BeMiller, J. N. and Colilla, W.,** Mechanism of corn indole-3-acetic acid oxidase *in vitro, Phytochemistry,* 11, 3393, 1972.

254. **Ahmad, S.,** Enzymatic adaptations of herbivorous insects and mites to phytochemicals, *J. Chem. Ecol.,* 12, 533, 1986.

255. **Ahmad, S.,** Roles of mixed-function oxidases in insect herbivory, in *Proc. 5th Int. Symp. Insect-Plant Relat., Wageningen, 1982,* Pudoc, Wageningen, The Netherlands, 1982, 41.

256. **Jones, C. G.,** Microorganisms as mediators of plant resource exploitation by insect herbivores, in *A New Ecology: Novel Approaches to Interactive Systems,* Price, P. W., Slobodchikoff, C. N., and Gaud, W. S., Eds., John Wiley & Sons, New York, 1984, 53.

257. **Dowd, P. F.,** Symbiont-mediated detoxification of plant substances by insects, in *Multitrophic-Level Interactions Among Microorganisms, Plants and Insects,* Barbosa, P., Krischik, V. A., and Jones, C. G. Eds., John Wiley & Sons, New York (in press).

258. **Kukor, J. J. and Martin, M. M.,** Cellulose digestion in *Monochamus marmorator* Kby. (Coleoptera: Cerambycidae): role of acquired fungal enzymes, *J. Chem. Ecol.,* 12, 1057, 1986.

259. **Wiklund, C.,** Generalist versus specialist utilization of host plants among butterflies, in *Proc. 5th Int. Symp. Insect-Plant Relat., Wageningen, 1982,* Pudoc, Wageningen, The Netherlands, 1982, 181.

260. **Berenbaum, M. R. and Miliczky, E.,** Mantids and milkweed bugs: efficacy of aposematic coloration against invertebrate predators, *Am. Midl. Nat.,* 111, 64, 1984.

261. **Smith, D. A. S.,** Cardiac glycosides in *Danaus chrysippus* L. provide some protection against an insect parasitoid, *Experientia,* 34, 844, 1978.
262. **Rothschild, M.,** Secondary plant substances and warning colouration in insects, *Insect/Plant Relationships,* van Emden, H. F., Ed., John Wiley & Sons, New York, 1973, 59.
263. **Fox, D. F.,** *Animal Biochromes and Structural Colors,* University of California Press, Berkeley, California, 1976, 213.

APPLICATION OF MOLECULAR GENETICS TO INSECT CONTROL

Andrew F. Cockburn and J. A. Seawright

INTRODUCTION

Currently we are witnessing the beginning of a new era in insect control. During the past four decades conventional insect control practices have consisted of a combination of the application of broad spectrum pesticides and cultural practices, e.g., postharvest tillage, timing of planting, and source reduction. The emphasis has been on the use of pesticides, and although much of the success of modern agricultural practices can be attributed to the synthetic chemicals used in pest control, their indiscriminate use has led to negative effects in the form of the development of resistant pests and caused considerable environmental damage due to direct toxic effects and to the buildup of harmful residues. Alternative strategies for insect control have been advanced and include a variety of possible approaches to using biological organisms for implementing the demise of pests. Although there have been some outstanding examples of biological control through the use of parasites, predators, and microbial agents, for many of the important pests we have relied on spraying and digging.

Many genetic phenomena, e.g., heritable chromosomal aberrations, hybrid sterility, genetic incompatibility, meiotic drive, etc., have been proposed and investigated as potential mechanisms; however, in spite of the favorable theoretical considerations and some promising results of limited field tests, none of these heritable traits has been used to control insect pests on a routine basis.[1] The most successful use of genetics in insect control has been the employment of the sterile insect technique, a method whereby huge numbers of a species are produced in a factory and released into the native habitat after sterilization by exposure to ionizing radiation. Although dramatic success has been obtained for a limited number of pest species, the sterile insect technique is not suitable for many economically important species for logistical, economical, and biological reasons.[2] Because of these limiting factors, the progress being made in the *in vitro* manipulation of genes is of interest to researchers involved in the use of genetics for insect control.

During the past decade, tremendous strides have been made in the genetic engineering of a diversity of organisms through the use of recombinant DNA technology (molecular genetics). Molecular geneticists have created bacteria, plants, animals, and fungi that have useful new properties, and many of these are being used or tested for commercial use. One application that has received inadequate attention is the practical use of molecular genetics for the control of pest insects. This is somewhat surprising since an insect, *Drosophila melanogaster*, is one of the most widely used organisms for basic research in molecular genetics, and it would seem that many of the molecular techniques that are currently used for *D. melanogaster* could also be used for modification of the genomes of other insects. Techniques for cloning genes from insects and the use of the P element transposon for the transformation of insects with those genes have recently been proposed (Figure 1).[3-7]

In this chapter, we review the available techniques and some possible strategies for manipulating the genes of pest insects to cause the cessation of the noxious characteristics of the pest, either by total eradication or the lowering of population density below economical thresholds. We first describe some of the important results that have come from the study of the molecular genetics of *Drosophila,* and then propose some specific applications of this technology to entomological problems. Proposals for using molecular genetics for insect control are still largely speculative, primarily because of the current lack of knowledge about

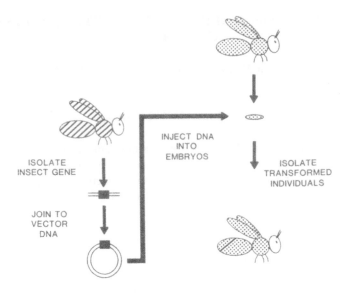

FIGURE 1.   Transfer of genes between insect species. DNA coding for an insect gene is isolated by cloning in *E. coli*. Following *in vitro* modification, if necessary, the DNA is introduced into another insect by injection into embryos. Incorporation of the injected DNA into the germ line might lead to a novel phenotype. See text for details.

pest species, but with the present interest in the development of genetic methods of control and the rapid progress being made in recombinant DNA methods, the futuristic schemes of today might become reality. The basics of insect molecular genetic technology have been described in detail before[3] and will be covered only briefly. We will not discuss genetic engineering of insect pathogens (bacteria, viruses, and fungi) because a recent review[8] is available for this important area.

BASIC TECHNOLOGY

OVERVIEW

Before any meaningful progress can be made in the genetic engineering of economically important species, the genetic and biochemical basis of useful traits must be established. Generally, a trait will only be useful if it is due to one or two genes, for at this time the manipulation of truly polygenic traits would be a formidable task. It may also be important to understand the physiological mechanism by which a gene exerts its effect before it can be effectively used for genetic manipulation of a pest insect. Once this basic information is available, molecular genetics can proceed through the following stages.

1.   The gene must be cloned and the DNA sequences coding for regulation and protein determined.
2.   Appropriate modifications must be made in the cloned gene. The most obvious type of change would be to connect the protein coding part of one gene to the regulatory part of another to create a gene that expresses the protein product in a novel tissue or stage of development.[3]
3.   The most important unsolved problem is that procedures must be developed for the introduction of a modified gene into the pest insect and for the identification of transformed individuals.[3]

These three stages will be treated in more detail below.

TABLE 1
Partial List of Genes Introduced into
Drosophila by P Element
Transformation

Gene	Ref.
Alcohol dehydrogenase	57
Chorion protein	79
Crooked neck	80
Dopa decarboxylase	58, 81
Fs(1)K10	80
Hsp26	15
Hsp27	67
Kurtz	80
Metallothionein	82
Pecanex	80
Sgs3	16
Sgs4	83
Transformer	84
Tropomyosin	85
White	86
Xanthine dehydrogenase	37, 39, 87, 88
Yolk protein 1	66

GENE ISOLATION

The first step in the use of a gene in genetic engineering is "cloning" it in bacteria. This involves fragmenting purified DNA of the insect into pieces about the size of a single gene, joining them to the DNA of a bacterial chromosome or virus, reintroducing the DNA into bacteria, and then isolating the bacteria containing the gene of interest.[3] Materials and techniques to perform all of these steps except the last are commercially available. Identifying and isolating a bacteria carrying a particular gene (out of tens of thousands of other genes) is trickier, which is why it is important to have as much genetic and biochemical information as possible about a gene before attempting to clone it.

Many genes have already been cloned from *Drosophila*,[9] so emphasis in this review will be given to the molecular genetics that could be done using these and other cloned and characterized genes. Many of these genes have been reintroduced into the *Drosophila* genome and function normally (Table 1). Generally, it is expected that genes from one species can also be used to effect phenotypic changes in a second species. Genes from other organisms have already been introduced into the genome of *D. melanogaster* (Table 2) and insect tissue culture cells,[10] so there are precedents for this type of genetic manipulation.

In some cases it might be desirable to work with a gene from the insect being manipulated, for example if there is no homologous gene in *Drosophila*. Cloned genes from other insects are becoming available as more molecular geneticists begin to work on economically important species. Since the same basic techniques work with any insect,[3] new technology will not have to be developed to clone genes from different species.

GENE STRUCTURE

A gene typically consists of two discrete parts: the DNA sequence which codes for the final protein product and the regulatory sequences responsible for ensuring that the gene makes the proper amount of protein in the proper cells at the proper time (Figure 2 and Figure 3).[11] Therefore it is possible, using *in vitro* molecular genetic techniques, to remove the regulatory sequences from one gene and join them to the protein coding sequences of a different gene. This hybrid gene will produce the protein specified by the second gene, but

TABLE 2
Partial List of Hybrid Genes Introduced into Drosophila by P Element Transformation

Promoter source	Coding sequence	Ref.
Fushi tarazu	Beta-galactosidase (*E. coli*)	19
Hsp26	Beta-galactosidase (*E. coli*)	89, 90
Hsp70	Neomycin phosphotransferase (*E. coli*)	61
Hsp70	Beta-galactosidase (*E. coli*)	89, 90
Hsp70	B2 (phage lambda)	91
Hsp70	White (*D. melanogaster*)	92
Larval serum protein	Chloramphenicol acetyltransferase (*E. coli*)	93
NinaE opsin	Chloramphenicol acetyltransferase (*E. coli*)	94
NinaE opsin	Beta-galactosidase (*E. coli*)	94
Sgs4	Alcohol dehydrogenase (*D. melanogaster*)	95
Yolk protein 1	(Phage m13)	18, 21

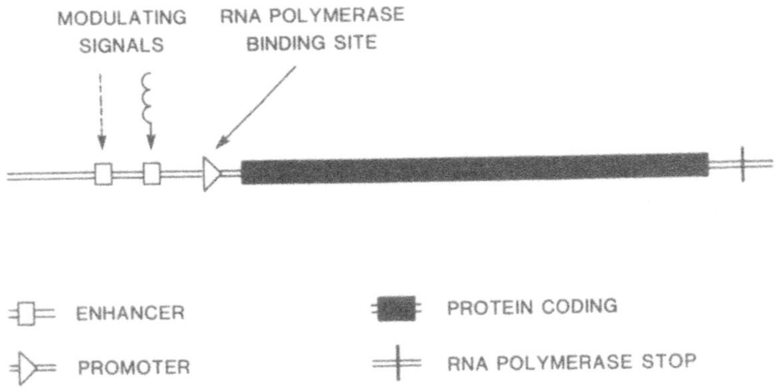

FIGURE 2.   Structure of a typical insect gene. Modulating signals (hormones and intracellular factors) act at enhancer sites to turn genes on or off. When a gene is on, RNA polymerase can bind at the promoter and transcribe the gene. A downstream stop site causes termination of transcription. See Freifelder[11] for details.

in an amount and at a time determined by the first gene.[12] Since the genetic code for protein is invariant (except in mitochondria, plastids, and a few protozoans), it is also possible to make functional "chimeric" genes by joining a protein coding sequence from one species to a regulatory sequence from another species. For example, the expression of eucaryotic coding sequences in bacteria has long been a standard technique in molecular genetics.[13] A list of some chimeric genes that have been used in *Drosophila* research is given in Table 2. Regulatory sequence do vary considerably from species to species, so a chimeric gene might only function properly in the species from which the regulatory part of the gene was taken.

The regulatory sequences of a gene are usually located in a region in front of the protein coding sequence.[14-16] The terminology used to designate eucaryotic regulatory sequences has not been standardized, hence the appearance of several terms, including "promoters" and "enhancers", in the literature on *Drosophila* research.[17-23] The nuances of these terms are not important for this discussion, so for simplicity we will refer to all of the regulatory sequences as "promoters", which is the term most often used. Other regulatory sequences, e.g., protein glycosylation sites,[24] leader peptides than direct the nascent protein to be secreted,[25] and protein translation controls,[26] are found in some genes and may be important when using these genes.

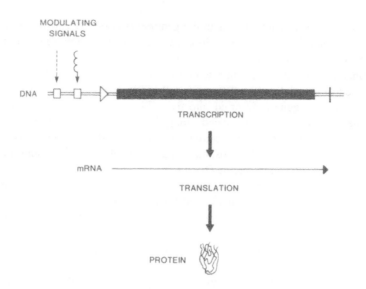

FIGURE 3. Molecular processes in gene expression. When a gene is expressed, RNA polymerase transcribes the DNA to produce RNA. The transcript is processed and exported to the cytoplasm as messenger RNA (mRNA). The mRNA is translated by the ribosomes to produce protein. See Freifelder[11] for details.

"Conditional genes" are expressed only in particular tissues or in response to particular stimuli, in contrast to "constitutive genes", which are expressed in all tissues at all times. "Conditional promoters", which are the regulatory sequences responsible for the tissue or stimuli-specific expression, will be important in turning on genes selectively. Tissue or developmental stage specific conditional genes can usually be cloned easily. For example, a simple screening procedure was used to isolate *Drosophila* cuticle protein genes,[27] and a similar differential hybridization approach was used to isolate genes expressed during *Drosophila* embryogenesis.[28] Among insect conditional genes that are activated by environmental stimuli are genes that respond to heat shock,[29] ecdysterone,[30] and the yolk protein genes which respond to both ecdysterone and juvenile hormone.[31] Other environmental events probably have similar effects, e.g., feeding is likely to result in expression of gut enzymes, exposure to heavy metal ions activates the metallothionein genes in other organisms,[32] and light conditions can affect pigmentation deposition.[33]

TRANSFORMATION

"Transformation" is the introduction of DNA into a cell or organism and the subsequent expression of the genes located on that DNA. In general, it has proven to be easier to clone genes than to develop reliable transformation systems to return them to the genome, and this has proven to be the case for insects. Techniques must be devised to transform cells in the germ line so that the genes will be passed on to future generations. There is a reliable transformation procedure that is being used in *Drosophila* research (Tables 1 and 2), so the developers of transformation systems have focused on adapting that procedure to other insects.[3]

Drosophila transformation involves injecting the cloned "P element" transposon[34-37] from *D. melanogaster* into embryos.[38,39] Transposons, also called transposable elements or selfish DNA, are short DNA sequences found in many genomes that have the ability to generate copies of themselves and insert the copies elsewhere in the genome.[40-43] The P element was originally identified by genetic analysis[44,45] and was later cloned.[34] It is capable of inserting itself at many different sites in the genome (frequently causing mutations in the process[35]), is terminated by the same 31 base pair sequence at each end,[36] and codes for a single protein that is necessary for transposition.[37]

When cloned P element DNA is injected into early embryos, the P element can transpose from the injected DNA molecule into a germ cell chromosome and be passed along to the progeny.[38,39] If another gene is inserted into the center of the P element DNA by genetic engineering the additional gene is carried along.[39] This technique has now been used to introduce many genes into the *D. melanogaster* genome and these genes usually function normally despite being in new chromosomal locations (Table 1). The same procedure has also worked in *D. hawaiiensis*[46] and in *D. simulans*.[47]

Injection of DNA into embryos of other species of insects is generally not difficult. Modifications of the injection techniques used for *Drosophila* embryos[48] have been used in this laboratory and elsewhere to introduce DNA into embryos of several species of dipterans[49-52] and lepidopterans.[7] We are also testing a shotgun injection system, which was originally developed to transform a large number of cells at once,[53] as a mass injection system for insect embryos.

Ensuring the proper functioning of the injected genes, especially the P transposon, is a more difficult problem. The P element protein is made in mammalian cells[54] and the P element enhances transformation in mammalian cells,[55] which would suggest that it should also work in other insect species. However, a sensitive *in vivo* assay for P element function has shown that there is no evidence of transposition in several species of nondrosophilid dipterans.[52] None of the many laboratories trying P element transformation with various nondrosophilid insect species has reported P element movement, which also argues that the P element will not be of general use.

If the *Drosophila* P element transformation system does not work in other insects, the most promising alternative is the isolation of analogous transposons from pest species. Transposons have been identified in many types of organisms[40-42] and probably occur in most species. Certainly, they are very common in *D. melanogaster*, where about 50 different types of elements are found.[43] Genetic phenomena similar to those caused by transposons, e.g., meiotic drive, hybrid sterility, recurrent breaks at specific chromosomal locations, and spontaneous mutations, may indicate the presence of such elements in other insects.[56] Other strategies for transforming insects have also been suggested.[3]

IDENTIFICATION OF TRANSFORMED INSECTS

A major difficulty in transformation experiments with any organism is identifying transformed individuals, since these are usually a small fraction of the total. One approach is to insert the normal gene for a specific enzyme into the DNA molecule to be transformed, and to inject the DNA into a mutant strain that does not produce the enzyme. Transformants can then be identified by assaying for the product of the gene in question. This approach has been very successful in *D. melanogaster*, where many cloned genes and mutant strains are available; for example, the genes coding for xanthine dehydrogenase,[39] alcohol dehydrogenase (ADH),[57] and dopa decarboxylase[58] have been used. These three genes are attractive for use as transformation markers because insects lacking the enzyme product are killed by exposure to a poisonous substrate. However, strains lacking the appropriate enzyme would have to be isolated from each species to be transformed. A selection system which can be used to select for null ADH mutants is available; it uses a chemical that is converted into a poisonous product by wild type enzyme.[59] Even with such a scheme it would be a major project to isolate a null mutant from most pest insects.

Identification of transformants could also be accomplished by using a pesticide resistance gene, such as the gene recently cloned from mosquitoes.[60] This approach has the advantage that pesticide susceptible strains are readily available. However, mass release of resistant pest insects is unlikely to be politically or environmentally feasible, so these genes will be primarily useful in laboratory research and development.

An alternative to the use of insect genes is to use bacterial or fungal enzyme genes which do not have analogs in insects. These would have to be joined to an insect promoter to ensure that they would be expressed. For example, a kanamycin resistance protein coding sequence from bacteria joined to a *D. melanogaster* promoter has been used to render *D. melanogaster*[61] and *Anopheles gambiae*[51] resistant to the antibiotic G418.

SPECIFIC PROPOSALS

In this section we propose some specific projects in which molecular genetics could have practical applications in entomology. We have chosen these examples to demonstrate the range of possible applications and the different types of gene constructions that could be used, starting with simple proposals and moving to the more complex.

IMPROVED PREDATORS AND PARASITES

One of the problems with the use of pesticides is that beneficial arthropods, viz, predators, parasites, etc., are usually adversely affected because of their susceptibility to chemical compounds. When the compound breaks down the pest can potentially increase to a higher level than before due to the reduced population of biological control agents. Molecular genetics can be employed to introduce pesticide resistance into beneficial predators, parasitoids, and pollinators. For example, pesticide-resistant predatory mites have been selected using classical genetic procedures.[62] To accomplish the same thing using molecular genetics will require the development of suitable transformation systems and the isolation of the appropriate pesticide resistance genes. Resistance would be selected for in the field, so the problem of introducing the modification into the environment is much easier than with genes that would be selected against.

STERILE INSECT TECHNIQUE

The sterile insect technique[63] would be generally more effective and efficient if only males are released. The selective elimination of females, either as eggs or neonate larvae, would save rearing resources, in some cases simplify sterilization, and eliminate damage that sterile females cause by feeding or oviposition.[64] Classical genetic sexing techniques have involved using reciprocal translocations to create pseudolinkage between a selectable gene and sex. Classical genetics has been criticized because, although it has good potential to enhance the sterile insect technique, it does not lead to alternative methods of control.[2] However, the molecular genetic technology for genetic sexing would probably be applicable to other types of genetic control.

Radiation or chemical sterilization can cause somatic damage which reduces the vigor and competitiveness of released insects.[65] Thus, higher costs are incurred because more insects must be reared and released to overcome the lack of competitiveness. One solution is to develop insects that can be sterilized by milder treatments such as heat shock, diet, or light regimens, which would induce cell lethal genes in the germ line. The promoter from the *D. melanogaster* hsp70 heat shock gene has been widely used to construct temperature-activated genes, so it is the most likely candidate for combination with a testes-specific promoter to activate a gene that would cause sterility in the male. Such sterility genes would have to be repressed while the colony is growing. If they contained the heat shock controlling elements and the insects were grown at normal temperature, the genes would not be expressed and the insects would be fertile. Before release, the insects could be heat-shocked to induce sterility. Other constructions could be devised using heavy metal responsive genes,[32] light regulated genes,[33] diet regulated genes (e.g., digestive enzymes), or hormone regulated genes.[18,66,67]

The promoter of choice would be attached to an appropriate protein coding sequence to sterilize the insect. Sex specific sterile mutants are common in *D. melanogaster*. It has been estimated that approximately 750 genes can be mutated to result in male sterility and a somewhat smaller number of genes can be mutated to result in female sterility.[68] Unfortunately, most of these mutants are recessive, the defects of most are not understood,[69] and the genes for even fewer have been cloned.[9] These genes may someday be used in genetic sterilization, but a more practical alternative is the construction of genes having the proper properties using molecular genetics.

The construction of a protein coding sequence that would selectively destroy the cells in which it was expressed could be accomplished by modifying genes that have already been cloned, such as those for important cellular structures like microtubules. Several *D. melanogaster* genes have been cloned by using corresponding genes of other eucaryotes as probes,[9] so genes for these structures are apparently highly conserved. In general, mutants in these genes are recessive lethals, but in several cases dominant mutant gene produce protein that copolymerizes with wild-type protein and kills the cell. An interesting example is a dominant male sterile gene in *D. melanogaster*, caused by a defective tubulin gene which is expressed only in testes.[70] These "poison subunits", if expressed in the cells of a specific tissue or sex, would efficiently kill those cells without damaging others. It might be possible to construct poison subunit genes by specifically modifying parts of the coding subunit to inactivate an active site without eliminating polymerization sites.

INSECT VECTORS OF DISEASE

Hematophagous insects are generally more important as vectors of disease than as pests in themselves. One alternative to eradication of a disease vector population is replacement with a nonvector (refractory) strain. For decades it has been known that populations of vector species are heterogeneous for vectorial capacity, which is genetically mediated and is often due to a single gene difference.[71] There are several ways this resistance might occur; toxic factors in the hemolymph have been implicated,[72] salivary glands can be impermeable to parasites,[73] and other mechanisms can be imagined.

Recently it has been shown that insects carrying disease organisms are less fit than uninfected individuals.[71] If this is generally true, then refractoriness would spread through an insect population by natural selection, but the degree and speed of such selection would vary depending on whether refractoriness is selected against in the absence of infection, and the extent to which the vector species contacts the disease organism. The disease organism will also be under selective pressure and might overcome the refractoriness, so modifications would probably be necessary continuously to maintain the refractoriness. Genes for refractoriness have not been cloned, but several laboratories are working on this difficult problem.

A similar approach is modification of the life history strategy of the vector to make it less likely to bite several hosts. If an insect takes fewer blood meals it will be less likely to transmit disease. Autogenous mosquitoes lay their first batch of eggs without a blood meal, so they feed one less time during their life. This effect might be achieved by turning on the gene coding for egg development neurosecretory hormone. This hormone controls the timing of egg production in mosquitoes.[74]

INTRODUCTION OF GENES FOR INCREASING THE GENETIC LOAD INTO WILD POPULATIONS

The current approach in autocidal control is to try to swamp the wild population with radiation-sterilized individuals. This approach should be feasible when dealing with a small population, such as an accidental introduction of a pest into a new environment. It is more complicated, difficult, and expensive when the target is a widespread population, although the success of the screwworm program in North America demonstrates that it is possible.

In terms of genetic control of pests, a mechanism that will spread in a population has always been considered as the most desirable situation because of the relatively inexpensive input of making only a small inoculative release of genetically altered insects. Currently, there are two fairly well-known mechanisms (viz, transposons and meiotic drive) that appear to have the desirable attributes for increasing the genetic load by spreading disadvantageous genes vertically in a pest population through the germ line.

Transposons can spread through a population by making sufficient copies so that more than 50% of the gametes of a host get at least one copy by chance, even if it confers a selective disadvantage on its host.[75] The process of transposition frequently causes deleterious mutations; about half of all spontaneous mutations studied in *D. melanogaster* have been shown to be due to transposon insertion.[43] Transposition can also cause chromosome breakage, chromosome rearrangements, inappropriate expression of genes, and cell death.[35,40,41,44,76] Families of transposons in *D. melanogaster* have between 5 and 50 copies per haploid genome; hence, the chance that a heterozygous individual would fail to transmit such an element would range from 10^{-3} to 10^{-6}. Most transposons that have been studied stop transposing when the number of copies per genome of that family reaches a certain level.[41] The identification and deletion of this repression system would generate a transposon that would continue to increase in copy number until the genetic load eliminated the individuals carrying it. Whether transposons can be used is still hypothetical, but this potential method should receive considerable attention in the future.

In meiotic drive systems, which are known from many species, the chromosome carrying the driving gene is preferentially passed on to the next generation.[77] The best characterized system is *Segregation distorter (Sd)* from *D. melanogaster*,[78] in which the presence of the Sd gene causes the homologous chromosome to break, resulting in inviable gametes.

CONCLUSIONS

We have described some of the techniques that are currently being used in molecular genetics and give a few examples of how we believe they may be applied to economically important arthropods. Molecular technology will probably be easier to transfer from *Drosophila* to other insects than standard genetic technology; this is critical, since genetic control has usually been a disappointment.[1,2] The techniques for cloning and manipulating genes *in vitro* are directly applicable to other insects. Transformation techniques for the genetic manipulation of other insects using cloned genes must be developed before molecular genetics has important practical applications. It will not be necessary to generate mutants or chromosome rearrangements in pest species, since the modified genes will be transferred. Once a transformation system is developed for any species, it can be used to introduce rapidly a large number of genetic variants into that species, including those that have controlled other species.

ACKNOWLEDGMENTS

We would like to thank Dr. S. Miller, Dr. D. O'Brochta, and Dr. A. Undeen for critical reading of the manuscript.

REFERENCES

1. **Whitten, M. J.,** The conceptual basis for genetic control, in *Comprehensive Insect Physiology, Biochemistry, and Pharmacology,* 1983.
2. **Curtis, C. F.,** Genetic control of insect pests: growth industry or lead balloon?, *Biol. J. Linn. Soc.,* 26, 359, 1985.
3. **Cockburn, A. F., Howells, A. J., and Whitten, M. J.,** Recombinant DNA technology and genetic control, *Biotechnol. Gen. Eng. Rev.,* 2, 69, 1984.
4. **Levin, B. R.,** Problems and promise in genetic engineering in its potential applications to insect management, in *Genetics in Relation to Insect Management,* Hoy, M. A. and McKelvey, J. J. J., Eds., The Rockefeller Foundation, New York, 1979, 170.
5. **Courtright, J. B. and Kumaran, A. K.,** A genetic engineering methodology for insect pest control: female sterilizing genes, *Biotechnol. Solving Agric. Probl., Beltsville Symp.,* 10, 325, 1985.
6. **Handler, A. and O'Brochta, D. A.,** Molecular genetic approaches to genetic sexing, in *Fruit Flies of Economic Importance,* Cavalloro, R., Ed., A. A. Balkema, 1988.
7. **Shirk, P. D., O'Brochta, D. A., Roberts, P. E., and Handler, A. E.,** Sex-specific selection using chimeric genes: applications to sterile insect release, in *Biotechnology in Crop Protection,* Hedin, P. A., Menn, J. J., and Hollingworth, R., Eds., ACS Books, Washington, D.C., 1987.
8. **Kirschbaum, J. B.,** Potential implication of genetic engineering and other biotechnologies to insect control, *Annu. Rev. Entomol.,* 30, 51, 1985.
9. **Merriam, J.,** Cloned DNA by chromosome location, *Drosophila Inf. Serv.,* 61, 9, 1985.
10. **Smith, G. E., Ju, G., Ericson, B. L., Moschera, J., Lahm, H.-W., Chizzonite, R., and Summers, M. D.,** Modification and secretion of human interleukin, *Proc. Natl. Acad. Sci. U.S.A.,* 82, 8404, 1985.
11. **Freifelder, D.,** Molecular Biology, Jones & Bartlett, Boston, 1983.
12. **Kelly, J. H. and Darlington, G. J.,** Hybrid genes: molecular approaches to tissue-specific gene regulation, *Annu. Rev. Genetics,* 19, 273, 1985.
13. **Maniatis, T., Fritsch, E. F., and Sambrook, J.,** Molecular Cloning. A Laboratory Manual, Cold Spring Harbor Labs, Cold Spring Harbor, New York, 1982.
14. **Serfling, E., Jasin, M., and Schaffner, W.,** Enhancers and eucaryotic gene transcription, *Trends Genet.,* 1, 224, 1985.
15. **Khoury, G. and Gruss, P.,** Enhancer elements, *Cell,* 33, 313, 1983.
16. **Roger, B. L. and Saunders, G. F.,** Transcriptional enhancers play a major role in gene expression, *BioEssays,* 4, 62, 1985.
17. **Corces, V., Pellicer, A., Axel, R., and Meselson, M.,** Integration, transcription, and control of a Drosophila gene in mouse cells, *Proc. Natl. Acad. Sci. U.S.A.,* 78, 7038, 1981.
18. **Garabedian, M. J., Hung, M. C., and Wensink, P. C.,** Independent control elements that determine yolk protein gene expression in alternative Drosophila tissue, *Proc. Natl. Acad. Sci. U.S.A.,* 82, 1396, 1985.
19. **Hiromi, Y., Kuroiwa, A., and Gehring, W. J.,** Control elements of the Drosophila segmentation gene fushi tarazu, *Cell,* 43, 603, 1985.
20. **Pelham, H. R. B.,** A regulatory upstream promoter element in the Drosophila hsp70 heat-shock gene, *Cell,* 30, 517, 1982.
21. **Garabedian, M. J., Shepherd, B. M., and Wensink, P. C.,** A tissue-specific transcription enhancer from the Drosophila yolk protein 1 gene, *Cell,* 45, 859, 1986.
22. **Cohen, R. S. and Meselson, M.,** Separate regulatory elements for the heat-inducible and ovarian expression of the Drosophila hsp26 gene, *Cell,* 43, 737, 1985.
23. **Bourouis, M. and Richards, G.,** Remote regulatory sequences of the Drosophila glue protein gene sgs3 as revealed by P-element transformation, *Cell,* 40, 349, 1985.
24. **Hubbard, S. C. and Ivatt, R. J.,** Synthesis and processing of asparagine-linked oligopolysaccharides, *Annu. Rev. Biochem.,* 50, 555, 1981.
25. **Walter, P., Gilmore, R., and Blobel, G.,** Protein translocation across the endoplasmic reticulum, *Cell,* 38, 5, 1984.
26. **McGarry, T. and Lindquist, S.,** The preferential translation of Drosophila hsp70 mRNA requires sequences in the untranslated leader, *Cell,* 42, 903, 1985.
27. **Snyder, M., Hirsh, J., and Davidson, N.,** The cuticle genes of Drosophila: a developmentally regulated gene cluster, *Cell,* 25, 165, 1981.
28. **Scherer, G., Telford, J., Baldari, C., and Pirrotta, V.,** Isolation of cloned genes differentially expressed at early and late stages of Drosophila embryonic development, *Dev. Biol.,* 86, 438, 1981.
29. **Ashburner, M. and Bonner, J. J.,** The induction of gene activity in Drosophila by heat shock, *Cell,* 17, 241, 1979.

30. **Morganelli, C. M., Berger, E. M., and Pelham, H. R. B.,** Transcription of Drosophila small hsp-tk hybrid genes is induced by heat shock and by ecdysterone in transfected Drosophila cells, *Proc. Natl. Acad. Sci. U.S.A.,* 82, 5865, 1985.

31. **Barnett, T., Pachl, C., Gergen, J. P., and Wensink, P. C.,** The isolation and characterization of Drosophila yolk protein genes, *Cell,* 21, 729, 1980.

32. **Karin, M.,** Metallothioneins: proteins in search of function, *Cell,* 41, 9, 1984.

33. **Benedict, M. Q. and Seawright, J. A.,** Changes in pigmentation in mosquitoes in response to color of environment, *Ann. Entomol. Soc. Am.,* 80, 55, 1987.

34. **Bingham, P. M., Kidwell, M. G., and Rubin, G. M.,** The molecular basis of P-M hybrid dysgenesis: the role of the P element, a P-strain-specific transposon family, *Cell,* 29, 995, 1982.

35. **Rubin, G. M., Kidwell, M. G., and Bingham, P. M.,** The molecular basis of P-M hybrid dysgenesis: the nature of induced mutations, *Cell,* 29, 987, 1982.

36. **O'Hare, K. and Rubin, G. M.,** Structure of P transposable elements of Drosophila melanogaster and their sites of insertion and excision, *Cell,* 34, 25, 1983.

37. **Karess, R. E. and Rubin, G. M.,** Analysis of P transposable element functions in Drosophila, *Cell,* 38, 135, 1984.

38. **Spradling, A. C. and Rubin, G. M.,** Transposition of cloned P elements into Drosophila germline chromosomes, *Science,* 218, 341, 1982.

39. **Rubin, G. M. and Spradling, A. C.,** Genetic transformation of Drosophila with transposable element vectors, *Science,* 218, 348, 1982.

40. **Kleckner, N.,** Transposable elements in procaryotes, *Annu. Rev. Genet.,* 15, 341, 1981.

41. **Calos, M. P. and Miller, J. H.,** Transposable elements, *Cell,* 20, 579, 1980.

42. **Green, M. M.,** Transposable elements in Drosophila and other Diptera, *Annu. Rev. Genet.,* 14, 109, 1980.

43. **Spradling, A. C. and Rubin, G. M.,** Drosophila genome organization: conserved and dynamic aspects, *Annu. Rev. Genet.,* 15, 219, 1981.

44. **Engels, W. R.,** Extrachromosomal control of mutability in Drosophila melanogaster, *Proc. Natl. Acad. Sci. U.S.A.,* 76, 4011, 1979.

45. **Kidwell, M. G.,** Intraspecific hybrid sterility, in *The Genetics and Biology of Drosophila,* Ashburner, M., Carson, H. L., and Thompson, J. N. J., Eds., Academic Press, Orlando, FL, 1984, 125.

46. **Brennan, M. D., Rowan, R. G., and Dickinson, W. J.,** Introduction of a functional P element into the germline of Drosophila hawaiiensis, *Cell,* 38, 147, 1984.

47. **Scavarda, N. J. and Hartl, D. L.,** Interspecific DNA transformation in Drosophila, *Proc. Natl. Acad. Sci. U.S.A.,* 81, 7515, 1984.

48. **Zalocar, M.,** A method for injection and transplantation of nuclei and cells, *Experientia,* 37, 1354, 1981.

49. **Grace, T. D. C. and Cockburn, A. F.,** unpublished data.

50. **Benedict, M. Q., Tarrant, C. A., and Cockburn, A. F.,** unpublished data.

51. **Miller, L. H., Sakai, R. K., Romans, P., Gwadz, R. W., Kantoff, P., and Coon, H. G.,** Stable integration and expression of a bacterial gene in the mosquito Anopheles gambiae, *Science,* 237, 779, 1987.

52. **O'Brochta, D. A.,** personal communication.

53. **Klein, T. M., Wolf, E. D., Wu, R., and Sanford, J. C.,** High-velocity microprojectiles for delivering nucleic acids into living cells, *Nature,* 327, 70, 1987.

54. **Rio, D. C., Laski, F. A., and Rubin, G. M.,** Identification and immunochemical analysis of biologically active Drosophila P element transposase, *Cell,* 44, 21, 1986.

55. **Clough, D. W., Lepinske, H. M., Davidson, R. L., and Stroti, R. V.,** Drosophila P element-enhanced transfection in mammalian cells, *Mol. Cell. Biol.,* 5, 898, 1985.

56. **Green, M. M.,** Mutable and mutator loci, in *The Genetics and Biology of Drosophila,* Ashburner, M. and Novitski, E., Eds., Academic Press, Orlando, FL, 1976, 929.

57. **Goldberg, D. A., Posakony, J. W., and Maniatis, T.,** Correct developmental expression of a cloned alcohol dehydrogenase gene transduced into the Drosophila germ line, *Cell,* 34, 59, 1983.

58. **Scholnick, S. B., Morgan, B. A., and Hirsh, J.,** The cloned dopa decarboxylase gene is developmentally regulated when reintegrated into the Drosophila genome, *Cell,* 34, 37, 1983.

59. **O'Donnell, J., Gerace, L., Leister, F., and Sofer, W.,** Chemical selection of mutants that affect ADH in Drosophila II. Use of 1-pentene-3-ol, *Genetics,* 79, 73, 1974.

60. **Mouches, C., Pasteur, N., Berge, J. B., Hyrien, O., Raymond, M., Robert de Saint Vincent, B., de Silvestri, M., and Georghiou, G.,** Amplification of an esterase gene is responsible for insecticide resistance in a California Culex mosquito, *Science,* 233, 778, 1986.

61. **Steller, H. and Pirrotta, V.,** A transposable P vector that confers selectable G418 resistance to Drosophila larvae, *EMBO J.,* 4, 167, 1985.

62. **Hoy, M. A.,** Recent advances in genetics and genetic improvement of the phytoseiidae, *Annu. Rev. Entomol.,* 30, 345, 1985.

63. **Baumhover, A. H., Graham, A. J., Bitter, B. A., Hopkins, D. E., New, W. D., Dudley, F. H., and Bushland, R. C.,** Screwworm control through release of sterilized flies, *J. Econ. Entomol.,* 48, 462, 1955.

64. **Waterhouse, D. F., LaChance, L., and Whitten, M. J.,** Use of autocidal methods, in *Theory and Practice of Biological Control,* Huffaker, C. and Messenger, P. S., Eds., Academic Press, Orlando, 1974, 637.

65. **LaChance, L. E.,** Genetic strategies affecting the success and economy of the sterile insect release method, in *Genetics in Relation to Insect Management,* Hoy, M. A. and McKelvey, J. J. J., Eds., The Rockefeller Foundation, New York, 1979, 8.

66. **Tamura, T., Kunert, C., and Postlewait, J.,** Sex- and cell-specific regulation of yolk polypeptide genes introduced into Drosophila by P-element mediated gene transfer, *Proc. Natl. Acad. Sci. U.S.A.,* 82, 7000, 1985.

67. **Hoffman, E. P., Gerring, S. L., and Corces, V. G.,** The ovarian, ecdysterone, and heat-shock-responsive promoters of the Drosophila melanogaster hsp27 gene react very differently to perturbations of DNA sequence, *Mol. Cell. Biol.,* 7, 973, 1987.

68. **Lindsley, D. L. and Lifshitz, E.,** Genetic control of spermatogenesis in Drosophila, in *Edinburgh Symp. Genetics Spermatogenesis,* M. Glucksohn-Welsh, Ed., R. A. Beatty, Copenhagen, 1971, 203.

69. **Lindsley, D. L. and Grell, E. H.,** Genetic Variations of Drosophila Melanogaster, Carnegie Inst. Wash. Publ., 1968, 471.

70. **Kemphues, T. C., Raff, E. C., Raff, R. A., and Kaufman, T. C.,** Mutation in a structural gene for a B-tubulin specific to testis in Drosophila melanogaster, *Proc. Natl. Acad. Sci. U.S.A.,* 76, 3991, 1979.

71. **Curtis, C. F. and Graves, P. M.,** Genetic variation in the ability of insects to transmit filariae, trypanosomes, and malaria parasites, *Curr. Top. Vector Res.,* 1, 31, 1983.

72. **Weathersby, A. B. and McCroddan, D. M.,** The effects of parabiotic twinning of susceptible and refractory mosquitoes on the development of Plasmodium gallinaceum, *J. Parasitol.,* 68, 1081, 1982.

73. **Rosenberg, R.,** Inability of Plasmodium knowlesi sporozoites to invade Anopheles freeborni salivary glands, *Am. J. Trop. Med. Hyg.,* 34, 687, 1985.

74. **Fuchs, M. S. and Kang, S.,** Ecdysone and mosquito vitellogenesis. A critical appraisal, *Insect Biochem.,* 11, 627, 1981.

75. **Hickey, D. A.,** Selfish DNA: a sexually transmitted parasite, *Genetics,* 120, 33, 1982.

76. **McGinnis, W. and Beckendorf, S.,** Association of a Drosophila transposable element of the roo family with chromosomal deletion breakpoints, *Nucl. Acids Res.,* 11, 737, 1983.

77. **Zimmering, S., Sandler, L., and Nicoletti, B.,** Mechanisms of meiotic drive, *Annu. Rev. Genet.,* 4, 409, 1970.

78. **Ganetzky, B.,** On the components of segregation distortion in Drosophila melanogaster, *Genetics,* 86, 321, 1977.

79. **DeCicco, D. V. and Spradling, A. C.,** Localization of a cis acting element responsible for the developmentally regulated amplification of Drosophila chorion genes, *Cell,* 38, 45, 1984.

80. **Haenlin, M., Steller, H., Pirrotta, V., and Mohier, E.,** A 43 kilobase cosmid P transposon rescues the fsD K10 morphogenetic locus and three adjacent Drosophila developmental mutants, *Cell,* 40, 827, 1985.

81. **Chen, Z.-Q. and Hodgetts, R. B.,** Functional analysis of a naturally occurring variant dopa decarboxylase gene in Drosophila melanogaster using P element mediated germ line transformation, *Mol. Gen. Genet.,* 207, 441, 1987.

82. **Otto, E., Allen, J. M., Young, J. E., Palmiter, R. D., and Maroni, G.,** A DNA segment controlling metal-regulated expression of the Drosophila melanogaster metallothionein gene Mtn, *Mol. Cell. Biol.,* 7, 1710, 1987.

83. **Krumm, A., Roth, G. E., and Korge, G.,** Transformation of salivary gland secretion protein gene Sgs-4 in Drosophila: stage- and tissue-specific regulation, dosage compensation, and position effect, *Proc. Natl. Acad. Sci. U.S.A.,* 82, 5055, 1985.

84. **Butler, B., Pirrotta, V., Irminger-Finger, I., and Nothiger, R.,** The sex-determining gene traits of Drosophila: molecular cloning and transformation studies, *EMBO J.,* 5, 3607, 1986.

85. **Fyrberg, E. A. and Karlik, C. C.,** Genetic rescue of muscle defects associated with a mutant Drosophila melanogaster tropomyosin mutant, *Mol. Cell. Biol.,* 7, 2977, 1987.

86. **Hazelrigg, T., Levis, R., and Rubin, G. M.,** Transformation of white locus DNA in Drosophila: dosage compensation, zeste interaction, and position effects, *Cell,* 36, 469, 1984.

87. **Rubin, G. M. and Spradling, A. C.,** Vectors for P element mediated gene transfer in Drosophila, *Nucl. Acids Res.,* 11, 6341, 1983.

88. **Spradling, A. C. and Rubin, G. M.,** The effect of chromosomal position on the expression of the Drosophila xanthine dehydrogenase gene, *Cell,* 34, 47, 1983.

89. **Lis, J. T., Simon, J. A., and Sutton, C. A.,** New heat shock puffs and beta-galactosidase activity resulting from transformation of Drosophila with an hsp70-lacZ hybrid gene, *Cell,* 35, 403, 1983.

90. **Simon, J. A. and Lis, J. T.,** A germline transformation analysis reveals flexibility in the organization of heat shock consensus elements, *Nucl. Acids Res.,* 15, 2971, 1987.

91. **Cohen, R. S. and Meselson, M.,** Inducible transcription and puffing in Drosophila melanogaster transformed with hsp70-phage hybrid heat shock genes, *Proc. Natl. Acad. Sci. U.S.A.,* 81, 5509, 1984.

92. **Klemenz, R., Weber, U., and Gehring, W. J.,** The white gene as a marker in a new P-element vector for gene transfer in Drosophila melanogaster, *Nucl. Acids Res.,* 15, 3947, 1987.
93. **Davies, J. A., Addison, C. F., Delaney, S. J., Sunkel, C., and Glover, D. M.,** Expression of the prokaryotic gene for chloramphenicol acetyltransferase in Drosophila under the control of larval serum protein 1 gene promoters, *J. Mol. Biol.,* 189, 13, 1986.
94. **Mismer, D. and Rubin, G. M.,** Analysis of the promoter of the ninaE gene opsin gene in Drosophila melanogaster, *Genetics,* 116, 565, 1987.
95. **Shermoen, A. W., Jongens, J., Barnett, S. W., Flynn, K., and Beckendorf, S. K.,** Developmental regulation by an enhancer from the Sgs-4 gene of Drosophila, *EMBO J.,* 6, 207, 1987.

INDEX

A

Printed and bound by CPI Group (UK) Ltd, Croydon, CR0 4YY

22/10/2024

01777630-0012